W9-AOV-825

TREATMENT AND DISPOSAL OF WASTEWATER SLUDGES

Revised Edition

P. AARNE VESILIND
Department of Civil Engineering
Duke University
Durham, North Carolina

ANN ARBOR SCIENCE
PUBLISHERS INC
P.O. BOX 1425 • ANN ARBOR, MICH. 48106

P.O. Box 1425, 230 Collingwood, Ann Arbor, Michigan 48106

Library of Congress Catalog Card Number 78-71431
ISBN 0-250-40290-4

Manufactured in the United States of America

PREFACE TO THE SECOND EDITION

Sequels are seldom as satisfying as the originals, and it was thus with some trepidation that I undertook writing "Son of Sludge." The need for the second edition became strikingly clear, however, when I surveyed the material on sludge published since 1974. Tremendous progress has been made in the five intervening years, especially in the area of ultimate disposal. Accordingly, this second edition has two chapters on ultimate disposal–one devoted to disposal in the marine environment and one that discusses disposal on land–none of which was covered in the earlier edition. In Europe, multinational cooperative programs in sludge management have significantly advanced the art. In addition, research in Canada, South Africa and other countries has contributed to the knowledge of sludge technology.

All this activity has not, however, solved our problems of sludge disposal. On balance, the problem seems to hinge on the significant lack of basic research on sludge treatment and disposal. The reasons for this are, I believe, twofold: the seemingly unsavory nature of the research, and the undefined and transient nature of the material in question. I am optimistic, however, that as more scientists and engineers recognize the challenges in this research area, further progress will be made, and that the next five years will see an even greater output of literature in research and operating experience.

I gratefully acknowledge the assistance of Dr. Yakir Hasit of Cornell University in the review of the original publication and in the co-authorship of Chapter 12, Regional Sludge Management. Dr. Miguel Medina of Duke University reviewed the chapter on ultimate disposal into the marine environment, and Dr. William Clarkson of Cornell reviewed the chapter on ultimate disposal on land. Their suggestions are gratefully acknowledged.

The manuscript for the second edition was typed by Kathy Worrell and Judy Edwards, two of the finest people anyone could wish to work with.

P. Aarne Vesilind
Durham, North Carolina
1979

PREFACE TO THE FIRST EDITION

Most people consider sludge to be a four-letter word. In fact, a book about sludge does lack a certain glamour and appeal other publications might offer. Perhaps this explains why no previous volume has been devoted to the engineering aspects of sludge handling and disposal.

This slight is unjust. With increasing urban development and industrial capacity, wastewaters will be carrying more and more solids to treatment plants. Problems associated with the treatment and disposal of these solids can't help but multiply. In certain areas, these problems have already reached critical levels. It thus seems appropriate that environmental engineers should pay more attention to the treatment and disposal of wastewater sludges.

This book emanated from a graduate engineering course in Solid Waste Engineering taught at Duke University. The first half deals with sludge (a solid waste) and the remainder is devoted to refuse collection and disposal.

This arrangement allows us to take the entire sludge treatment section out of the unit operations course, leaving more time for other topics while still providing the students with a comprehensive background in wastewater sludge technology.

Basically this volume is a textbook for graduate students and upper level undergraduates. It is also a reference source for practicing engineers and researchers. I assume the reader has some knowledge of wastewater treatment, but nevertheless I have introduced the necessary jargon in Chapter 1.

As I began writing this book, I soon discovered that a set of class notes is not a manuscript. The transition from notes to manuscript occurred during the summer of 1973 while I was a visiting lecturer at the Danish Technological University in Lyngby, and a research associate at the Norwegian Institute for Water Research in Oslo. The opportunity to study, reflect and write, coupled with the friendly atmosphere and Scandinavian milieu, contributed significantly to the otherwise trying project. Specifically, the assistance of Paul Harremoes and Jens Hansen of the Danish Technological University, and Kjell Baalsrud, Peter Balmér and Arild Eikum of the Norwegian Institute for Water Research is gratefully acknowledged.

The text was reviewed by Richard Cole, formerly of the University of North Carolina at Chapel Hill and presently of the Atomic Energy Commission, James E. Smith, Jr., Sanitary Engineer, and Stephen McCullers, Duke University. Their valuable comments are greatly appreciated.

Finally, special thanks go to my wife, who typed the first rough draft from my barely intelligible recordings. Her perseverence and encouragement were vital to the completion of the manuscript.

P. Aarne Vesilind
Durham, North Carolina
1974

P. Aarne Vesilind is Associate Professor of Civil Engineering at Duke University, Durham, North Carolina, and Director of the Duke Environmental Center. Born in Estonia, he holds BS and MS degrees in civil engineering from Lehigh University, and a Master's degree and a PhD in sanitary engineering from the University of North Carolina. Dr. Vesilind has also been a Fellow at the Norwegian Institute for Water Research in Oslo.

A member of several professional societies, the author has done research, design and consulting work for a number of U.S. firms. He has also made an extensive study of European sludge management practices.

For several years he has taught a graduate engineering course in wastewater solids management, and recently was the recipient of a Fullbright-Hayes Senior Lectureship to study in New Zealand.

Among his many honors and awards are the Duke University Outstanding Professor Award for 1971-1972 and the 1971 Collingwood Prize from the American Society of Civil Engineers.

Dr. Vesilind is the author of *Environmental Pollution and Control*, an Ann Arbor Science Publishers, Inc. publication, and more than three dozen related articles in national and international journals.

To My Father

CONTENTS

CHAPTER 1

SLUDGE SOURCES, QUANTITIES AND MANAGEMENT

The treatment of wastewaters invariably produces a residual which must be disposed of into the environment. Most often this residual is a semisolid, odiferous, unmanageable and dangerous material commonly termed *sludge*. Sludge, however, can also contain substantial nutrients and organics and these can be considered a replenishable natural resource.

The quantities of sludges produced in our modern society are staggering. Some present and future sludge production estimates for the United States are shown in Figure 1-1. The increase in sludge production is due mainly to the upgrading of wastewater treatment plants as mandated by federal legislation. The effect of such upgrading and new plant construction has been to effectively double the sludge production between 1972 and 1990. More importantly, this increase is not evenly distributed geographically, and the larger cities, with limited available land area for ultimate disposal, are facing critical problems.

In this chapter, the sources and quantities of wastewater sludges are discussed first, followed by comments on sludge management within a treatment plant.

SLUDGE SOURCES

Wastewater treatment systems are designed to reduce adverse effects on the environment. As wastewaters differ greatly, so do their treatment systems. Historically, three basic systems have evolved for domestic wastewater treatment.

The first diagram in Figure 1-2 shows a typical *primary treatment plant*. This was the first widely used treatment method, and still is the only method

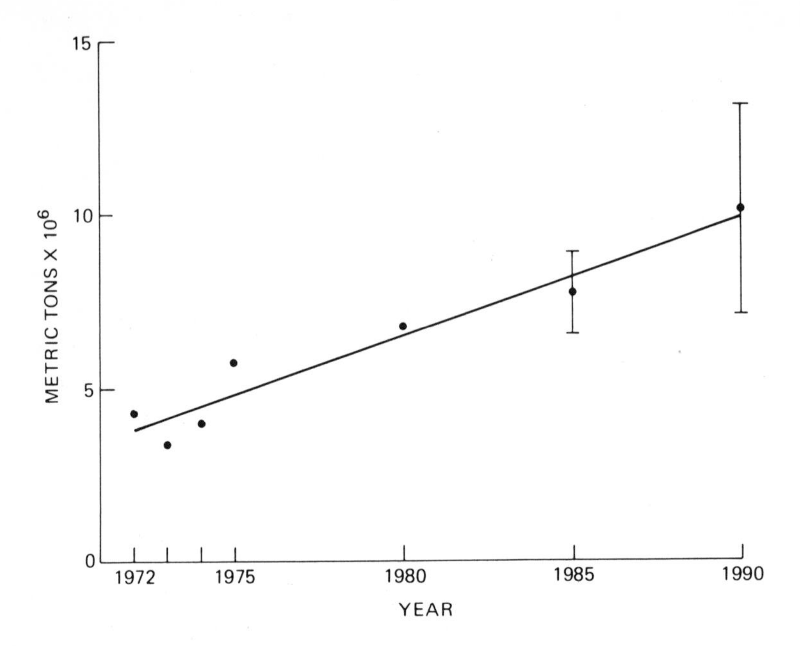

Figure 1-1. Estimates of present and future total sludge production in the U.S. (National Academy of Sciences, 1978).

of treatment in some large cities. The principal objective of primary treatment is to remove the settleable solids from wastewater and thus make the water less objectionable. With primary treatment systems, while no attempt is deliberately made to remove oxygen-demanding materials, some of the biochemical oxygen demand (BOD) is removed as a result of removing solids.

The primary clarifier (also known as a settling tank or, less appropriately, a sedimentation tank) operates simply by allowing heavier solids to settle to the bottom and lighter solids to float to the top. The quantity of floated material, called scum, is usually not very great and is either treated with the settled solids or conveniently disposed of on nearby land without further treatment. The solids removed from the bottom of the clarifier are known as *raw primary sludge*.

Raw primary sludge is quite objectionable and has a high percentage of water, two characteristics which make further handling difficult.

This sludge is often *digested* to make it less objectionable and is then known as *primary digested sludge*. Many digesters operate to allow the lighter fraction, or *supernatant*, in the digester to be pumped back to the head of the treatment plant. Supernatant often has a high solids concentration, and this operation can result in the recirculation of solids and eventual operating problems.

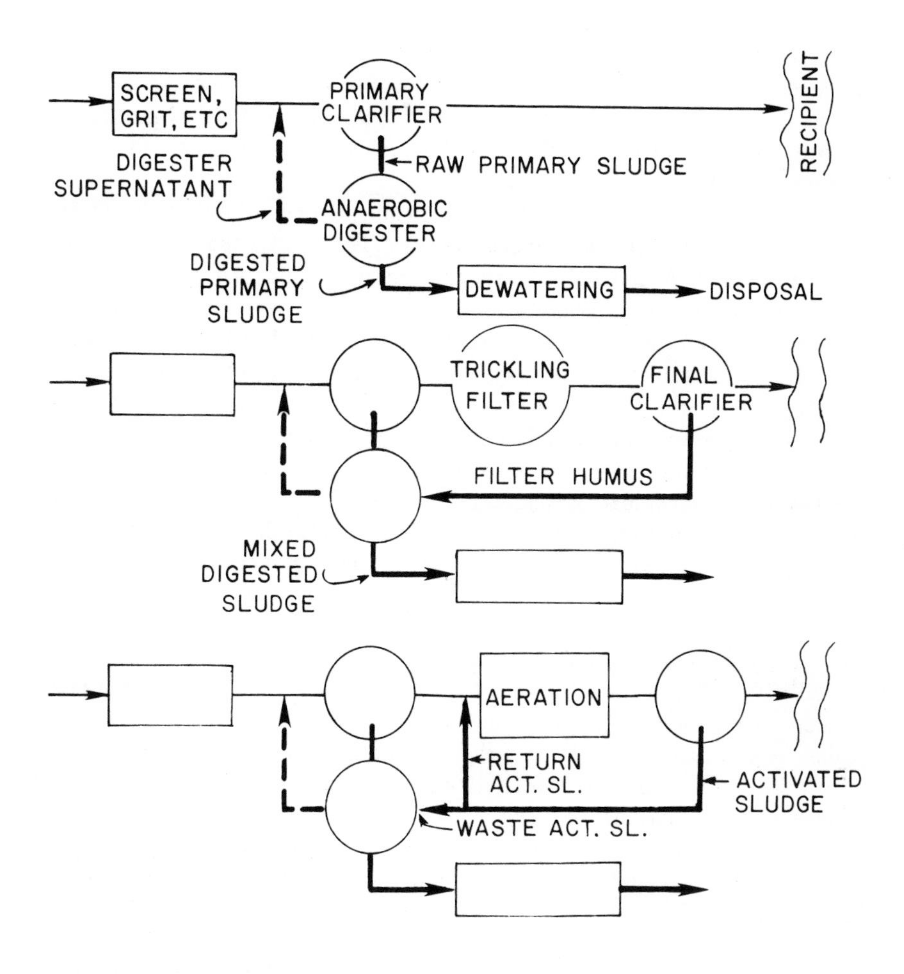

Figure 1-2. Typical wastewater treatment systems.

Anaerobic digestion results in a 50% reduction of volatile solids, elimination of objectionable odor, and a significant destruction of pathogenic organisms. Digested sludge can be disposed of directly on land, dewatered on drying beds, or by mechanical means before ultimate disposal.

The second and third illustrations in Figure 1-2 show the operation of a typical *secondary treatment plant* (those designed to remove BOD as well as solids). BOD is usually removed by a biological means, but recently physical-chemical operations also have been used for this purpose.

Trickling filters (a bed of rocks covered by zoogleal mass over which wastewater is trickled) are widely used as a biological treatment method for BOD. Their major drawback is that they require too much land for most large-scale applications. The solids that slough off the filter rocks are captured in the *final clarifier*. This sludge, called *filter humus*, is limited in quantity (except at certain times of the year when the flora on the filter rocks changes).

More commonly, modern secondary treatment plants follow some variation of the basic activated sludge system. The biomass that degrades the oxygen-demanding material is suspended in the liquid. Air is driven into this *mixed liquor* by porous diffusers, mechanically by surface aerators or by some other means such as brushes or aspirators. The biomass cultured in the aeration tank must be settled out in the final clarifier and returned to the head of the aeration system to be reused. In the process the microorganisms consume the dissolved organic material producing CO_2, water and more microorganisms, and thus reduce the oxygen demand of the liquid.

This is an almost ideal system to remove and dispose of the dissolved high-energy material found in wastewater. Unfortunately, the amount of microorganisms produced in the activated sludge process exceeds the amount required by the system; thus some of these solids must be wasted. This waste biological material, referred to as *waste activated sludge*, represents one of the true headaches of wastewater treatment.

Both filter humus and waste activated sludge are often mixed with raw primary sludge and digested by anaerobic digesters. The resulting material, called *mixed digested sludge*, usually is dewatered before its final disposal.

An alternative to anaerobic digestion, sometimes employed in activated sludge plants, is aerobic digestion, which is simply an extension of the aeration system. Waste activated sludge is aerated in a separate tank for several days and thus is stabilized in terms of its oxygen demand and fraction of volatile solids. The resulting sludge is referred to as *aerobically digested sludge*.

Now that removal of nutrients such as nitrogen and phosphorus is often considered just as important as removal of the oxygen-demanding material and solids, physical-chemical treatment has received wide attention, and several such plants have gone into operation. The addition of chemicals such as iron or aluminum salts to primary clarifiers results in the production of sludges that have some characteristics of secondary biological sludges. The addition of chemicals to secondary effluents results in sludges similar to water plant sludge. In both cases, the sludges are difficult to handle and dewater.

An additional source of sludge in sanitary engineering is the waste from water treatment. Aluminum sulfate (alum), the most widely used chemical for coagulation and flocculation in water treatment, produces a sludge known as *waste alum sludge*.

Table 1-1 summarizes the kinds of sludges encountered in water and wastewater treatment as well as some of their characteristics. Raw primary sludge, commonly found at solids concentrations of 4 or 8% dry solids, is quite a nuisance in wastewater treatment due to its very strong odor and poor drainability. The odor problem is aggravated by allowing solids to become septic either in the sewers or in the bottom of primary clarifiers. Raw primary sludge does not drain well on sand drying beds but can be dewatered with mechanical equipment.

Although anaerobically digested sludge will dewater well on drying beds, it is more difficult to dewater by mechanical means. Anaerobically digested sludge is dark in color and has a musty odor that is not at all objectionable.

Filter humus is a light fluffy sludge as is waste activated sludge. The latter is a very wet material somewhere between 0.5 and 1.5% solids, and its color can range anywhere from yellow to almost black, although a deep brown is considered desirable. Obviously, the kind of sludge produced in an activated sludge plant depends both on the method of operation and

Table 1-1. Common Water and Wastewater Sludges

Sludge	Concentration % Solids	Characteristics
Raw Primary	4-8	Vile, bad odor; gray-brown; does not drain well on drying beds; but can be dewatered mechanically
Anaerobic Primary Digested	6-10	Dewaters well on drying beds; black; musty; produces gas
Filter Humus	3-4	Fluffy; brown
Waste Activated	0.5-1.5	Little odor; yellow-brown; fluffy; difficult to dewater; very active biologically
Mixed Digested (Primary + Waste Activated)	2-4	Black-brown; produces gas; musty; not as easy to dewater as digested primary
Aerobic Digested	1-3	Yellow-brown, sometimes difficult to dewater; biologically active
Waste Alum	0.5-1.5	Gray-yellow; odorless; very difficult to dewater

the substrate, the food fed to the microorganisms. The entire ecology of the aeration basin is, therefore, a function of operation and feeding.

Mixed digested sludge which is a combination of primary and waste activated (or filter humus) is light brown and not too odoriferous, although it is not as easy to dewater as anaerobic primary digested sludge. Aerobically digested sludge has a low solids concentration, and its dewatering and disposal are complicated by its relatively high biological activity. Extended periods of anaerobiosis (lack of dissolved oxygen) will also affect its characteristics, such as making it more difficult to dewater.

Waste alum sludge from water treatment also can differ in color depending on the type of material removed from the water, although very often it will be a grayish-yellow color. It is very difficult to dewater, but its saving grace is its relative biological inactivity.

It must be emphasized that vastly different types of sludges are encountered in water and wastewater treatment. The above list includes only some of the more common sludges, and such troublesome sludges as those resulting from some industrial waste treatment operations have not even been mentioned. This does not mean they are unimportant, or easy to handle.

If all sludges were the same, there would be no need for sophisticated methods of sludge characterization and analysis, or for the development of methods for designing sludge handling and disposal operations—or, for that matter, this book. If any one generalization can be made about sludges, it is that they are different.

SLUDGE QUANTITIES

Figure 1-3 is a generalized schematic of a secondary wastewater treatment plant adapted from Kormanik (1972). The secondary part of the plant could be either for aeration or trickling filtration. The purpose here is to develop a generalized plant. The symbols are defined as follows:

S_0 = influent BOD kg/hr (5-day 20°C)
X_0 = influent suspended solids (SS), kg/hr
h = fraction of BOD not removed in primary clarifier
i = fraction of BOD not removed in aeration of trickling filter
X_f = plant effluent SS, kg/hr
k = fraction of X_0 removed in primary clarifier
j = fraction of solids not destroyed in digestion
ΔX = net solids produced by biological action, kg/hr
Y = yield = $\Delta X/\Delta S$, where $\Delta S = hS_0 - ihS_0$

For example, incoming BOD (S_0) can be calculated as some reasonable BOD concentration such as 250 mg/l x the flow rate (Q) as m^3/hr x 10^{-3} = kg/hr. In English units S_0 is calculated as 250 mg/l x 8.34 (which is a conversion factor) x the flow rate in million gallons per day (mgd). S_0 thus

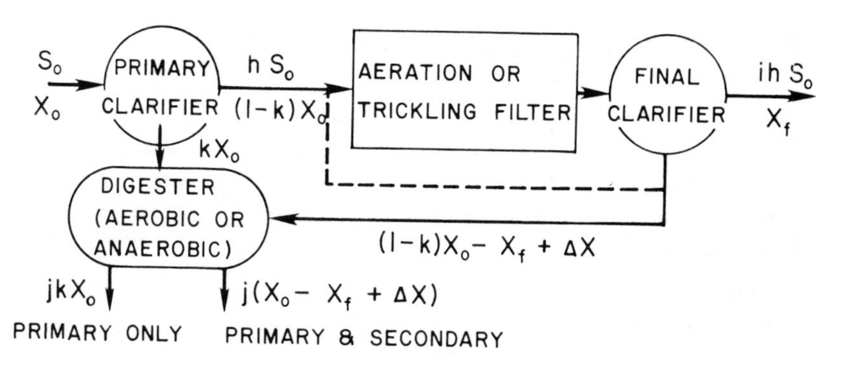

Figure 1-3. A generalized wastewater treatment system.

is expressed in lb BOD/day. Similarly, X_0 can be calculated as 225 mg/l x 8.34 x flow rate (Q). The value for k can be assumed to be about 0.6. The value for h is about 0.7, indicating that primary clarification is not a very efficient method for removing BOD (which is quite understandable since it is not designed to be). The amount of solids escaping the plant, about 20 mg/l, is denoted as X_f. The value for j is about 0.8, although solids reduction in aerobic digestion can exceed this, as shown by Lindstedt *et al.* (1971). When designing a plant, however, it might be wise to allow for only limited total SS reduction in aerobic digestion.

The fraction of BOD not removed in aeration or trickling filters, denoted by the symbol i, is equal to about 0.1 for a well-operated activated sludge plant and about 0.2 for a trickling filter plant—indicating for activated sludge a 90% or better BOD removal (common in a well-operated plant). The sludge yield, or pounds of solids produced per pound of BOD used in the aeration system, is about 0.5 for activated sludge and about 0.2 for trickling filters. This yield is the waste biological sludge produced in the secondary system. BOD reduction is obviously equal to hS_0-ihS_0. Knowing the yield and the destruction of BOD, the net solids produced by biological action (ΔX) can be readily calculated.

To summarize, some typical values for domestic wastewater are:

$$S_0 = \begin{cases} 250 \times 10^{-3} \times Q = \text{kg/hr if Q is m}^3\text{/hr} \\ 250 \times 8.34 \times Q = \text{lb/day if Q is in mgd} \end{cases}$$

$$X_0 = \begin{cases} 225 \times 10^{-3} \times Q = \text{m}^3\text{/hr} \\ 225 \times 8.34 \times Q = \text{lb/day} \end{cases}$$

$$k = 0.6$$

$$h = 0.7$$

$$X_f = \begin{cases} 20 \times 10^{-3} \times Q = \text{m}^3\text{/hr} \\ 20 \times 8.34 \times Q = \text{lb/day} \end{cases}$$

$j \approx$ 0.8, assuming no supernatant withdrawal

$i =$ 0.1 for well-operated activated sludge

$i =$ 0.2 for trickling filters

Y = 0.5 for activated sludge
Y = 0.2 for trickling filters

Example 1-1: Using the above typical values, calculate the various solids produced in a 10-mgd wastewater treatment plant (standard activated sludge, anaerobic digestion).

a. Raw Primary Sludge

Total solids in influent, X_0 = 225 mg/l x 10 mgd x 8.34
= 18,800 lb/day

Efficiency of primary clarifier, $k = 0.6$
Raw primary sludge = 0.6 x 18,800 = 11,200 lb/day
Assume raw primary sludge is 4% solids, the volume of sludge is

$$\frac{11{,}200}{0.04} \times \frac{1 \text{ gal}}{8.34 \text{ lb}} = 34{,}000 \text{ gal/day}$$

b. Waste Activated Sludge

BOD removed,

$$\begin{aligned} \Delta S &= hS_0 - ihS_0 \\ &= 0.7\,(250) - 0.1\,(0.7)\,(250) \\ &= 157 \text{ mg/l} \\ \text{or } 157 \times 8.34 \times 10 &= 13{,}100 \text{ lb/day} \end{aligned}$$

The net solids production,

$$\begin{aligned} \Delta X &= (\Delta S)\,Y \\ &= 13{,}100 \times 0.5 \\ &= 6{,}550 \text{ lb/day} \end{aligned}$$

Total waste activated sludge

$$\begin{aligned} &= (1-k)\,X_0 - X_f + \Delta X \\ &= (0.4 \times 18{,}800) - (20 \times 10 \times 8.34) + 6{,}550 \\ &= 15{,}720 \text{ lb/day} \end{aligned}$$

Assume waste activated sludge is 1% solids, the volume of sludge is

$$\frac{15{,}720}{0.01} \times \frac{1}{8.34} = 188{,}000 \text{ gal/day}$$

c. Mixed Digested Sludge

Total influent to digester = 11,200 + 15,720
= 26,920 lb/day
Mixed digested sludge = 0.8 x 26,920
= 21,536 lb/day

Obviously, it is possible to calculate the quantities of sludges produced in different types of treatment plants by inserting appropriate constants and values from real life into a generalized scheme as above. However, it must be emphasized that the designer should be very careful in using such a generalized system in the calculation of sludge quantities. Special circumstances and special wastes might make the use of such an approach inappropriate and even dangerous.

For example, some communities still use septic tanks, and the resulting sludge can significantly affect sludge production. A small amount of septic tank sludge can be easily assimilated into an existing wastewater treatment plant without any problems. In some locations, however, a large portion of residences might have septic tanks, and the quantity might even be great enough to require a separate treatment plant.

Septic tank sludge quantities are difficult to estimate as they vary with frequency of cleaning. Some cities require cleaning at least twice a year and the sludge quantities can thus be calculated.

According to Kolega (1971) septic tank sludge has a high chemical oxygen demand (COD) (26,000 mg/l) and a low BOD (4,800 mg/l). Total solids can vary from about 2 to 10%, with over half being volatile.

For purposes of comparison, Table 1-2 is a listing of some typical average sludge production rates as proposed by a number of authors. Again caution in the application of these numbers is urged. With increased contributions from industry, the sum of such average numbers for design purposes can lead to serious error.

Table 1-2. Typical Sludge Volumes Produced in Wastewater Treatment

	Gallons (or m^3) of Sludge Produced per Million Gallons (or million m^3) of Wastewater Treated		
	Raw Primary Sludge	Trickling Filter Humus	Waste Activated Sludge
Keefer (1940)	2,950	745	19,400
Fair and Imhoff (1965)	3,530	530	14,600
Babbitt (1953)	2,440	750	18,700
McCabe and Eckenfelder (1963)	3,000	700	19,400

Also not included in the above discussion is the effect on chemical sludges by the addition of chemical flocculants. The quantities produced will vary almost directly with the kinds and amounts of chemicals used for precipitation, and thus can be easily calculated from simple stoichiometry (see Chapter 9). Rough figures for design guidance can be obtained from actual operating experience in adding chemicals. Tables 1-3, 1-4 and 1-5 show the data obtained in one study of 13 treatment plants (Adrian and Smith, 1972).

Table 1-3. Additional Sludge to be Handled with Chemical Treatment Systems: Primary Treatment for Removal of Phosphorus (Adrian and Smith, 1972)

Sludge Production Parameter		Conventional Primary	Low Lime Addition to Primary Influent	High Lime Addition to Primary Influent	Aluminum Addition to Primary Influent	Iron Addition to Primary Influent
Level of Chemical Addition (mg/l)		0	350-500	800-1,600	13-22.7	25.8
Percent Sludge Solids	Mean	5.25	11.1	4.4	1.2	2.25
	Range	5.0-5.5	3.0-19.5	2.1-5.5	0.4-2.0	1.0-4.5
lb/mil gal	Mean	788	5,630	9,567	1,323	2,775
	Range	600-950	2,500-8,000	4,700-15,000	1,200-1,545	1,400-4,500
gal/mil gal	Mean	4,465	8,924	28,254	23,000	21,922
	Range	3,600-5,000	4,663-18,000	16,787-38,000	10,000-36,000	9,000-38,000

Table 1-4. Additional Sludge to be Handled with Chemical Treatment Systems: Phosphorus Removal by Mineral Addition to Aerator (Adrian and Smith, 1972)

Sludge Production Parameter		Al^{+++} Addition to Aerator		Fe^{+++} Addition to Aerator	
		Conventional Secondary	With Al^{+++} Addition	Conventional Secondary	With Fe^{+++} Addition
Level of Chemical Addition (mg/l)		0	9.4-23	0	10-30
Percent Sludge Solids	Mean	0.91	1.12	1.2	1.3
	Range	0.58-1.4	0.75-2.0	1.0-1.4	1.0-2.2
lb/mil gal	Mean	672	1,180	1,059	1,705
	Range	384-820	744-1,462	918-1,200	1,100-2,035
gal/mil gal	Mean	9,100	13,477	10,650	18,650
	Range	7,250-12,300	7,360-20,000	10,300-11,000	6,000-24,000

Table 1-5. Additional Sludge to be Handled with Chemical Treatment Systems: Phosphorus Removal by Mineral Addition to Secondary Effluent (Adrian and Smith, 1972)

Sludge Production Parameters		Lime Addition	Aluminum Addition	Iron Addition
Level of Chemical Addition (mg/l)	–	268-450	16	10-30
Percent Sludge Solids	Mean	1.1	2.0	0.29
	Range	0.6-17.2	–	–
lb/mil gal	Mean	4,650	2,000	507
	Range	3,100-6,800	–	175-781
gal/mil gal	Mean	53,400	12,000	22,066
	Range	50,000-63,000	–	6,000-36,000

Whenever possible, it is desirable to use pilot plant equipment to calculate the amount of sludge produced. Dry calculations, such as the generalized system presented, are a poor substitute for wet testing.

RATIONALE OF SLUDGE MANAGEMENT IN A WASTEWATER TREATMENT PLANT

A wastewater treatment plant exists for the purpose of producing an essentially clear effluent which can be discharged to the environment without adverse impact. The design engineer and plant operator therefore are mostly concerned with the liquid processing within the plant, and solids management becomes important only when the plant malfunctions. Often, these upsets can be avoided by following a few fairly simple guidelines for sludge management within a plant.

Stated concisely, these laws of sludge management are:

1. Don't hold sludges.
2. Don't mix sludges.
3. Don't recirculate sludges.

As is the case with any generality, these are not always applicable. They have been found to be true at enough plants, however, to warrant discussion.

Holding sludge (saving it, storing it, etc.) is usually the doing of a plant operator, and the procedure is often dictated by the vagaries of the labor force. It should nevertheless be the aim of an operator not to hold sludge before a subsequent operation such as dewatering. Although this principle seems to be especially applicable to raw primary sludges, the holding of digested sludge (and the resulting cooling) has been shown to cut the capacity of mechanical dewatering to one third of the original (Vermehen, 1978).

One clear outward sign of impending trouble is bubbling primary clarifiers, due to gas formation within the settled raw sludge. This will decrease the efficiency of solids removal, as well as place an extra load on the alkalinity of the anaerobic digesters. Sludge should be pumped from the primary clarifiers at sufficiently frequent intervals so as to avoid septicity.

The second axiom, "Don't mix it," is usually violated by design engineers. For example, experience seems to indicate that the discharge of excess waste activated sludge to the head of the plant results in deterioration of the primary clarifier performance, significantly increases the volume of sludge pumped, and necessitates more frequent pumping due to increased septicity problems.

It has been known for some time that when a raw primary sludge is mixed with waste activated sludge, the mixture assumes many of the undesirable characteristics of waste activated sludge. It is, for example, very difficult to dewater a 1:1 mixture of raw primary:waste activated on a vacuum filter.

It is, however, necessary to achieve the highest solids concentrations possible before a process such as anaerobic digestion, because the governing operational parameter in anaerobic digestion is solids retention time. One means of achieving this has been to blend the raw primary and waste activated sludges in a gravity thickener (Torpey, 1954). This has at times caused operational problems such as floating sludge and odors, mainly due to the oxygen-starved nature of the sludges (EPA, 1974). A more reasonable solution seems to be the gravitational thickening of primary sludges and the use of flotation thickeners for aerobic secondary sludges, with blending following the separate thickening operations (Figure 1-4). In plants where this type of scheme has been instituted, the solids dewatering has improved markedly and operational problems have been reduced.

Another example of improved operation by not mixing sludges is the scheme of using anaerobic digestion for raw primary sludge only, and employing aerobic digestion for the waste activated sludge. These stabilized sludges can then be blended and dewatered at a cost significantly lower than the alternative of mixing prior to stabilization (Cohen and Puntenny, 1973).

The final axiom, "Don't recirculate sludge," is probably the most difficult of the three laws to promote, since the recirculation of various sundry process streams to the head of the plant is almost a knee-jerk reaction for design engineers. Yet these secondary streams can impose a significant excess solids load on the process, as shown below.

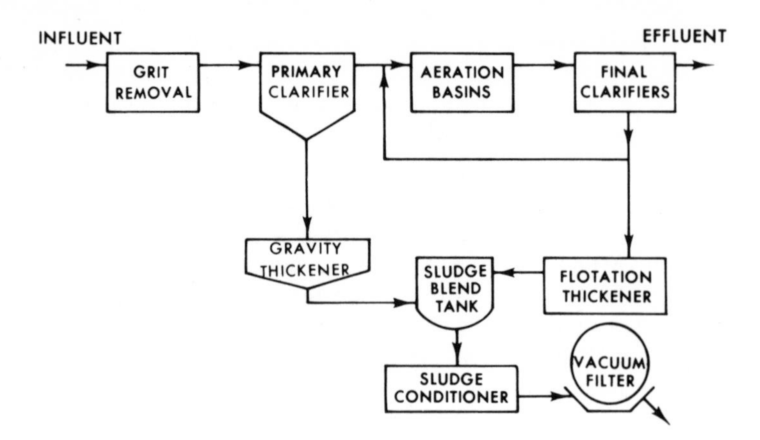

Figure 1-4. Suggested scheme for sludge management (EPA, 1974).

Example 1-2. Consider a 10-mgd plant, as described in Example 1-1, but with a thickener before the anaerobic digester. Assume the plant has a gravity thickener for the primary and waste activated sludges and a centrifuge to dewater the anaerobically digested sludge. The solids recovery of the thickener and centrifuge is 90% and 80%, respectively, and the supernatant from the secondary digester has a solids concentration of 2%, with a digester influent and effluent solids of 5% and 8%, respectively. What fraction of the total solids load on the treatment plant is produced by the return flows from the thickener, centrifuge and secondary digester?

As calculated in Example 1-1, the total solids flows are

Raw primary	11,200 lb/day at 4% solids
Waste activated	15,720 lb/day at 1% solids

a. Thickener overflow, with 90% solids capture, is

$$0.10\,(11{,}200 \text{ lb/day} + 15{,}720 \text{ lb/day}) = 2{,}692 \text{ lb/day}$$

b. The total solids flow into the anaerobic digester is thus

$$0.9\,(11{,}200 \text{ lb/day} + 15{,}720 \text{ lb/day}) = 24{,}228 \text{ lb/day}$$

Of this solids load, assume 20% is lost (gas plus dissolved solids) so that the digester effluent solids (supernatant + underflow) is 19,383 lb/day, while 4,846 lb/day is lost in the digestion. Mass balance on the digester is

$$\text{Inflow} = \text{Underflow} + \text{Supernatant} + \text{Loss in Digestion}$$
$$24{,}228 \text{ lb/day} = x \text{ lb/day} + y \text{ lb/day} + (-4{,}846 \text{ lb/day})$$

The volume balance is

$$\text{Inflow} = \text{Underflow} + \text{Supernatant}$$

$$\frac{24{,}228}{0.05} \text{ lb/day} = \frac{x}{0.08} \text{ lb/day} + \frac{y}{0.02} \text{ lb/day}$$

Solving the two simultaneous equations:

$$x = \text{solids in underflow} = 6{,}461 \text{ lb/day}$$
$$y = \text{solids in supernatant} = 8{,}076 \text{ lb/day}$$

c. The centrifuge operates at 80% solids capture on the underflow, so that the centrate contains

$$0.2\ (6{,}461 \text{ lb/day}) = 1{,}292 \text{ lb/day}$$

The total contribution of recycled solids is thus

$$2{,}692 \text{ lb/day} + 8{,}076 \text{ lb/day} + 1{,}292 \text{ lb/day} = 12{,}060 \text{ lb/day}.$$

Since the plant influent carries 18,800 lb/day, the recycled solids increase the total solids load on the plant by

$$\frac{12{,}060}{18{,}800} \times 100 = 64\%.$$

In addition to the total magnitude of the additional solids load on the plant, the *types* of solids recirculated are almost exclusively the fine, difficult to handle solids. More than one treatment plant operator has found himself inundated with these fine solids, at the eventual expense in treatment efficiency. The only solution in such cases is to clean the plant out. In the design and operation of a plant, therefore, one should not recirculate solids if at all possible. In the example above, the 2% solids supernatant clearly has an impact on operation, and the practice of supernatant withdrawal is of highly questionable value. Many good operators prefer to dewater all of the sludge, and to do this at a high rate of solids recovery, recognizing that this investment is highly profitable for future good plant operation.

CONCLUSIONS

The types and quantities of sludges produced in wastewater treatment can vary tremendously from plant to plant and even, from time to time, in the *same* plant. Sludge handling is responsible for about 30-40% of the capital cost of a treatment plant, and about 50% of the operating cost. Most treatment plant operators will agree, however, that in terms of the headaches and trouble caused, sludge handling is worth about 90% of the wastewater treatment system. As a result, the importance of sludge handling and disposal cannot be overestimated.

REFERENCES

Adrian, D., and R. Smith (1972). "Dewatering Physical-Chemical Sludges," *Applications of New Concepts of Physical Chemical Treatment* (New York: Pergamon Press, Inc.).

Babbitt, H. E. (1953). *Sewerage and Sewage Treatment*, 7th ed. (New York: John Wiley & Sons, Inc.).

Cohen, D. B., and J. L. Puntenny (1973). "Metro Denver's Experience in Large-Scale Aerobic Digestion of Waste Activated Sludge," Water Poll. Control Fed. Conference, Denver

EPA (1974). "Process Design Manual for Sludge Treatment and Disposal," U.S. EPA Technology Transfer, 625/1-74-006, Washington, D.C.

Fair, G. M., and K. Imhoff (1965). *Sewage Treatment*, 2nd ed. (New York: John Wiley & Sons, Inc.).

Keefer, C. E. (1940). *Sewage Treatment Works* (New York: McGraw-Hill Book Company).

Kolega, J. J. (1971). "Design Curves for Septage," *Water Sew. Works* 118:5.

Kormanik, R. A. (1972). "Estimating Solids Production for Sludge Handling," *Water Sew. Works* 119:12.

Lindstedt, K. D., E. R. Bennett and T. Pontenney (1971). "Aerobic Digestion for Waste Activated Sludge Solids Reduction," *Water Sew. Works* 118:6.

McCabe, J., and W. W. Eckenfelder (1963). *Advances in Biological Waste Treatment* (New York: Pergamon Press, Inc.).

National Academy of Sciences (1978). *Multimedium Management of Municipal Sludge*, Washington, D.C.

Torpey, W. N. (1954). "Concentration of Combined Primary and Activated Sludges in Separate Thickening Tanks," *Proc. ASCE,* 80, September, No. 443.

Vermehen, P. I. (1978). "Chemical Removal of Nutrient Salts from Plant Effluent," Sixth Nordic Symposium on Water Research, Copenhagen.

PROBLEMS

1. Using reasonable values, estimate the excess solids loading, both dissolved as well as suspended, on a 10-mgd activated sludge plant due to supernatant return (from anaerobic digesters).

2. Estimate the sludge production in a modified aeration activated sludge plant (no primary clarifier) treating a dairy waste which has the following influent characteristics: Flow = 3,000 m^3/day, SS = 10 mg/l, BOD = 4,000 mg/l. Assume effluent limits are 40 mg/l of BOD and SS.

3. A 300-mgd primary plant is to be upgraded to a secondary activated sludge plant. The incoming BOD is 140 mg/l on the average (heavy infiltration), and the influent SS are 200 mg/l. A common 20 mg/l effluent standard is imposed on both BOD and SS. Estimate the percent increase in raw sludge produced, both in weight and volume.

CHAPTER 2

SLUDGE CHARACTERISTICS

The characteristics of sludges produced in wastewater treatment can vary greatly, due to the tremendous difference in types of wastewaters, and in the design and operation of wastewater treatment plants. It is important to be able to measure some of the characteristics that can be used in design and operation. Because of the complex nature of sludge, basic parameters are of only limited value, and it has been necessary to develop some operational parameters—a number are discussed in this chapter.

PHYSICAL CHARACTERISTICS

Specific Gravity

Specific gravity is defined as the ratio of the weight of the material to that of an equal volume of water. Most sludges have a specific gravity of almost 1.0. In other words, they are almost equal to the weight of water.

If one liter of sludge weighs 1,010 g, the specific gravity is calculated as:

$$S = \text{Sludge Specific Gravity} = \frac{1{,}010 \text{ g}}{1{,}000 \text{ g}} = 1.01$$

Sludges are seldom composed of only one solid and only one liquid; thus the specific gravity must be calculated as

$$\frac{1}{S} = \sum_{i=1}^{n} \left(\frac{W_i}{S_i}\right)$$

where S = specific gravity of the sludge
W_i = weight fraction of the *ith* component in the sludge
S_i = specific gravity of the *ith* component

Example 2-1. A sludge contains the following components:

	% by weight
Volatile Solids (S = 1.0)	5
Fixed Solids (S = 2.5)	5
Water (S = 1.0)	90

The sludge specific gravity is calculated as

$$\frac{1}{S} = \frac{0.05}{1.0} + \frac{0.05}{2.5} + \frac{0.90}{1.0}$$

$$S = 1.03$$

Reported values of mixed sludge dry solids' specific gravities range from 1.4 to 2.1, while the overall sludge specific gravities range from 1.0032 to 1.054 (Campbell *et al.*, 1975). Dick (1972) found the specific gravity of dry activated sludge solids to be 1.08.

SOLIDS CONCENTRATION

The relative solid and liquid fractions of a slurry are most commonly described by the *solids concentration*, expressed as mg/l or percent solids. If we assume that the slurry has a specific gravity of 1.0,

$$10{,}000 \frac{\text{mg}}{\text{l}} = 1\% \text{ solids}$$

and the percentage is expressed in terms of weight/weight. Sludges composed of heavy solids (*e.g.*, lime or iron flocs) may have higher specific gravities, hence the mg/l to percent relationship no longer would be valid.

Total solids in a sludge equals the sum of dissolved and suspended solids, the latter defined as that fraction that does not pass through a fiber filter. Each of these three types of solids (total, dissolved, suspended), can be further subdivided into fixed and volatile, the latter defined as the fraction that oxidizes at 600°C. It is commonly assumed that the volatile solids measurement is a reasonable approximation of the organic material in the sludge.

In this text, the following notations are used:

TS = total solids
DS = dissolved solids
SS = suspended solids*

---

*Some authors use the term Total Suspended Solids (TSS). This is unnecessary, can lead to misunderstandings, and should therefore be avoided.

VSS = volatile suspended solids
FSS = fixed suspended solids

The common method of measuring solids is by evaporation. For example, according to standard procedure, total solids is measured by pouring exactly 100 ml into an evaporating dish, driving off the moisture in a 103°C oven, and weighing the dish and dry matter. Subtraction of the tare weight yields the solids, and the concentration is calculated in mg/l.

Unfortunately, for concentrated slurries and especially for industrial sludges, the mg/l thus calculated cannot be translated to percent, the common way of reporting solids in sludge.

A better measuring method–that should be used whenever possible–is to place a reasonable quantity of sludge into a preweighed evaporating dish and weigh the dish plus the wet sludge. Following evaporation of the moisture, the dish plus the dry sludge is weighed. The difference equals the moisture driven off, and the solids concentration thus calculated is in terms of *percent solids*, expressed as weight/weight (wt/wt).

Incidentally, some authors express concentration as *percent moisture*. Obviously,

$$\% \text{ Solids} = 100 - \% \text{ Moisture}$$

Wastewater treatment engineers will recognize that a 1% sludge is a significant solids concentration, even though such a slurry might have a specific gravity of only 1.01. Many chemical sludges are much heavier of course, while many biological sludges have specific gravities of much less than 1.01.

Although the specific gravity of most wastewater sludges closely approximates that of water, it is nevertheless true (and convenient) that most are greater than 1.0 and thus will settle.

Settling

Sludges often can be characterized by how well they settle. Settling tests are commonly conducted in 1-liter cylinders. First the sludge is mixed to distribute the solids evenly. As the solids settle, a clear interface will emerge between the solids and the liquid, and the height of this interface can be recorded over time.

In such tests, it is assumed that the settling rate is independent of all test conditions, and that the settling velocity of a specific sludge is an inverse function of the sludge solids concentration only. In other words, if the concentration of solids is greater the sludge is likely to settle at a slower rate. Conversely, if the sludge is dilute, the settling will be rapid. This is quite easily understood if one considers that the solids can move to the bottom of the cylinder only if they are able to displace an equal volume of water to the top of the cylinder. This water must work its way through the sludge bed and reaches the top much more easily if a great many void spaces exist between the sludge solids; in other words, the sludge is dilute in its solids concentration. Thus a dilute sludge settles faster, and a more concentrated sludge settles slower.

Sludge settleability is an important parameter especially in the operation of an activated sludge plant. Because of the need for a simple and inexpensive measure of sludge settleability in operational plant control, the Sludge Volume Index (SVI) has been widely accepted (Mohlman, 1934). The SVI is measured by allowing sludge to settle in a liter cylinder for 30 min. Sludge depth is measured in ml and divided by the sludge suspended solids concentration in mg/l. Thus the SVI is calculated as

$$\text{SVI} = \frac{\text{ml sludge} \times 1{,}000}{\text{mg/l suspended solids}}$$

and can also be defined as the volume of sludge in ml occupied by a gram of solids.

Example 2-2. If the initial concentration of solids in the sludge is 2,000 mg/l and after 30 min the sludge height is 200 ml, the SVI is calculated as

$$\frac{200 \times 1{,}000}{2{,}000} = 100 \text{ ml/g}$$

In activated sludge plants, a sludge with an SVI less than 100 is considered a very well-settling sludge, whereas one with an SVI greater than 100 is often troublesome.

Although SVI was developed as an operational tool for plant control, it has acquired a much broader meaning than was originally intended. Its use in research is especially questionable (Dick and Vesilind, 1969), as discussed in Chapter 5.

Particle Size

Particles in sludges vary not only in size but also in consistency and shape. Thus it is extremely difficult to characterize sludges by particle size, although this has been attempted by several researchers. Not only is sludge composed of many different-sized particles, but these sizes change with time and test conditions. It is thus not proper to measure the size of such particles unless the type of test and immediate history of the sludge are also defined. The process of sludge agglomeration is shown in the before and after photographs of Figure 2-1 which dramatically illustrates the change in sludge texture with time. The two photographs show the same transparent settling cylinder—one at the start of a test and one about 5 minutes later. Particle size thus varies with time making measurement quite difficult.

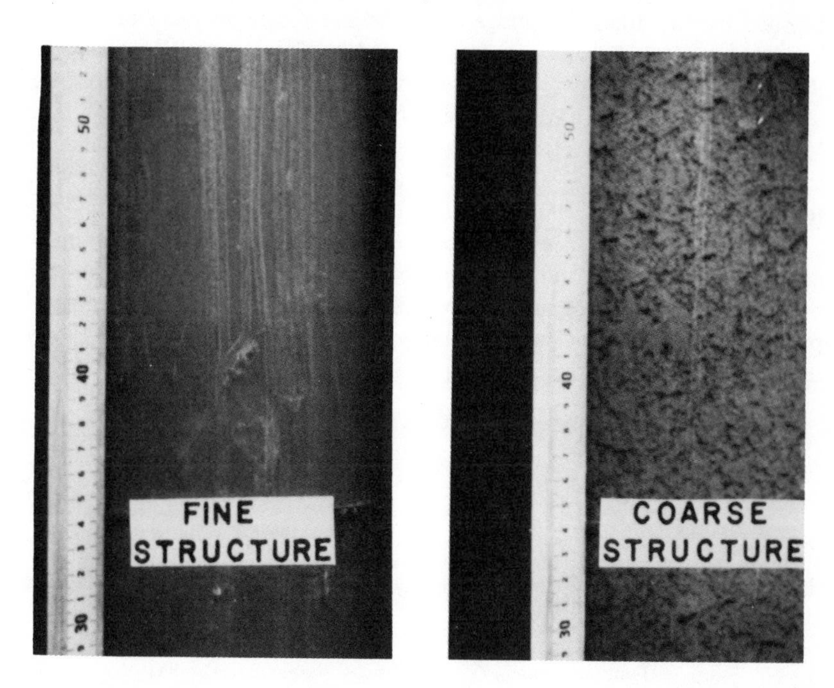

Figure 2-1. Sludge texture before and after agglomeration has occurred (courtesy of Richard F. Cole).

Recognizing this difficulty, Javaheri and Dick (1969) proposed an *aggregate volume index*, defined as the ratio between the volumetric aggregate concentration and the volumetric solids concentration. Aggregates contain a great deal of water that acts as part of the particle during settling. They proposed that this water would be squeezed out during compression, and that the amount of squeezing is a characteristic of the sludge.

The size of sludge aggregates (or flocs) is of major interest in thickening and dewatering operations. The proportion of water and solids in a floc, if it can be measured, can be of immediate use in sludge characterization. Accordingly, Thomas (1964) proposed the use of the *alpha value* of a floc as

$$\alpha = \frac{\text{water volume}}{\text{sludge solids volume}}$$

Although this concept has merit, it is hampered by its difficult (and very questionable) method of measurement. One requirement is the visual measurement of floc volume.

Schroeper *et al.* (1955) used photographs to calculate floc volume, while Finstein (1965) dried the sludge on slides and projected them on a screen for measurement. Alba *et al.* (1965) used the Stokes settling technique. Randall *et al.* (1973) employed a series of micron-sized filters and weighed the dry residue. This, however, gave only a percentage distribution by mass. Mueller *et al.* (1966), after trying several methods, concluded that "it is extremely difficult to obtain a meaningful floc dimension by visual techniques."

Nevertheless, particle size has a significant effect on sludge treatment, especially dewatering, and the value of particle size analysis can be demonstrated. For example, in a thorough investigation of the influence of particle size on sludge dewaterability, Karr (1976) fractionated several different sludges using the following scheme:

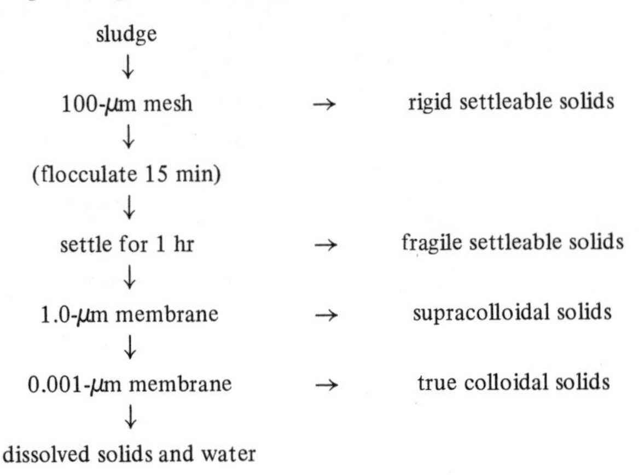

Capillary Suction Time (CST) and specific resistance to filtration tests* were run on sludge samples with spiked particle size distributions. It was found that the greatest effect on dewatering seems to be by the supracolloidal solids (solids which do not settle in 1 hr, but are removed in a 1.0-μm membrane filter). Table 2-1, for example, shows three sludges with roughly equal total and suspended solids concentrations, but quite different particle size distributions and dewatering characteristics. It is postulated that the adverse effect of these solids is due to their blinding the filter medium during filtration.

---

*This test is discussed fully in Chapter 6. For now, it is necessary to recognize that a sludge with a high specific resistance to filtration is difficult to dewater. Low specific resistance values are therefore desirable. Conversely, a low CST suggests that the sludge can probably be readily dewatered.

Table 2-1. Characteristics of Three Sludges: Particle Sizes and Dewaterability (Karr, 1976)

	Type of Sludge		
	Raw Primary Sludge	Activated Sludge	Mixed Digested (Anaerobic) Sludge
Specific Resistance (m/kg)	2.1×10^{14}	4.8×10^{13}	9.3×10^{14}
CST (sec)	17	14	144
Total Solids (mg/l)	9698	8841	10,266
Rigid Settleable (percent of total)	6452	1920	3374
Fragile Settleable (percent of total)	2320	6587	4054
Supracolloidal (percent of total)	355	84	1997
True Colloidal (percent of total)	45	7	301
Dissolved (percent of total)	526	243	540

The effect of polymer conditioning is to increase the particle size, as shown by typical data obtained by a Coulter Counter (Campbell *et al.*, 1975) shown in Figure 2-2.

Distribution of Water

Related to the above approach, it is possible to characterize sludge according to distribution of water. First attempted by Heukelekian and Weisberg (1956), they suggested that water in sludge is either in the free state with the characteristics of water, or is "bound" to the solids and therefore has completely different characteristics. They noted that one such difference is that bound water will not freeze at 0°C. It is possible to measure the bound water in a sludge by measuring volume changes with accurate (petroleum-filled) pipettes, and this has been related to sludge stability and other characteristics. Unfortunately, this is a difficult measurement technique and has thus found little practical use.

Another approach is to consider water in sludge in four categories:

1. *Free water* which is not attached to sludge solids in any way and can be removed by simple gravitational settling.
2. *Floc water* which is trapped within the flocs and travels with them. Its removal is possible by mechanical dewatering.
3. *Capillary water* which adheres to the individual particles and can be squeezed out only if these individual particles are forced out of shape and compacted.
4. *Bound water* which is chemically bound to the individual particles.

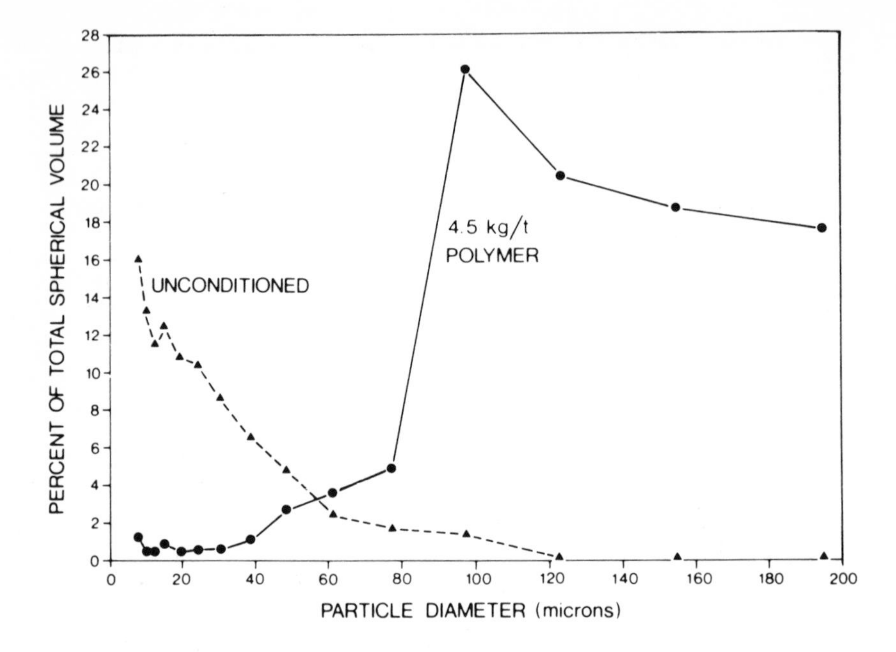

Figure 2-2. Effect of polymer addition on particle size distribution (Campbell *et al.*, 1975).

Such an hypothesis is supported by high-speed centrifuge test-tube data for activated sludge. A typical curve is shown in Figure 2-3. The free water was defined by a settling test. The volume of sludge after settling in a 1-liter cylinder in this case was 260 ml, defining the free water as (1,000 - 260) = 740 or 74%. The volume occupied by the sludge at higher centrifuge speeds levels off to a constant volume. This is thought to be due to the strength of the particle lattice resisting further compaction, and defines the floc water that was driven out. The capillary water was driven out only at very high centrifugal speeds, as the individual particles were compacted and rearranged.

The bound water consists of highly oriented water molecules and will travel through the bulk solution in association with whatever particle it is bound to. The layers of oriented water molecules on all hydrophilic surfaces such as polysaccharides and proteins may be about 10 to 20Å thick (Dugan, 1972).

For a typical activated sludge, the water is distributed as shown in Table 2-2. This method of classifying water in a sludge might eventually be used to evaluate the applicability of a specific thickening or dewatering operation to a sludge.

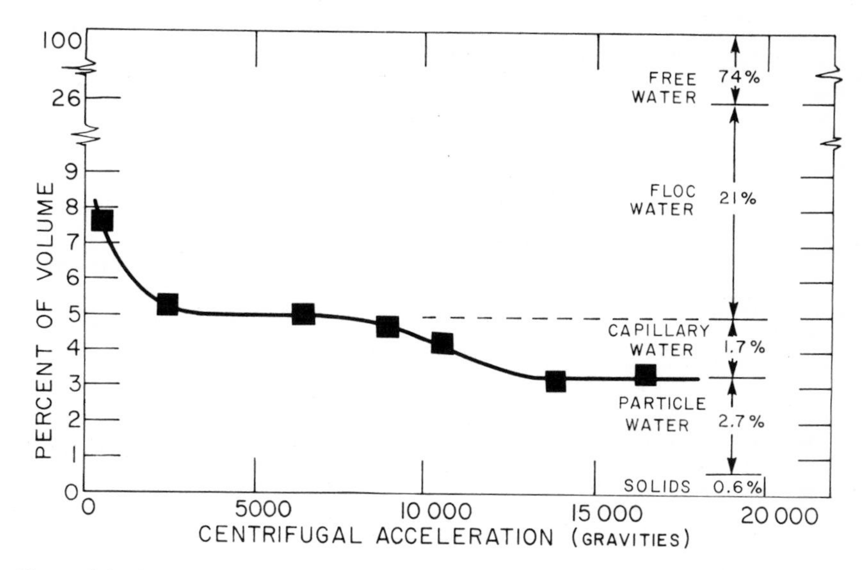

Figure 2-3. Classification of water in a waste activated sludge by gravitational and centrifugal settling.

Table 2-2. Distribution of Water in an Activated Sludge

	% Volume
Free Water	75
Floc Water	20
Capillary Water	2
Bound Water	2.5
Solids	0.5
Total	100%

Although the nomenclature is slightly different, Bjorkman (1969) has found that for a mixed digested sludge the water distribution is as shown in Table 2-3.

Rheology

All fluids can be classified in terms of their flow properties, or their rheology. One common rheological measurement is viscosity, classically defined as the rate of displacement of a fluid with a given shear force. Consider Figure 2-4 and note that the fluid is between a fixed lower boundary, and the upper boundary which moves at some velocity, u. This velocity

Table 2-3. Distribution of Water in a Mixed Digested Sludge (Bjorkman, 1969)

	Percent Water
Between Cells Water	70
Adhesion and Capillary Water	22
Adsorption and Intracellular Water	8
Total	100

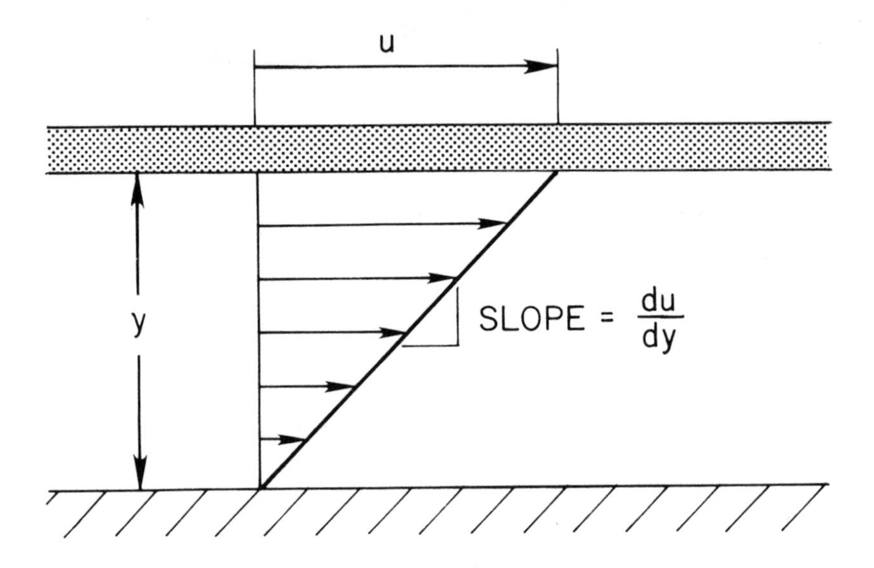

Figure 2-4. Velocity profile in a fluid with an imposed shear stress.

is the result of a force τ imposed upon the upper boundary. The velocity profile thus formed is linear for many fluids, with a slope of du/dy. This slope relates the shear force and the displacement as

$$\tau \propto \frac{du}{dy} \text{ or } \tau = \mu \frac{du}{dy}$$

where μ is the proportionality constant called the viscosity.

Most fluids can be classifed as Newtonian fluids which behave like the fluid above—that is, the shear stress is proportional to the displacement. However, many fluids do not behave like this. For example (Figure 2-5), the *plastic* fluid requires a certain finite shear before it can be displaced. This shear, called *yield stress*, is another rheological property of that particular type of fluid. The shear rate-shear stress relationship for a plastic

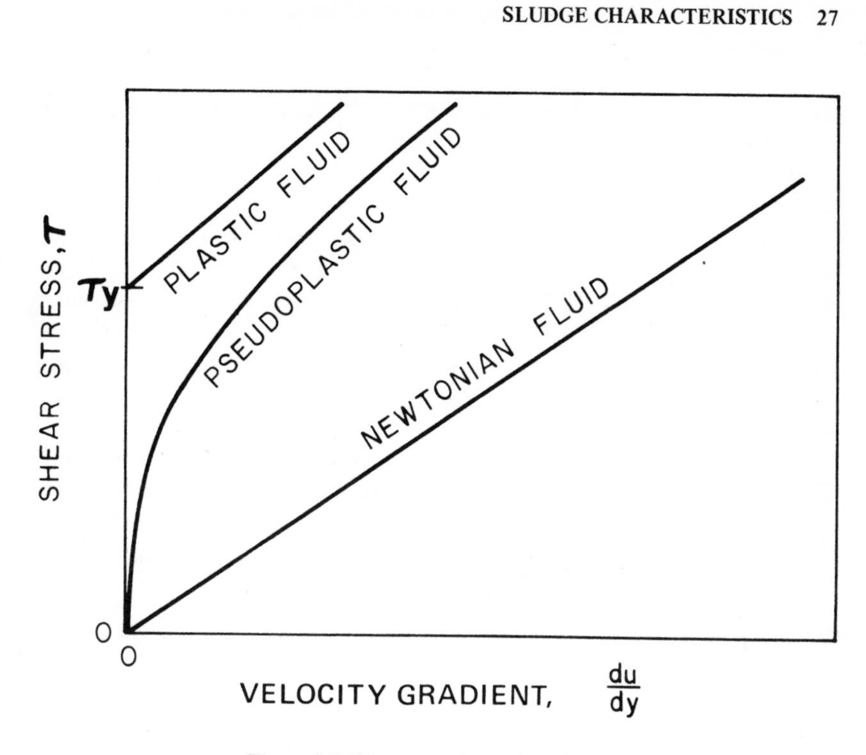

Figure 2-5. Rheograms for various fluids.

fluid may be expressed as

$$\tau = \tau_y + \eta \frac{du}{dy}$$

where τ_y = yield stress
η = plastic viscosity

Most wastewater sludges are neither Newtonian nor plastic but fall somewhere in between. These fluids are called *pseudo-plastic* fluids and can be approximated by an equation known as the power law,

$$\tau = \eta \left(\frac{du}{dy}\right)^n$$

where η = plastic viscosity
n = a constant, known as the flow behavior index

Note that Newtonian flow can be considered merely a special case of the power law, where n = 1 and $\eta = \mu$.

The constants in the power law equation can be determined by plotting the shear stress divided by the shear rate versus the shear rate, on log-log coordinates. The slope of this line would be n - 1, and the intercept at du/dy = 1 would be read as the apparent viscosity, η.

Sludge has another rather annoying characteristic from the rheological standpoint; it is thixotropic, meaning that its viscosity decreases with the duration of the applied stress. This occurs most probably due to a destruction of floc structures. When the stress is removed, the floc structure reforms, increasing the viscosity. This reversibility gives rise to a hystersis curve as shown in Figure 2-6, which can be used to measure the extent of thixotropy. This curve is developed by starting at zero shear and zero displacement, (zero speed on the rotating viscometer) and increasing the speed slowly, thus transcribing the upper curve. If the test is then continued from the highest back to the lowest speed, the lower line might be transcribed. This hysteresis loop is typical for sludges and indicates that the characteristic of the sludge was changed during the viscosity determination. Thus one cannot readily denote the rheological characteristic of the sludge without first also specifying the immediate prior history of the sludge.

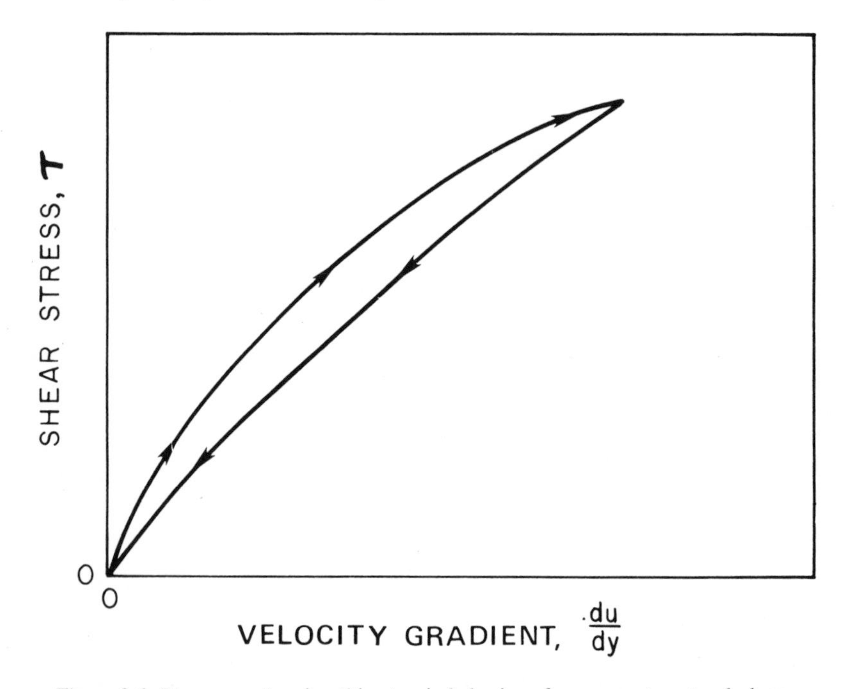

Figure 2-6. Rheogram showing thixotropic behavior of many wastewater sludges.

Measurement of viscosity and rheological properties is rather difficult for many reasons. Dick and Ewing (1967) started with a spindle viscometer to estimate sludge rheological properties. It was determined that the standard spindles did not give the required reproducibility and shear, and also that sludge solids tended to move away from the spindle. As a result they could only measure the viscosity of the water. The spindle was replaced with a roughened drum and, to rectify the problem of particle movement outward, the outside container was made to rotate instead of the inside spindle. In addition to all these problems, sludge tends to settle during such tests, and they found it necessary to place the whole contraption on a shaking tray to prevent settlement of solids. It should be clear that present methods of sludge viscosity measurement are inadequate, and better methods are required.

Sludge rheology varies widely. Rheological data of some sludges are described in Table 2-4. Obviously, the sludge temperature and solids concentration will greatly influence the rheological properties.

Table 2-4. Typical Sludge Rheological Properties

Sludge	Temp. (°C)	Solids Concentration (%)	Yield Strength (dynes/cm^2)	Plastic Viscosity (g/cm sec)	Source
Water	20	0	0	0.01	–
Raw Primary	12	6.7	43	0.28	Babbit & Caldwell (1939)
Mixed Digested	17	10	15	0.92	Merkel (1934)
Activated A	20	0.4	0.1	0.06	Dick & Ewing (1967)
Activated B	20	0.4	0.07	0.05	Dick & Ewing (1967)
Activated C	20	0.2	0.2	0.07	Dick & Ewing (1967)

It is interesting that there are very little rheological data available in the literature, and that what does exist shows a wide variation in rheological properties among sludges. Thus it seems possible that rheological properties could be used increasingly in the future in both plant control and research.

Atterberg Limits

Soils engineers use Atterberg limits to define the behavior of a soil at various moisture levels. The two limits are the plastic limit at which the material exhibits plastic behavior and the liquid limit at which the material acts like a liquid. A third value, the plasticity index, is defined as the difference between the liquid and plastic limits, expressed as a percentage.

The plastic limit is the moisture content at which the sludge begins to crumble when rolled into a 1/8-in. thread under the palm of the hand. The test is performed on progressively drier samples until the thread begins to break. The moisture content (percent moisture) at that point is the plastic limit.

The liquid limit is determined by placing a sample of sludge on a plate, cutting a 1/2-in. groove into the sludge, and observing how the sludge flows together under the influence of hammer blows using a standard device. The moisture content at which the sludge is sufficiently dry to no longer flow together under the impact of 25 blows is defined as the liquid limit.

Both of these tests, standardized by ASTM as D424-59 and D423-66, are useful if the sludge is expected to have structural properties. Other than for fixed industrial sludge, this possibility is fairly remote.

Fuel Value

Many wastewater sludges contain high concentrations of organic material, and thus have a fuel value. Dry sludge has a fuel value of about 5,500 cal/g (10,000 Btu/lb) of dry volatile solids. This compares favorably with coal, for example, which has a fuel value of about 7,700 cal/g (14,000 Btu/lb). Unfortunately, the sludges are wet and only partially volatile—reducing the fuel value to only about 550 cal/g (1,000 Btu/lb) of sludge. Usually auxiliary fuel (*e.g.*, oil) is required in sludge combustion.

The fuel value of sludge is discussed further in Chapter 8.

Fertilizer Value

Most fertilizers require the presence of nitrogen, phosphorus and potassium. A common fertilizer is 8-8-8, which has 8% nitrogen, 8% phosphorus (as P_2O_5) and 8% potassium (as K_2O). Sludges from wastewater treatment plants seldom contain such high percentages of these nutrients, as shown by some typical values in Table 2-5. Part of the problem is that much of the nitrogen is lost as ammonia in the digester supernatant, and much of the phosphorus is resolubilized in the digester. Almost all sludges are low in potassium.

Another concern of using sludges as fertilizer is the presence of unwanted materials such as heavy metals and chlorinated hydrocarbons. Table 2-6 lists some typical values for some of these materials. It is difficult to say what standards might be used to estimate whether a sludge is fit for fertilizer use based on such heavy metals and other unwanted materials. However,

Table 2-5. Typical Nutrient Concentration of Domestic Wastewater Sludges

Sludge	% Nitrogen (N)	% Phosphorus (P_2O_5)	% Potassium (K_2O)	Source
Raw Primary	2.4	1.1	–	Burd (1968)
Raw Primary	2.9	1.6	–	Burd (1969)
Raw Primary	3.0	1.6	0.4	Metcalf & Eddy (1972)
Trickling Filter	2.9	2.8	–	Anderson (1959)
Trickling Filter	3.0	3.0	0.5	Metcalf & Eddy (1972)
Activated	5.6	7.0	0.56	Anderson (1959)
Activated	3.5	2.8	–	Burd (1968)
Activated	3.0	3.6	–	Burd (1968)
Activated	5.6	5.7	0.4	Metcalf & Eddy (1972)
Mixed Digested	5.9	3.5	–	Burd (1968)
Mixed Digested	2.0	1.4	0.14	Burd (1968)
Mixed Digested	2.5	1.2	0.2	Burd (1968)
Mixed Digested	4.6	1.4	0.38	Jansson (1972)
Mixed Digested	1.8	3.5	0.18	Lynam *et al.* (1972)
Mixed Digested	2.5	3.3	0.40	Anderson (1959)
Mixed Digested	3.7	1.7	0.4	Metcalf & Eddy (1972)

Table 2-6. Concentrations of Various Chemicals in Sewage Sludge (Sommers, 1977)

Component	Sample		Range	Median	Mean
	Type[a]	Number		(%)	
K	Anaerobic	86	0.02-2.64	0.30	0.52
	Aerobic	37	0.08-1.10	0.38	0.46
	Other	69	0.02-0.87	0.17	0.20
	All	192	0.02-2.64	0.30	0.40
Na	Anaerobic	73	0.01-2.19	0.73	0.70
	Aerobic	36	0.03-3.07	0.77	1.11
	Other	67	0.01-0.96	0.11	0.13
	All	176	0.01-3.07	0.24	0.57
Ca	Anaerobic	87	1.9-20.0	4.9	5.8
	Aerobic	37	0.6-13.5	3.0	3.3
	Other	69	0.1-25.0	3.4	4.6
	All	193	0.1-25.0	3.9	4.9
Mg	Anaerobic	87	0.03-1.92	0.48	0.58
	Aerobic	37	0.03-1.10	0.41	0.52
	Other	65	0.03-1.97	0.43	0.50
	All	189	0.03-1.97	0.45	0.54

Table 2-6. Continued

	Sample		Range	Median	Mean
Component	Type[a]	Number		(%)	
Ba	Anaerobic	27	<0.01-0.90	0.05	0.08
	Aerobic	10	<0.01-0.03	0.02	0.02
	Other	23	<0.01-0.44	<0.01	0.04
	All	60	<0.01-0.90	0.02	0.06
Fe	Anaerobic	96	0.1-15.3	1.2	1.6
	Aerobic	38	0.1-4.0	1.0	1.1
	Other	31	<0.1-4.2	0.1	0.8
	All	165	<0.1-15.3	1.1	1.3
Al	Anaerobic	73	0.1-13.5	0.5	1.7
	Aerobic	37	0.1-2.3	0.4	0.7
	Other	23	0.1-2.6	0.1	0.3
	All	133	0.1-13.5	0.4	1.2
	Sample		Range	Median	Mean
Component	Type[a]	Number		(mg/kg)	
Mn	Anaerobic	81	58-7,100	280	400
	Aerobic	38	55-1,120	340	420
	Other	24	18-1,840	118	250
	All	143	18-7,100	260	380
B	Anaerobic	62	12-760	36	97
	Aerobic	29	17-74	33	40
	Other	18	4-700	16	69
	All	109	4-760	33	77
As	Anaerobic	3	10-230	116	119
	Aerobic	–	–	–	–
	Other	7	6-18	9	11
	All	10	6-230	10	43
Co	Anaerobic	4	3-18	7.0	8.8
	Aerobic	–	–	–	–
	Other	9	1-11	4.0	4.3
	All	13	1-18	4.0	5.3
Mo	Anaerobic	9	24-30	30	29
	Aerobic	3	30-30	30	30
	Other	17	5-39	30	27
	All	29	5-39	30	28
Hg	Anaerobic	35	0.5-10,600	5	1,100
	Aerobic	20	1.0-22	5	7
	Other	23	2.0-5,300	3	810
	All	78	0.2-10,600	5	733

Table 2.6. Continued

Metal	Sludge No.	Range	Median	Mean
		(mg/kg)[b]		
Cd	1	109-372	170	210
	2	4-39	15	19
	3	483-1,177	806	846
	4	3-150	40	53
	5	24-756	663	503
	6	12-163	12	43
	7	22-256	154	136
	8	11-32	11	16
Cu	1	4,083-7,174	6,525	6,079
	2	5,741-11,875	8,386	8,381
	3	2,081-3,510	2,390	2,594
	4	452-802	683	662
	5	391-6,973	476	1,747
	6	300-1,800	682	778
	7	422-1,392	894	871
	8	979-1,475	1,144	1,154
Ni	1	1,932-4,016	3,543	3,184
	2	663-1,351	1,053	1,015
	3	468-812	651	649
	4	75-219	95	119
	5	40-797	86	252
	6	46-92	88	81
	7	47-547	367	349
	8	65-93	79	80

[a] "Other" includes lagooned, primary, tertiary and unspecified sludges. "All" signifies data for all types of sludges.
[b] Oven-dry basis.

if sludges contain concentrations much greater than those shown in the table, some concern should be voiced in their use as fertilizers. This problem is discussed further in Chapter 11.

Food Value

Sludge can sometimes be used as a food source, such as adding dried activated sludge to cattle feed. Rudolfs (1940) reported that domestic sludges contain quite high concentrations of protein.

Experiments are also continuing in which yeast is grown on wastes, then recovered as a protein. Church *et al.* (1972) produced 0.5 g of *Trichoderma viride* mycelium containing 54% protein per g of BOD removed. The product was found comparable to soy meal.

Dean and Bouthilet (1972) have developed a technique for producing animal feed from sludge. Sulfur dioxide is used to solubilize sludge, and the filtrate is evaporated to produce a molasses-like material containing 20% protein.

Obviously, sludges do have food value and may be used more extensively in the future. At present, however, information characterizing sludges by the food value is quite scarce.

Electrical Charge of Solid Particles

The net electrical charge on particles is best measured as the "zeta potential." If the charge on the particles is large, this tends to inhibit flocculation, and these sludges tend to be difficult to dewater. Typically, municipal sludges exhibit zeta potentials in the range of -10 to -20 millivolts, and there does not seem to be much difference in the various types of municipal plant sludges. Some industrial sludges, on the other hand, have been found to have zeta potentials exceeding -80 millivolts (Campbell *et al.*, 1975).

BIOLOGICAL CHARACTERISTICS

The two main biological characteristics of interest are taxonomy (the classification of organisms) and the presence of pathogenic organisms.

Raw primary sludge has a mixed taxonomy because it contains a tremendous variety of organisms both from the intestinal tract and other sources. Activated sludge has a taxonomy dependent entirely on the feed and environmental conditions. It is almost impossible to classify all organisms responsible for the biological reactions in activated sludge. In fact, van Gils (1964) stated that during a study on the bacteriology of activated sludge, "identifying all of the bacteria isolated from the activated sludge seemed an almost impossible task."

Not only are we dealing with bacteria in activated sludge, but also molds, yeasts, protozoa, crustaceans, and even rotifiers are often present in activated sludge tanks. Further, it is not only a complex ecology, but also a dynamic one (Cassell, 1964), because the types of organisms are changing continually even when the substrate is constant.

For mixed digested sludges, the taxonomy is of course determined entirely by the type of anaerobic or aerobic reaction involved. Both are reviewed further in Chapter 5.

Raw primary sludge has a high concentration of pathogenic organisms. In fact, raw primary sludge has been known to concentrate the pathogens in sewage (Kabler, 1959).

Activated sludge also contains pathogenic organisms, as demonstrated by van Gils (1964).

Anaerobic digestion reduces the number of, but does not totally destroy, pathogenic organisms. Pathogens which survive normal digestion times of 30 days are *Salmonella typhosa* (McKinney *et al.*, 1958), *Entamoeba histolitica, Ascaris* eggs and helminth cysts (Peterson *et al.*, 1973; McWhorter, 1974). Most enteric (pathogenic) viruses are inactivated during digestion, but positive coliphage tests indicate the strong possibility of the persistence of enteric viruses in digested sludge (Maline *et al.*, 1975).

Table 2-7 is a listing of some typical values for pathogen concentrations in various sludges. Coliforms, also listed in this table, are of course normally not pathogenic but are indicators of pathogens. Similarly, enteric viruses are estimated based on plaque-forming units (PFU) which are actually coliphages (viruses that attack coliforms) but are themselves not pathogenic to humans (Hanson, 1972).

Table 2-7. Pathogens in Sludge

Sludge	Virus (PFU/100 ml)	Coliform (10^6/ml)	*Salmonella* (per 100 ml)	*Pseudomonas* (per 100 ml)
Raw Primary	7.9	11.0-11.4	460	46,000
Activated	–	2-2.8	74-23,000	1,100-24,000
Trickling Filter	–	11.5	93	11,000
Mixed Digested	0.85	0.4	29	34
Source	Palfi (1973)	Farrell (1974)	Farrell (1974)	Farrell (1974)

The fate of pathogens in sludge is discussed further in Chapters 3 and 11.

BIOCHEMICAL CHARACTERISTICS

ATP (Adenosine Triphosphate)

ATP is a specific indicator of cell viability because it exists only in living cells. The quantity of ATP can thus be used as an indicator of life in activated or any other kind of sludge (Ford *et al.*, 1966). The difficulty of using ATP as an indicator of biological activity is that, although accurate, the test is expensive and complicated to run.

DNA (Dioxyribonucleic Acid)

DNA is a normal constituent of all cells and remains fairly stable with age. Domestic biological sludge contains about 3% DNA. Although DNA appears to be a valuable parameter, it is again difficult to measure.

Enzymes

There seems to be some correlation between the activity of the dehydrogenase enzymes and sludge drainage characteristics (Randall *et al.*, 1971). Such a relationship suggests that it might be possible to add a certain enzyme to a sludge and thus make it easier to dewater or, alternatively, make it more stable.

CONCLUSIONS

It should be clear that characterization of any sludge is not simple. The identifying characteristics of interest must be specified relative to the problem or method of disposal. For example, if a sludge is to be thickened in a gravitational thickener, the characteristics of interest might be those of settling; whereas, if the sludge is to be utilized as fertilizer, its characteristics of interest might be those of nutrient concentrations.

It is also necessary to mention that not all known ways of characterizing sludges have been discussed above. New methods are being developed for sludge handling and disposal, and it will be necessary to continue the development of more sophisticated and useful tools for sludge characterization.

REFERENCES

Alba, S., A. E. Humphrey and N. F. Millis (1965). "Biochemical Engineering," U. of Tokyo Press (from Mueller *et al.*, 1966).

Anderson, M.S. (1959). "Fertilizing Characteristics of Sewage Sludge," *Sew. Ind. Wastes* 31:6.

Babbit, H. E., and D. H. Caldwell (1939). "Laminar Flow of Sludges in Pipes with Special Reference to Sewage Sludge," Univ. Ill. Eng. Exp. Sta. Bull. 319.

Berggren, B., and S. Oden (1971). "Analycretultat Rörande Fundmetaller och Klorerade Kolväten i Rötslam Fråm Svenska Reningsverk 1968-1971," Institutionen for Marketenskab, Lantbrukshögskolan, Uppsala, Sweden.

Bjorkman, A. (1969). "Heat Processing of Sewage Sludge," *Pro. Congress of the IRGR*, Basle.

Burd, R. S. (1968). "A Study of Sludge Handling and Disposal," FWPCA (EPA) Pub. WP-20-4, Washington, D.C.

Campbell, H. W., R. J. Rush and R. Tew (1975). "Sludge Dewatering Design Manual," Wastewater Technology Centre, Environmental Protection Service, Burlington, Ontario.

Cassell, E. A., F. T. Sulzer and J. C. Lamb (1966). "Population Dynamics and Selection in Continuous Mixed Culture," *J. Water Poll. Control Fed.* 38:9.

Church, B. D., E. E. Erickson and W. Brosz (1972). "It's Go, Go, Go, Instead of Ho, Ho, Ho, from the Valley," *Water Wastes Eng.* 9:3.

Dean, R. B., and R. J. Bouthilet (1973). "Treating Sewage Sludge with Sulfur Dioxide for Ultimate Feed Manufacture," German Patent 2,052,667 (after Dick, 1973).

Dick, R. I. (1972). "Sludge Treatment," in *Physiochemical Processes for Water Quality Control*, W. J. Weber, Jr., Ed. (New York: John Wiley & Sons, Inc.).

Dick, R. I., and B. B. Ewing (1967). "The Rheology of Activated Sludge," *J. Water Poll. Control Fed.* 39:4.

Dick, R. I. (1973). "Sludge Treatment, Utilization and Disposal," Annual Literature Review, *J. Water Poll. Control Fed.* 45:6.

Dick, R. I., and P. A. Vesilind (1969). "The Sludge Volume Index: What Is It?," *J. Water Poll. Control Fed.* 41:7.

Dugan, P. R. (1972). *Biochemical Ecology of Water Pollution* (New York and London: Plenum Press).

Ewing, B. B., and R. I. Dick (1970). "Disposal of Sludge on Land, Water Quality Improvement by Physical Processes," Center for Research in Water Resources, Univ. Texas, Austin.

Farrell, J. (1974). "Overview of Sludge Handling and Disposal," in *Municipal Sludge Management* (Rockville, MD: Information Transfer, Inc.)

Farrell, J. B., and B. V. Solatto (1973). "The Effect of Incineration on Metals, Pesticides, and Polychlorinated Biphenyls in Sewage Sludge, Ultimate Disposal of Wastewaters and their Residuals," Wat. Res. Res. Inst., N.C. State Univ., Raleigh.

Finstein, M. S. (1965). "Gross Dimensions of Activated Sludge Floc," *Bact. Proc. Am. Soc. of Microbiology*, 13 (from Mueller *et al.*, 1966).

Ford, D. L., J. T. Yang and W. W. Eckenfelder (1966). "Dehydrogenase Enzyme as a Parameter of Activated Sludge Activities," *Proc. 21st Ind. Waste Conf.*, Purdue Univ.

Hansen, J. A., and J. C. Tjell (1978). "Guidelines and Sludge Utilization Practice in Scandinavia," *Proc. Conference on Utilization of Sewage Sludge on Land* (Medmenham, England: Water Research Centre).

Hanson, J. B. (1972). "Viral Contamination as Measured by Coliphages," Duke Univ. Environmental Center Publication, Durham, NC.

Heukelekian, H., and E. Weisberg (1956). "Bound Water and Activated Sludge Bulking," *Sew. Ind. Wastes* 28:4.

Jansson, S. (1972). *Slambehandling*, SIF Ingenjör Utbildning, Stockholm (in Swedish).

Javaheri, A. R., and R. I. Dick (1969). "Aggregate Size Variations During Thickening of Activated Sludge," *J. Water Poll. Control Fed.* 41:5.

Kabler, P. (1959). "The Removal of Pathogenic Organisms by Sewage Treatment Processes," *Sew. Ind. Wastes* 31:12.

Karr, P. R. (1976). "Factors Influencing the Dewatering Characteristics of Sludge," PhD Dissertation, Clemson Univ., SC.

Lynam, B. T., B. Sosewitz and T. D. Hinesly (1972). "Liquid Fertilizer to Reclaim Land and Produce Crops," *Water Res.* 6:545.

Malina, J. F., Jr., K. Ranganathan, B. P. Sagik and B. E. Moore (1975). "Poliovirus Inactivation by Activated Sludge," *J. Water Poll. Control Fed.* 47:8.

McKinney, R. E., H. E. Langley and H. D. Tomlinson (1958). "Survival of *Salmonella typhosa* During Anaerobic Digestion," *Sew. Ind. Wastes* 30:12.

McWhorter, D. B. (1974). "Environmental Monitoring of the On-Land Disposal Site," *Proc. Conference on On-Land Disposal of Municipal Sewage Sludge by Subsurface Injection* (Fort Collins, CO: Colorado State University).

Merkel, W. (1934). "Die Fliesseigenschaften von Abwerschlamm," *Beihef. Gesundheits-Ingenieur* 11:14 (from Dick, 1972).

Metcalf and Eddy (1972). *Wastewater Engineering* (New York: McGraw-Hill).

Mohlman, F. W. (1934). "The Sludge Index," *Sew. Works J.* 6:1.

Mueller, J. A., K. G. Voelkel and W. C. Boyle (1966). "Nominal Diameter of Floc Related to Oxygen Transfer," *J. San Eng. Div. ASCE* 92:SA2.

Palfi, A. (1973). "Survival of Enteroviruses During Anaerobic Sludge Digestion," *Proc. Sixth International Conference*, Jerusalem (Oxford and New York: Pergamon Press).

Pauly, H. (1972). *"Kommuneundersøgelse,"* Mineralogisk Institutt, Danish Technical University, Lyngby (to be published).

Peterson, J. R., C. Lue-Hing and D. R. Zenz (1973). "Chemical and Biological Quality of Municipal Sludge," in *Recycling Treated Municipal Wastewater and Sludge through Forest and Cropland*, W. E. Sopper and L. T. Kardos, Eds. (University Park, PA: Pennsylvania State University Press).

Randall, C. W., D. G. Parker and A. Rivera-Cordero (1973). "Optimal Procedures for Processing of Waste Activated Sludge," Virginia Water Resources Research Center, Bull. 61, Blacksburg.

Randall, C. W., J. K. Turpin and P. H. King (1971). "Activated Sludge Dewatering Factors Affecting Drainability," *J. Water Poll. Control Fed.* 43:1.

Rudolfs, W. (1940). "Fertilizer and Fertility Value of Sewage Sludge," *Water Sew. Works* 87:12.

Schroepfer, G. J., A. S. Johnson and N. R. Ziemke (1955). "Effects of Various Factors on Hydraulic Separation in the Anaerobic Contact Process," San. Eng. Rep. 101s, Univ. Minnesota, Minneapolis.

Sommers, L. E. (1977). "Chemical Composition of Sewage Sludges and Analysis of Their Potential Use as Fertilizers," *J. Environ. Quality* 6:225.

Thomas, D. G. (1964). "Turbulent Disruption of Flocs in Small Particle Size Suspensions," *J. Am. Inst. Chem. Eng.* 10:517.

Ulmgren, L. (1972). *Slambehanding*, STF Ingenjör Utbildning, Stockholm (in Swedish).

van Gils, H. W. (1964). "Bacteriology of Activated Sludge," Research Inst. for Pub. Health Eng., Delft, Netherlands.

PROBLEMS

1. A liter cylinder is used to measure the settleability of a 0.5% suspended solids sludge. After 30 min, the settled sludge solids occupy 600 ml. Calculate the SVI.
2. A raw sludge has a plastic viscosity of 0.3 g/cm sec and a flow behavior index of 0.8. Plot a rheogram for this sludge. Construct a second line if this sludge behaved as a Newtonian fluid (flow behavior index = 1.0).
3. How much dried (40% solids) mixed digested sludge is necessary to have the same fertilizer value as 1 tonne (1,000 kg) of 8-8-8 inorganic fertilizer?
4. How much mercury might the above sludge contain (give range)?

CHAPTER 3

SLUDGE STABILIZATION

The term "stabilization" is at once widely used and understood but surprisingly difficult to define. Sludge disposal guidelines (see Chapter 11) often speak of stabilized sludges without making any attempt to define the term, perhaps assuming that the meaning is intuitively obvious. Unfortunately, because of the absence of a strict definition, considerable misunderstanding can result.

In this chapter, an effort is made to logically define what we should mean by stabilization, followed by a description of various methods and devices used to achieve that end.

MEASUREMENT OF SLUDGE STABILITY

Sludge stabilization is often practiced without any consideration of what is meant by stable sludge and how and why this must be obtained.

Although there can be many definitions for stabilization, it is not unreasonable to suggest that a stable sludge is one that can be disposed of without damage to the environment, and without creating nuisance conditions. If such a definition is accepted, it is necessary to define what we mean by "damage" or "nuisance."

Damage can be defined as an undesirable rate or method of degradation, or undesirable effect on existing ecology. This effect may be either toxic or simply the accumulation of an undesirable quantity of an inert material. Nuisance can be considered simply an affront to the senses of sight or smell.

Based on these definitions, the following parameters are proposed for defining sludge stability. The list is divided into three sections: those considered necessary in the measurement of stability, those considered valuable, and several proposed by others but that, in the author's opinion, are not as useful.

Important Parameters for Measuring Sludge Stability

Odor Production. In addition to hydrogen sulfide, many other chemicals either in or produced by sludge may create odors. The large number of potential odor-causing chemicals makes it extremely difficult to measure quantitatively the potential or the amount produced of an existing odor. In addition, there is also the problem of defining the strength and type of odor produced.

People often disagree on the desirability or undesirability of an odor. For example, familiarity with an odor makes it less perceptible and more acceptable. But, association of odors with their sources can also lead to misunderstandings.

Odor can be measured at the present time only by two methods–the panel and the dilution. The panel method involves submitting a number of people to various odors, and recording and analyzing their opinions. It is felt this can indicate an *average* opinion of the strength and nuisance value of a certain odor.

The dilution method is utilized in waterworks practice to determine the odors of drinking waters. It is possible to obtain a quantitative value for odor in water simply by diluting the water with equal quantities of odorless distilled water until the odor is no longer detectable by the average nose. The number of these dilutions would then be recorded as the *odor number*. The dilution method has also been used for atmospheric odors, but with limited success.

One semiquantitative test for odor measurement in sludge is based on the evolution of deadly hydrogen sulfide, the end product of the anaerobic sulfur cycle. At low concentrations H_2S has the characteristic rotten egg odor. However, at higher levels H_2S takes on a rather pleasant, though lethal, smell. The olfactory threshold of H_2S is 1.3×10^{-3} ppm, with the danger level at 10 ppm (Summer, 1971). The percentage of hydrogen sulfide in olfactory gas from anaerobic digestion is very small but, because of its low threshold, contributes greatly to the potential odor problems.

Rüffer (1966) reported a simple test to determine the amount of hydrogen sulfide being emitted. A strip of lead acetate paper is put in a bottle with some sludge in it. The time required for this paper to turn from white to brown due to the hydrogen sulfide evolution is recorded, and this indicates the amount of hydrogen sulfide emitted.

Some typical data, obtained during an aerobic digestion study, are shown below in Table 3-1.

Lack of hydrogen sulfide emission does not necessarily mean the sludge is odorless. However, the presence of a fairly substantial quantity of hydrogen sulfide definitely means the sludge will prove to be a nuisance.

Table 3-1. Results of Lead Acetate Test (Eikum, 1972)

	Time of Aerobic Digestion (days)				
	0	5	10	15	35
	Time for Paper to Turn Brown				
Primary Sludge	10 min	4 days	4 days	NC	NC
Mixed Primary and Chemical	3 hr	NC[a]	NC	NC	NC

[a]NC: no change in color.

Toxicity. Undoubtedly, toxicity is one of the most important criteria for determining whether a sludge is sufficiently stable for disposal into the environment. The difficulty in measuring toxicity is in determining "toxicity to what?"

A small quantity of common table salt, for example, may be harmful to some plants, and not to others. Most plants are sensitive to such toxic materials as chromium, lead, zinc, and some chlorinated hydrocarbons. All of these could be found in sludge and in fairly high quantitites.

Disposal into the marine environment can also cause a great deal of damage–for example, near New York City where wastewater sludge has been dumped for almost 50 years into an area which is now effectively a dead sea.

The measurement of toxicity is complicated by the fact that sludges are often transformed from one environment to another, sometimes accidentally. For example, septic tank sludge was stored for many years in lagoons and behind dikes on the banks of one of the best salmon rivers in Norway. A few years ago one of these dikes broke during a period of heavy rainfall, causing the contents of several lagoons to flow into the river, thereby destroying the salmon population for several years.

The stability of a sludge, as measured by its toxicity, is a function of the method of disposal, and each case must be evaluated separately. If a sludge is to be disposed of on farmland, for example, we would normally look for such characteristics as high concentrations of heavy metals, high oxygen demand, abnormally high or low pH, or unsafe levels of pathogenic organisms. Disposal in the marine environment, on the other hand, would dictate a different set of parameters.

Other Potentially Valuable Parameters for Measuring Stability

Reduction in Volatile Suspended Solids. In aerobic and anaerobic digestion, as might be suspected, a strong correlation exists between the drop in percent VSS and aeration or detention time (Table 3-2). After some time, the percent VSS often reaches a constant value, and many investigators have suggested that a steady percent VSS indicates that the sludge has been stabilized.

Table 3-2. Reduction of Volatile Suspended Solids During Aerobic and Anaerobic Stabilization

Detention Time (days)	Sludge	Reduction VSS (%)	Type of Plant	Source
Aerobic				
4	Raw Primary	8	Pilot	Krauth (1969)
8		13		
16		13		
32		13		
5	Raw Primary	12	Pilot	Eikum (1972)
10		22		
15		44		
25		52		
35		56		
Anaerobic				
5	Mixed	40	Pilot	Eckenfelder (1967)
10		50		
15		56		
20		58		
25		58		

Kempa (1967) suggests the following equation for calculating the degree of stability:

$$M = 100\left(1 - \frac{V_1 m_0}{V_0 m_1}\right)$$

where V_0 and V_1 = the volatile solids as percent of total solids in raw and treated sludge

m_0 and m_1 = the fixed solids in raw and treated sludge

Some engineers automatically equate stability with volatile solids reduction. This is understandable, since the organic fraction in the sludge is the most likely cause of odor and also makes sludge handling and dewatering difficult. However, if we accept the definition of stability (toxicity and nuisance), a sludge with a high percent VSS is not necessarily unstable, unless the high volatile fraction causes environmental problems. A reduction of VSS is at best an indirect indicator of stability and may at times be misleading.

Change in Qo_2 as an Indication of Aerobic Activity

The rate of oxygen removal by aerobic organisms is referred to as the Qo_2, or the milligrams of oxygen used by the microorganisms per hour per gram VSS in a mixed flask or respirometer (Lamb *et al.*, 1964). This has been used as a method both of calculating the BOD of a substance as well as determining the respiration rate of an activated sludge. High oxygen use per gram of solids would tend to indicate a very active, viable sludge. Thus a decrease in the Qo_2 would indicate a sludge has been either poisoned in some way or has attained a state of respiration where its biological activity is minimal.

The Qo_2 has been suggested as a method of measuring sludge stability. Mudrack (1966) found that for a primary sludge a rapid Qo_2 occurs in the first 6 hours then levels off to a value of 6.7 mg O_2/g volatile solids per hour. After 4 days, this value drops even further, to about 5. Accordingly, Mudrack suggested 5 as the limit where sludge would be stable.

Eikum (1972), however, obtained Qo_2 values below 2 mg O_2/g VSS/hr for mixed sludge (primary plus chemical precipitated). Even after only 5 days aeration Qo_2 values were less than 5. Eikum suggests this may have been due to the high chemical solids concentration but, nevertheless, the Qo_2 value of 5 suggested by Mudrack was not applicable in this case.

Obviously, different organisms have different oxygen uptakes. For example, protozoa in aerobic sludge produce a high Qo_2, but their presence is an indication of a fully digested hence *stable* sludge.

Change in Gas Production as an Indication of Anaerobic Activity. Gas production is to anaerobic biodegradation what oxygen uptake is to aerobic. The end products of anaerobic decomposition are still high in energy, but most can be disposed of into the environment without undue damage or nuisance. For most wastewater treatment plants 10-20 days of anaerobic digestion usually are required to stabilize metabolic activity. In most laboratory studies, this time is much less. In both cases, the decrease in metabolic activity is indicated by a reduction in gas production. Biological stability can thus be indicated by the absence of gas. Obviously, some toxic chemicals or other inhibitors may prevent the methane-forming bacteria from producing gas, even though the volatile solids concentration is still high. Once the inhibition is removed, biological activity can continue, and thus the use of gas production as an indicator of stability is of dubious value.

Nitrification. In aerobic stabilization, the nitrogen cycle follows from organic nitrogen to ammonia to nitrite to nitrate. This same type of nitrification occurs when waste activated sludge is aerated in an aerobic stabilization unit. The presence of a high percentage of nitrate (NO_3^-) would indicate that most of the nitrogen has been converted to the fully oxidized form. Eikum (1972) found that at 25°C 30 days aeration was sufficient to completely convert most organic and ammonia nitrogen to nitrate form. The degree of nitrification may thus be a good indicator of stability under aerobic conditions. Under anaerobic conditions, the end product of the nitrogen cycle is still the potentially harmful ammonia, and nitrification as a measure of stability is not of much value.

ATP (Adenosine Triphosphate). ATP is a specific indicator of cell viability (Patterson *et al.*, 1970). ATP has been found to increase markedly with aerobic stabilization and, after 5 days, gradually decrease. It is possible the ATP level will stabilize when most organisms either have formed spores or are in a very low-energy endogenous respiration state. This would obviously correspond to a low Q_{O_2}.

ATP levels are sensitive to pH, heavy metals and other environmental conditions, and are good indicators of a sludge's viability. One drawback to using ATP is that organic materials need not be alive to cause eventual problems in the environment, and thus the measurement of ATP may give a misleading reading of sludge stability.

In addition, ATP is difficult to measure. The standard method is by luminescence of a firefly enzyme, which is unique among bioluminescent phenomena. This can be accomplished only with a fairly expensive measuring device and with skilled technicians.

Enzymes. Enzyme activity, especially of the dehydrogenase enzymes, has been correlated with the drainability of a sludge (Randall *et al.*, 1971). However, it is again possible to have a sludge low in dehydrogenase activity be a serious nuisance, thus this parameter has little value as an indicator of sludge stability. Further, enzyme activity is difficult to measure and holds little promise for on-line monitoring.

Total Organic Carbon. Total organic carbon (TOC) is a fairly good measurement of the potential degradability of sludge. However, it is rather difficult to measure in sludge by known means because most available analyzers are easily clogged and quite expensive.

BOD and COD of the Liquor. The filtrate from raw primary sludge has a BOD of approximately 1,000 mg/l (Swanwick, 1972). On the other hand, digested sludge filtrate has a BOD of around 100 mg/l. It might thus be possible to estimate the stability of a sludge by measuring the BOD (or COD) of the sludge liquor.

Microbiology. It is known that bacteria, viruses, and higher forms of pathogens such as worm eggs can be discharged from the treatment plant with the sludge. If the sludge is to be disposed of where it might come into human contact, the concentration of pathogens in the sludge should be measured. Other than the coliform test, there are no acceptable methods for estimating the concentration of pathogens in sludge. In some places in Europe, sludge is presently being pasteurized to ensure that pathogens are no longer viable.

Other Suggested Parameters of Sludge Stability

Dewaterability. Dewaterability has been suggested as a measurement of sludge stability. The belief is that if a sludge dewaters well it must be stable. This is absolutely untrue and can be terribly misleading. Dewaterability has nothing to do with the stability of a sludge and, although it may be a desirable attribute, it is not a necessary condition for sludge stabilization.

Viscosity. Viscosity might be a good method of control for estimating the stability of a specific sludge. Viscosity measurement cannot apply to all sludges because sludge viscosities are inherently different, and it would be extremely difficult to use such measurements in determining general stability values.

Caloric Value. The fuel value of sludge is directly related to the VSS and to the TOC as mentioned before. Both are more easily measured than the caloric value.

Stability Defined

Based on the above argument, we can then define a stable sludge as one which does not create, upon its disposal into the environment, a significant detrimental impact. This detrimental impact may take many forms, but most of the effects can be categorized as either toxic or nuisance.

Following is a discussion of the various methods which have been used for sludge stabilization.

BIOLOGICAL STABILIZATION

Anaerobic Digestion

Anaerobic stabilization of sludge is an old process. Septic tanks, for example, use the anaerobic decomposition process in the partial stabilization of domestic sludge. The old Imhoff tanks also utilize anaerobic decomposition as the method of partially stabilizing sludge.

The *conventional* digester is a large tank with either a floating or fixed cover. The sludge is pumped into the middle, and the stabilized solids are removed from the bottom (Figure 3-1). Since the sludge is not mixed, it separates into layers of scum, supernatant, and active and stabilized solids. The gas is collected in the top.

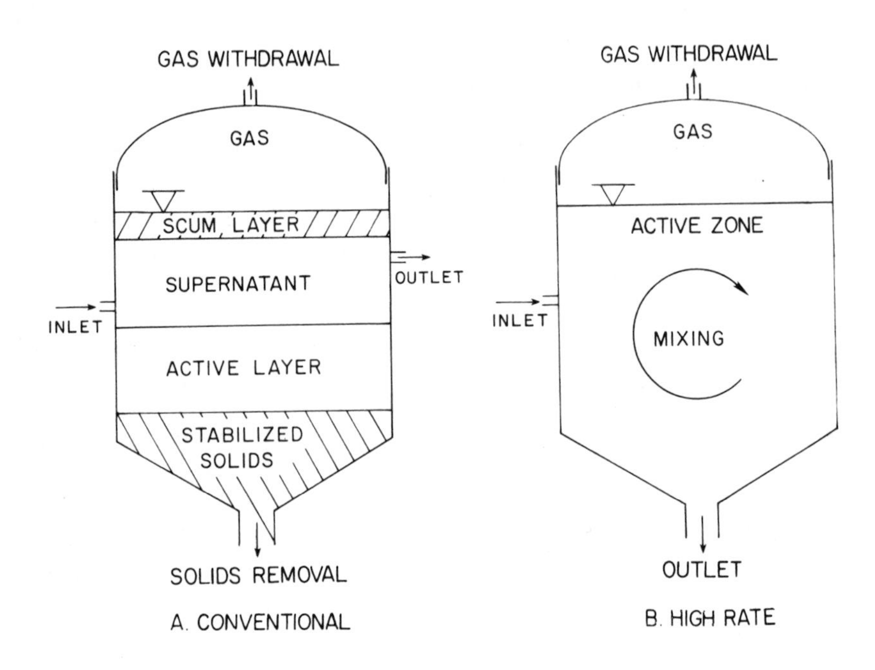

Figure 3-1. Schematics of conventional and high-rate anaerobic digesters.

Although these digesters will stabilize sludge, the required retention time is quite long, and thus the process is inefficient. The modern digester, or *high-rate* tank, is mixed either mechanically or by the recycling of compressed digester gas. This creates a large zone of active decomposition and the stabilization reaction is speeded up significantly.

A number of treatment plants are designed with two digesters in series, the first used for mixing and heating (the primary digester) and the second used for gas, supernatant, and stabilized sludge separation and storage. Very little actual stabilization occurs in the secondary digester, and this volume is not used in the calculation of digester volume requirements.

Anaerobic digestion is a series operation in that specific groups of organisms use, in series, the end products of a previous decomposition step, as illustrated in Figure 3-2. The first step is the solubilization of the organics by extracellular enzymes. The soluble organics are in turn degraded by a group of organisms known as acid formers or acid producers. The organic (volatile) acids produced are mostly propionic and acetic. These acid-forming organisms are hardy and plentiful and are not rate-limiting.

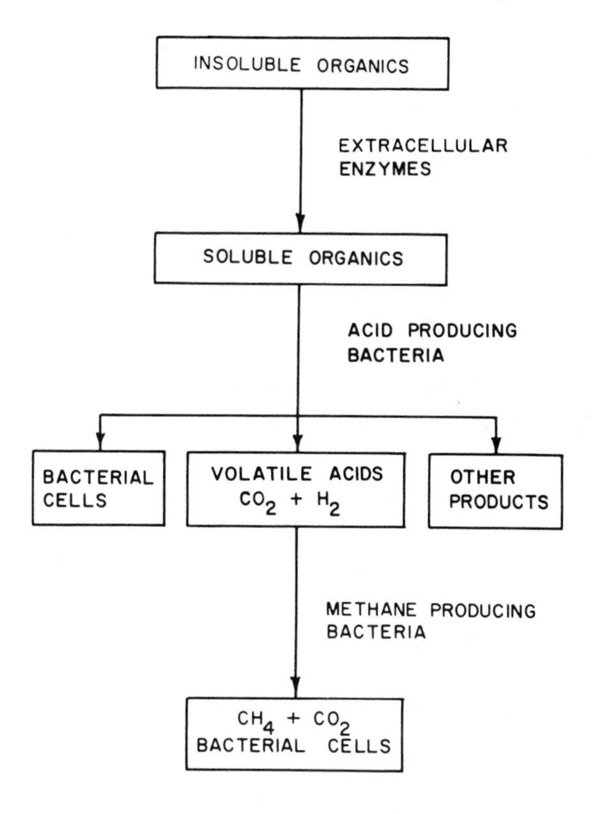

Figure 3-2. Schematic representation of the anaerobic digestion of organic waste (Andrews and Graef, 1970).

The last group of bacteria are known as methane producers or methane formers. These strict anaerobes are so sensitive to environmental change that they are often referred to as the "prima donnas" of wastewater treatment. For example, they are sensitive to such things as pH changes, presence of heavy metals, detergents, changes in alkalinity, ammonia, sulfides and temperature. In addition, methane formers are slow growing, and there is a real danger of flushing out these microorganisms. A number of investigators have suggested recycling the concentrated sludge from the secondary digester. This operation is similar to the activated sludge system and has been tagged the "anaerobic contact process' (Schroepfer and Ziemke, 1959). It helps maintain a high concentration of methane formers in the primary digester and increases the solids retention time in the digestion process. Dague (1968) has suggested that all digestion systems be routinely operated in this manner.

The anaerobic decomposition of a specific carbohydrate such as glucose can be illustrated as

$$C_6H_{12}O_6 \xrightarrow{\text{Acid Formers}} 3\ CH_3COOH$$

$$3\ CH_3COOH + 3\ NH_4HCO_3 \longrightarrow 3\ CH_3COONH_4 + 3\ H_2O + 3\ CO_2$$

$$3\ CH_3COONH_4 + 3H_2O \xrightarrow{\text{Methane Bacteria}} 3\ CH_4 + 3\ NH_4HCO_3$$

The first equation refers to the breakdown of glucose to acetic acid by acid-forming bacteria. This acid must be neutralized, as shown by the second equation. If sufficient buffer (such as the bicarbonate shown) is not available, the pH will drop, and the third reaction will be inhibited. Note that the buffer consumed in the second reaction is reformed in the third.

The overall general equation for anaerobic decomposition of an organic material has been suggested by Buswell (1939) as

$$C_nH_aO_b + \left(n - \frac{a}{4} - \frac{b}{2}\right) H_2O \rightarrow \left(\frac{n}{2} + \frac{a}{8} + \frac{b}{4}\right) CO_2 + \left(\frac{n}{2} + \frac{a}{8} - \frac{b}{4}\right) CH_4$$

This formula, however, does not consider the fraction of substrate converted to microorganisms, which is critical in the application of growth kinetics.

Operation. Dague (1968) summarized the factors that require operational control in digesters and responses that can be expected from manipulating these controls. He lists these controls as

Food	Temperature
Contact	pH
Time	Toxins

Although the food used by the bacteria usually cannot be controlled by the operator, the manner of feeding the food to the microorganisms is controllable. Experience has shown that a uniform feed rate is mandatory for successful digestion, and twice a day should be the minimum frequency for sludge pumping.

Adequate contact between the food and microorganisms can be achieved by mixing, either mechanically or by recycling the digester gas.

The time required for digestion is a function of digestion temperature. Higher temperatures allow for shorter solids retention time.

Digesters can be heated by internal or external heat exchangers, and the source of energy is usually the methane generated during the digestion.

Two distinct temperature ranges of digestion are usually recognized: mesophilic (92 to 98°F or 33 to 37°C) and thermophilic (105 to 110°F or 40 to 43°C). Thermophilic digestion is much faster and achieves higher gas production, but it is expensive to operate due to fuel requirements and is more sensitive to upset than mesophilic digesters. Apparently the ecological system within a thermophilic process is more fragile, with fewer species occupying various niches. The conversion of digesters from mesophilic to thermophilic has been used as a fairly low capital-intensive alternative to building more digesters in overloaded plants.

One of the most important parameters in digester operation is the maintenance of an acceptable environment for the methane-producing bacteria. These bacteria seem to be especially sensitive to change in pH. Detecting changes in pH, however, is not an acceptable operating procedure because, once the pH changes, the digester is already in serious trouble, and major effort is needed to restart it. It is thus necessary to find other parameters which can foretell the oncoming upset.

In this regard the most important variable to monitor seems to be volatile acid concentration. Since the anaerobic digestion process is a series reaction, any problems with the methane-producing bacteria will immediately be indicated by an increase in the volatile acids. Volatile acids in well-operated digesters should be less than about 250 mg/l.

Alkalinity is another major operational parameter which must be monitored, because a low alkalinity (low buffering capacity against pH drop) indicates a probable instability in digester operation. The most useful measure of alkalinity, as proposed by Banta and Pomeroy (1934) would be a measure of bicarbonate alkalinity, because the normal total alkalinity measurement is to pH 4.3, while the pK of volatile acids is about 4.5; therefore, a substantial fraction of the titrant would be used to neutralize the volatile acids (Andrews and Graef, 1970). Typical values of alkalinity found in well-operated digesters are between 1,000 and 5,000 mg/l.

Another operating variable is the methane production rate and the composition of the gas. The production rate is, however, difficult to measure in practice, especially in the case of nonuniform feeding. Gas composition should theoretically be a sensitive parameter for detecting upset, but this is not commonly measured in most wastewater treatment plants. The normal CH_4 concentration should be about 65% and CO_2 about 30%. The normal gas production should be between 0.5 and 0.75 m^3/kg (8 and 12 ft^3/gas/lb) volatile solids added, or 0.75 and 1.1 m^3/kg (12 and 18 ft^3/lb) volatile solids destroyed. Digester gas has a heat value of about 5.3×10^6 calories/m^3 (600 Btu/ft^3).

Thus the volatile acids, alkalinity, gas production and composition all are precursors to pH drop. During the time that a digester is beginning to go sour, the pH may change very little if at all. Once the pH drops, however, it is too late, and major effort is needed to cure the sick digester.

The standard method of bringing a sour digester back to operation is to add lime, sometimes in massive quantities, as a sour digester (pH less 6.5) has high acidity. In some cases, lime addition is inadequate, and the digester must be cleaned out completely and restarted, using seed from another digester.

Recently, the advisability of adding lime has been questioned, as lime may have some undesirable side effects. One of these is that the addition of lime will result in the following reaction

$$Ca(OH)_2 + 2CO_2 \rightarrow Ca(HCO_3)_2$$

which is the removal of CO_2. Although this increases the pH, it also forces the reduction in bicarbonate alkalinity. Since

$$[H^+] + [HCO_3^-] \rightleftarrows [H_2CO_3] \rightleftarrows [CO_2] + [H_2O]$$

the reduction in soluble CO_2 drives these equations to the right, thus resulting in a reduction in the concentration of the bicarbonate ion, HCO_3^-. It seems to make much more sense to use lime only for gross pH adjustment, and then add bicarbonate for the final adjustment to neutrality (Barber, 1978).

Using a dynamic model of anaerobic digestion developed by Andrews and Graef (1970) it is possible to simulate digester failure given almost any adverse environmental effect. For example, if a digester is overloaded with a sudden step change in feed, the interaction of the variables discussed above can be predicted. The alkalinity will begin to drop, the volatile acids will increase and the production of CO_2 will increase, thus producing an increase in the total gas flow rate. The methane-forming organisms will at first increase the production of CH_4 due to the ready availability of feed, but will soon be overwhelmed by the excess of volatile acids, and cease to function properly, resulting in further reductions in alkalinity, pH and CH_4 and an increase in volatile acids. Given sufficient time, the digester will go sour and cease to function. This failure is graphically depicted in Figure 3-3.

Anaerobic digestion has had a very rocky history in the last 10-20 years because it has been considered more of a nuisance than a benefit in wastewater treatment. Many digesters presently in use have had some problems maintaining operation, mostly due to the inhibition of methane-forming bacteria by toxic substances. If the cause of the upset is a slug load of some inhibitory material that can be flushed out or assimilated, increasing the pH with lime and decreasing the feed may be sufficient to bring the digester back into operation. However, sometimes the cause for digester upsets is the constant and prolonged input of a toxin. As a result, lime would treat only the symptoms but would not cure the digester problem.

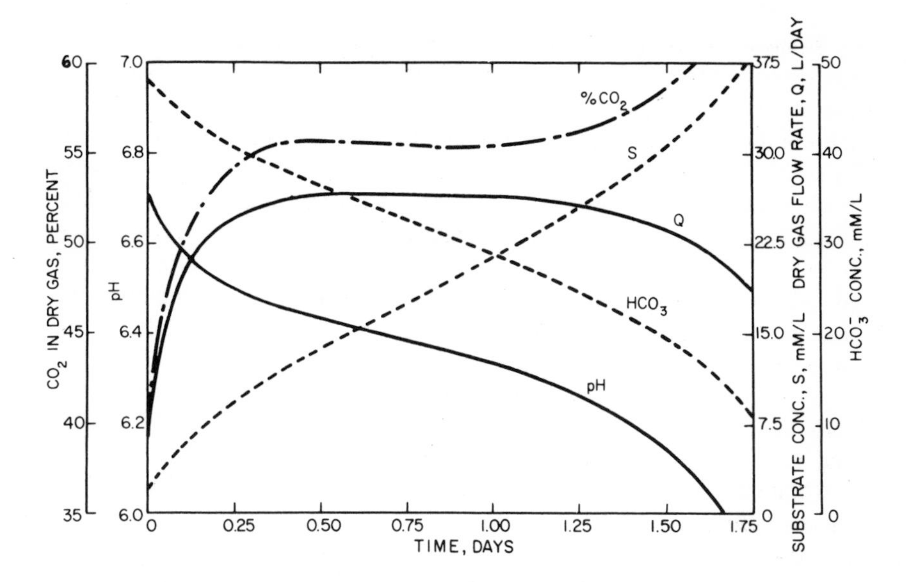

Figure 3-3. Anaerobic digestion process failure by a sudden increase in substrate concentration (Andrews and Graef, 1970). (Note: m*M*/l = millimoles per liter.)

Toxicity within a digester can be caused by an excessive amount of almost any material, even substances normally considered as necessary nutrients (*e.g.*, ammonia). The concentration at which a substance begins to be toxic to digestion is difficult to determine because it depends on many variables such as the digester operation, as well as the effects of antagonism, synergism and acclimation (Kugelman and Chin, 1970). Some typical toxic concentrations for common problem substances are shown in Table 3-3, but it must be emphasized that these values can be considered only as rough guides.

Table 3-3. Typical Toxicity Levels for Anaerobic Digestion (Lawrence, Kugelman and McCarty, 1964)

Substance	Toxic Concentration (mg/l)
Sulfides	200
Soluble Heavy Metals	>1
Sodium	5000-8000
Potassium	4000-10,000
Calcium	2000-6000
Magnesium	1200-3500
Ammonium	1700-4000
Free Ammonia	150

Swanwick *et al.*, (1972) suggested that it is necessary to look at what ails the digester and to cure it by first analyzing the problem. For example, if the problem is a high concentration of detergents or other types of surfactants in the sludge, one solution might be the addition of a sufficient quantity of stearine amine. On the other hand, if the problem is heavy metals, their toxicity can often be overcome by adding sufficient quantities of sulfides, which will precipitate the metals and take them out of solution. For example, in one laboratory analysis the addition of cadmium resulted in 80% reduction in gas production. However, if a sufficient amount of sodium sulfide was also added to the digestant simultaneously, the cadmium precipitated out as cadmium sulfide, and no drop in gas production occurred.

Swanwick agrees that there are some materials, such as chlorinated hydrocarbons, that are extremely toxic and difficult to remove in the digester and must be removed before the sludge enters the digestion process.

He concludes, however, that "there appears to be no reason to abandon the digestion process solely because of the possibility of inhibition."

Although anaerobic digestion reduces pathogenic bacterial and viral populations, anaerobically digested sludge is not free of pathogens. The temperature in mesophilic digesters is not sufficiently high to cause pathogen die-offs at rates much greater than would occur in nature. Table 3-4 lists some survival rates for several microorganisms.

Table 3-4. Survival of Microorganisms in Anaerobic Digestion (EPA, 1974)

Organism	Digestion Period (days)	Removal (%)
Endamoeba hystolytica	12	<100
Salmonella typhosa	20	92
Tubercle bacilli	35	85
Escherichia coli	49	<100

Design. The design of the anaerobic digestion process involves selection of the digester volume, digestion temperature and an estimate of gas production. The design can be based on experience with similar sludge or, if unusual conditions are expected, on pilot plant studies.

Design by Experience. During many years of building and operating anaerobic digesters, some rules of thumb have evolved. Burd (1968) summarized this experience as shown in Table 3-5.

The "Ten State Standards" (Recommended Standards for Sewage Works 1968) is still used by many regulatory agencies and a maximum digester

Table 3-5. Typical Design Criteria for Low-Rate and High-Rate Digesters (After Burd, 1968)

Parameter	Low-Rate	High-Rate
Solids Retention Time (days)	30 to 60	10 to 20
Solids Loading (lb VSS/ft^3/day)	0.04 to 0.1	0.15 to 0.40
(kg VSS/m^3/day)	(0.64 to 1.6)	(2.4 to 6.4)
Volume Criteria (ft^3/capita) (m^3/capita)		
Primary Sludge	2 to 3 (0.06 to 0.09)	1.3 to 2 (0.035 to 0.06)
Primary Sludge + Trickling Filter Sludge	4 to 5 (0.12 to 0.14)	2.6 to 3.3 (0.075 to 0.085)
Primary Sludge + Waste Activated Sludge	4 to 6 (0.12 to 0.17)	2.6 to 4 (0.075 to 0.12)
Combined Primary + Waste Biological Sludge Feed Concentration (% solids–dry basis)	2 to 4	4 to 6
Anticipated Digester Underflow Concentration (% solids–dry basis)	4 to 6	4 to 6

loading of 1.3 kg/m^3/day (0.08 lb VSS/ft^2/day) for high-rate digestion and 0.6 kg (0.04 lb) VSS loading for single-stage digestion is recommended. It is recognized that these loadings are conservative and process modifications can greatly accelerate digestion, thus making higher loadings possible.

Gas production will vary according to the efficiency of mixing, the temperature, and many other variables. A well-operated digester should produce gas at a rate of about 0.5 m^3/kg (8 ft^2/lb) of volatile solids added. It is advisable when designing an anaerobic digestion system not to depend solely on digester gas for heating, but make it possible to utilize other fuels if gas production is less than anticipated.

Average operating data for the Chicago high-rate digestion system, shown in Table 3-6, confirm the validity of the design criteria (Lynam *et al.,* 1967).

Design Based on Laboratory Studies. As long as waste sludges were of purely domestic origin, there was no need for more sophisticated design procedures. With increasing industrial waste components, however, rules of thumb become useless and even dangerous. Accordingly, an attempt has been made to analyze anaerobic digestion from the standpoint of growth kinetics; this would provide general models that would allow for more accurate design and analysis.

Andrews (1969) has shown that growth kinetics can be applied to the anaerobic digestion process, and Andrews and Graef developed a dynamic model for the analysis of digester operation (1970).

Table 3-6. Average Operating Results for a High-Rate Digestion System (Lynam *et al.*, 1967)

Feed Sludge	
Total Dry Solids (%)	3.76
Total Volatile Solids (%)	59.5
pH	6.6
Flow (mgd)	0.449 (71 m^3/hr)
Volatile Solids (tons/day)	41.2 (156 kg/hr)
Volatile Solids (lb/day/ft^3)	0.084 (0.056 kg/hr/m^3)
Hydraulic Detention Time (days)	17
Temperature (°F)	95 (35°C)
Drawoff Sludge	
Total Dry Solids (%)	2.91
Total Volatile Solids (%)	46.9
pH	7.2
Gas Produced (ft^3/day)	558,000 (660 m^3/hr)
Total Solids Reduced (tons/day)	18.6 (700 kg/hr)
Volatile Solids Reduced (%)	38.2

Lawrence and McCarty (1969) suggest that the net growth of microorganisms in a continuous-flow completely mixed anaerobic treatment system can be described as

$$\frac{dX}{dt} = Ym\ bX \qquad (3\text{-}1)$$

where X = microorganism concentration, mass/volume
t = time
Y = growth yield coefficient
b = microorganism decay coefficient, $time^{-1}$
m = rate of waste utilization per unit volume, mass/volume-time

This expression can be combined with a relationship similar to the well-known Monod (1942) formula derived empirically for enzyme kinetics,

$$m = \frac{kSX}{K_s + S} \qquad (3\text{-}2)$$

where k = maximum rate of waste utilization per unit weight of microorganisms, $time^{-1}$
S = waste concentration, mass/volume
K_S = half velocity coefficient equal to the waste concentration when $m = k/2$

Combining these expressions yields a model which describes a continuous steady-state completely mixed digester.

$$\frac{dX/dt}{X} = \frac{YkS}{K_s + S} - b = (\theta_c)^{-1} \qquad (3\text{-}3)$$

where θ_c = solids retention time, defined as the ratio of the total weight of active microbial solids in the system to the quantity of solids withdrawn daily.

The constants can be evaluated by conducting laboratory analyses with continuous bench-scale digesters similar to the schematic shown in Figure 3-4. The substrate concentration, S, is approximated by the COD or BOD of the sludge, while the microbial population is estimated from a measurement of VSS.

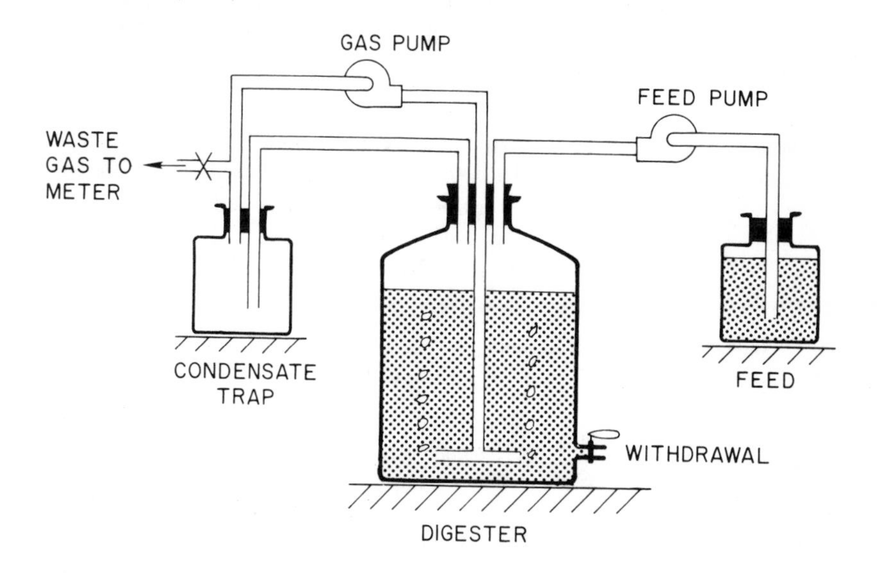

Figure 3-4. Laboratory apparatus for conducting anaerobic digestion studies.

The constants can be evaluated by recognizing that Equations 3-1 and 3-2 represent straight lines if they are written as

$$\left(\frac{dX}{dt}\right)\left(\frac{1}{M}\right) = Y\frac{m}{X} - b$$

and

$$\frac{1}{(m/X)} = \frac{K_s}{k}\left(\frac{1}{S}\right) + \frac{1}{k}$$

(dX/dt) 1/X is the reciprocal of the solids retention time and, for a continuous steady-state reactor, this is equal to the hydraulic retention time. Thus plotting the hydraulic retention time against the rate of substrate utilization divided by the microorganism concentration will allow the constants Y and b to read from the graph as the slope and intercept, respectively, Similarly, a plot of 1/(m/X) vs 1/S yields Ks/k as the slope and 1/k as the intercept. Typical curves are shown in Figure 3-5.

From Figure 3-5A, the intercept is 0.08 and the slope is 17.8, so that (Ks/k) = 17.8 and (1/k) = 0.08. Hence k = 12.5 and Ks = 222.

Similarly, from Figure 3-5B, the intercept is - b = - 0.03 and the slope is Y = 0.04.

Commonly, the yield (Y) anaerobic digestion ranges between 0.040 and 0.054, with an average value of 0.044 mg/mg, and the decay constant (b) ranges between 0.010 and 0.040, with an average value of 0.019 day^{-1} (Lawrence and McCarty, 1969).

Example 3-1. If the data in Figure 3-5 were obtained for a pilot study for which the feed (substrate) is 10,000 mg/l COD, and the desired efficiency of removal is 90%, what solids retention time (θ_c) is required?

$$\frac{1}{\theta_c} = \frac{YkS}{K_s + S} - b$$

Y = 0.04, k = 12.5, K_s = 222, and
b = 0.03 from Figure 3-5.

$$\text{Efficiency} = 90\% = \left(\frac{10{,}000 - S}{10{,}000}\right) 100$$

and hence S = 1,000 mg/l

$$\frac{1}{\theta_c} = \frac{(0.04)\,(12.5)\,(1{,}000)}{222 + 1{,}000} - 0.03 = 0.38$$

$$\theta_c = 2.6 \text{ days}$$

Using the above equations, it is also possible to develop the following relationships (Lawrence, 1971):

Concentration of microorganisms in the digester

$$X = \frac{Y(S_o\text{-}S)}{1 + b\theta_c}$$

Excess microorganism production rate (mass/time)

$$P_x = \frac{YQ(S_o\text{-}S)}{1 + b\theta_c}$$

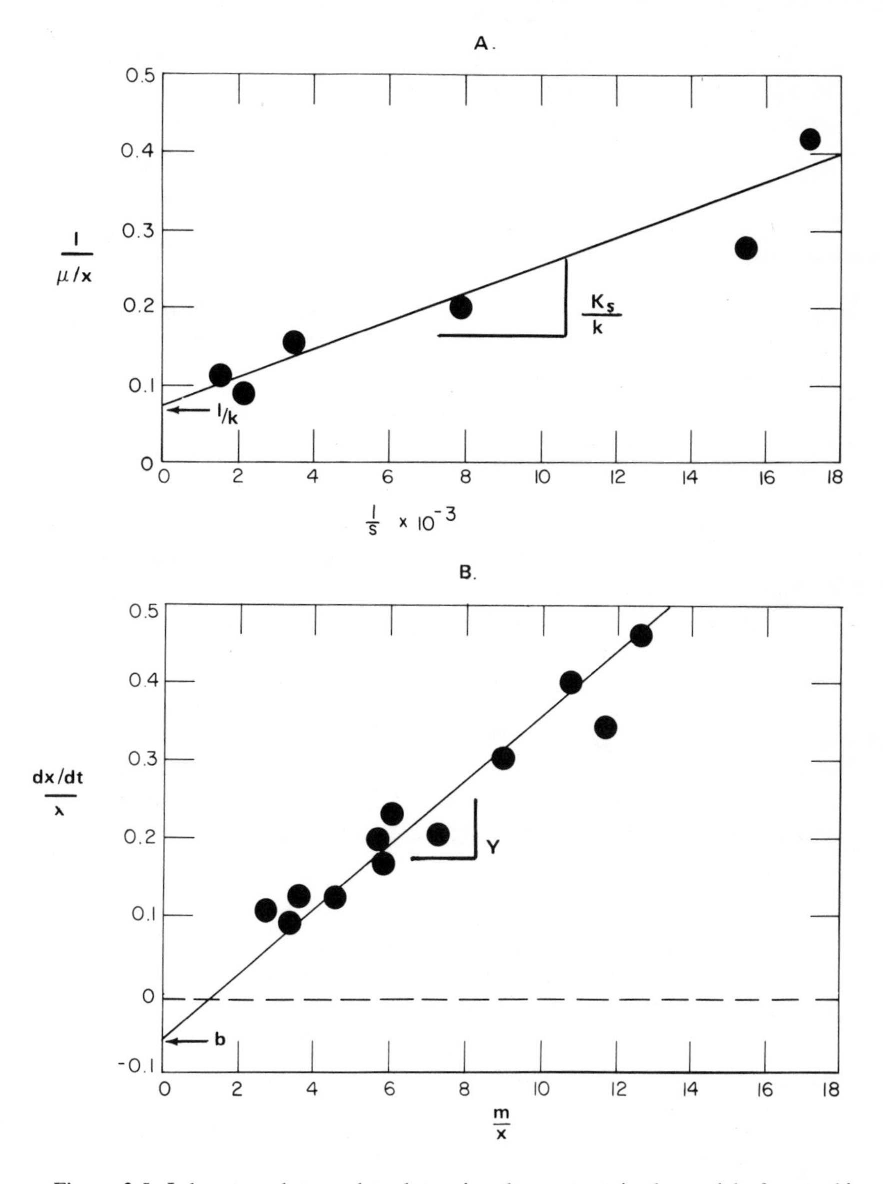

Figure 3-5. Laboratory data used to determine the contents in the model of anaerobic digestion. (Lawrence and McCarty, 1969)

and the efficiency of substrate utilization is of course

$$E_s = \frac{100(S_o\text{-}S)}{S_o}$$

Temperature has a significant effect on the process dynamics. O'Rourke (1968) has found that k and Ks vary with temperature as shown in Table 3-7. Table 3-7.

Table 3-7. Variation of Kinetic Coefficients for Anaerobic Digestion of a Complex Waste (O'Rourke, 1968)

Temperature (°C)	k day^{-1}	K_s (mg/l as COD)
35	6.67	2,224
25	4.65	5,790
20	3.85	10,610

The continuous anaerobic reactor can fail when the solids retention time (θ_c) is reduced to a value at which the microorganisms are washed from the system at a rate greater than their maximum net growth rate. At this point, the influent waste concentration is equal to the effluent, and efficiency drops to zero. Equation 3-3 then reduces to the following approximate form, with the θ_c at the minimum.

$$(\theta_c)_{lim} = (Y\ k - b)^{-1}$$

Lawrence (1971) has tabulated the minimum values of the solids retention time, as shown in Table 3-8. A strong effect of temperature is again noted, with thermophilic digestion requiring the shortest time.

Table 3-8. Minimum Values of Solids Retention Time (θ_c) for Anaerobic Digestion of Various Substances (Lawrence, 1971)

	$[\theta_c]_{lim}$ (days)			
Substrate	**35°C**	**30°C**	**25°C**	**20°C**
Acetic Acid	3.1	4.2	4.2	–
Propionic Acid	3.2	–	2.8	–
Butyric Acid	2.7	–	–	–
Long-Chain Fatty Acid	4.0	–	5.8	7.2
Hydrogen	0.95[a]	–	–	–
Sewage Sludge	4.2[b]	–	7.5[b]	10
Sewage Sludge	2.6	–	–	–

[a]For 37°C.
[b]Computed value.

Aerobic Digestion

It has long been known that if mixed liquor is aerated a long time, most organics will be destroyed. Extended aeration plants are designed specifically to eliminate all sludge disposal problems and to allow only inert solids to escape over the effluent weir. These plants are used extensively by the dairy industry, motels and small industries, among others. Unfortunately, of course, few extended aeration plants operate without producing net sludge, and this excess sludge must still be disposed of.

Waste biological sludge produced in activated sludge plants can also be stabilized by simply reserving several aeration basins for the aeration of sludge. A number of plants have been built in the last 10-15 years in which aerobic sludge stabilization is incorporated into the design.

Operation. The basic reaction of the aerobic sludge stabilization is:

$$\boxed{\begin{array}{c}\text{Complex}\\ \text{Organics}\end{array}} \xrightarrow[\text{Organisms}]{\text{Aerobic}} \boxed{\begin{array}{c}CO_2\\ H_2O\end{array}}$$

Of course, this is again greatly simplified and does not express the complex biochemical reactions involved. The aerobic stabilization process has the advantage of a complete food chain and a mixed and varied ecology including some anaerobic and facultative organisms. This results in a resilient ecosystem, and thus the aerobic stabilization process is much less susceptible to be upset than is the anaerobic process.

The sludge produced from the aerobic process during the first 5 days of aeration often shows a marked increase in the sludge volume index, and drainability is severely hampered. After about 10 days aeration, however, the situation improves (Bisogni *et al.*, 1971; Jaworsky and Rohlich, 1963).

Eikum (1972) showed that specific resistance to filtration of aerobically stabilized sludge drops from about 10^{14} m/kg to 10^{13} m/kg after 30 days aeration. However, the compressibility of the sludge increases markedly, thus making it more difficult to filter.

Volatile suspended solids reduction has been found to vary with hydraulic retention time and temperature. Recent experiments have shown that increasing the temperature of an aerobic digester to 60°C accomplishes 30% reduction in VSS in 18 hours. Studies are also continuing on the use of pure oxygen in aerobic digestion.

Design. As with anaerobic digesters, the design of aerobic stabilization units is based either on experience, in which the detention time is the important variable, or on pilot plant data.

Design Based on Experience. In a recent EPA manual on upgrading existing treatment plants (Weston, 1971), design parameters (Table 3-9) are suggested for aerobic digestion of mixed sludges.

Operating data for aerobic digesters are scarce, but indications are that for waste activated sludge solids loadings of 0.1-0.2 kg VSS/day/kg solids in the digester (for hydraulic detention times of about 10-15 days) 40-50% reduction in VSS can be achieved with the ammonia-N and BOD of the clarified effluent at 20 mg/l and 10 mg/l, respectively. The volatile solids reduction achieved depends on the sludge age (kg solids under aeration/kg solids withdrawn per day) and the temperature.

Design Based on Laboratory Studies. The dynamic model based on the Monod analysis, discussed under *Anaerobic Digestion*, can also be applied to aerobic digestion (Middlebrooks and Garland, 1968; Pearson, 1968). Data from continuous pilot digesters are required as before.

In some cases, adequate design information can be obtained from laboratory-scale batch digesters. One approach is to express the decay of biodegradable material in a batch aerobic digester as

$$\frac{dS}{dt} = K_d S$$

where S = concentration of biodegradable material, mg/l
t = time
K_d = rate of decay constant, $time^{-1}$

In other words, the rate of bidegradation is proportional to the concentration of the biodegradable material.

After integration,

$$\frac{S_t}{S_o} = e^{-K_d t}$$

where S_t = concentration of biodegradable material at time t

In aerobic digestion, S is conveniently selected as mixed liquor volatile suspended solids (MLVSS), a gross but acceptable measure of the biomass.

For a completely mixed flow condition that would normally exist in an aerobic digester, Reynolds (1967) calculates the materials balance as

Input - Output - Change in Reactor = Net Change

$$\frac{QS_o}{V} - \frac{QS_t}{V} - K_d S_t = 0$$

where Q = flow rate,
V = volume of reactor

Table 3-9. Aerobic Digestion Design Parameters (Weston, 1971)

Parameter	Value	Remarks	Reference
Hydraulic Detention Time	15-20 days	Waste activated sludge alone.	–
	20-25 days	Primary + waste activated sludge.	–
Air Requirements Diffuser System	20-35 cfm/1,000 ft^3	Enough to keep the solids in suspension and maintain a DO between 1-2 mg/l. Waste activated sludge alone.	Burd, 1968
	60 cfm/1,000 ft^3	Primary and waste activated sludge.	
Mechanical System	1.0-1.25 hp/1,000 ft^3	This level is governed by mixing requirements. Most mechanical aerators in aerobic digesters require bottom mixers for solids concentration greater than 8,000 mg/l, especially if deep tanks ($>$12 ft) are used.	–
Temperature	15°C	If sludge temperatures are lower than 15°C, additional detention time should be provided so that stabilization will occur at the lower biological reaction rates.	–
Volatile Solids Reduction	40-50%		
Tank Design		Aerobic digestion tanks are open and generally require no special heat transfer equipment or insulation. For small treatment systems (0.1 mgd) the tank design should be flexible enough so that the digester tank can also act as a sludge thickening unit. If thickening is to be utilized in the aeration tank, sock-type diffusers should be used to minimize clogging.	Burd, 1968

The net change is zero because steady state is assumed. Obviously, Q/V equals inverse of the residence time, θ. Hence

$$\theta = \frac{S_o - S_t}{K_d S_t}$$

For a given influent MLVSS (S_o) and a desired effluent MLVSS (S_t), it is possible to calculate the required residence time (θ) if the decay constant K_d is known.

Using batch pilot plant data K_d can be determined by measuring the slope of the line ($K_d/2.303$) of a plot of S_t/S_o vs time as a semilog plot, such as Figure 3-6.

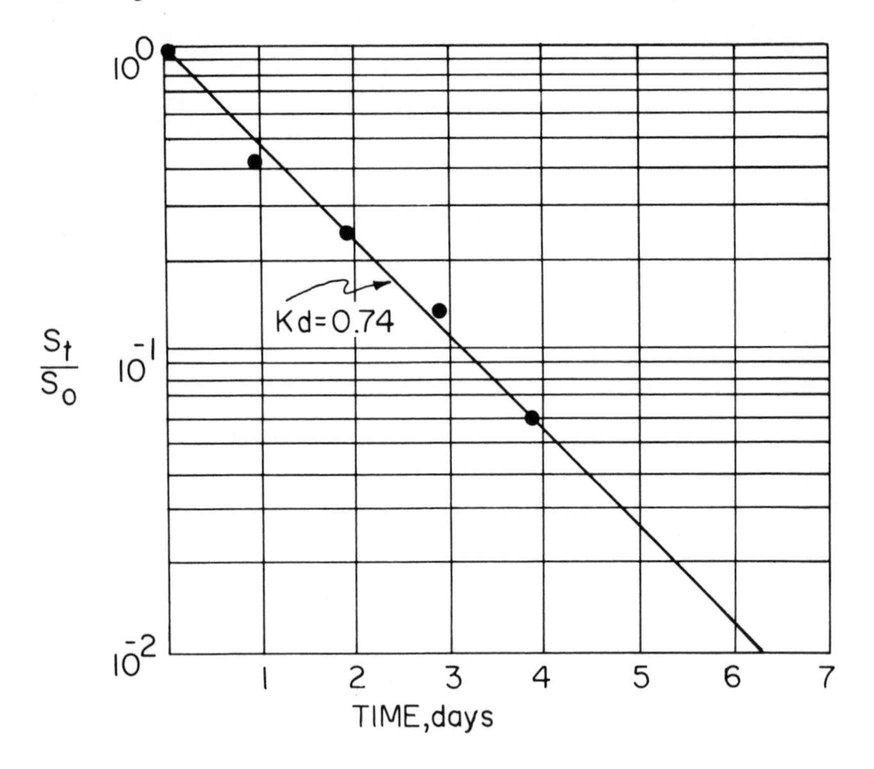

Figure 3-6. Determination of the decay constant in aerobic digestion experiments (after Reynolds, 1967).

Example 3-2. Batch data from the aeration of waste activated sludge results in MLVSS reduction as shown in Figure 3-6. The slope of the line is equal to

$$K_d/2.303 = 3.1 \text{ and } K_d = 0.74$$

If the influent MLVSS (S_o) are 10,000 mg/l, what is the required detention time for 75% reduction in MLVSS?

$$S_t = 10{,}000 \text{ mg/l} \times 0.25 = 2{,}500 \text{ mg/l}$$

$$\theta = \frac{S_o - S_t}{K_d S_t}$$

$$= \frac{10{,}000 - 2{,}500}{0.74\,(2{,}500)} = 4 \text{ days}$$

Lagoons

Lagoons are simply large holding basins with earth embankments. They are designed either as temporary holding basins or as a means for ultimate sludge disposal.

The latter, called *permanent lagoons*, tend to fill up about the time the design engineer retires from active practice. In other words, very few such lagoons are permanent. Lagooning is almost always a temporary holding, given cheap land and space available for the construction of big holes. In some cities lagoons are used for dewatering digested sludge, but the odors from such lagoons can be atrocious if the sludge is not well digested.

Some lagoons are designed as digesters. The requirement for thorough digestion in lagoons is generally about three years detention time, one of the years being required for resting without sludge. Obviously, these lagoons are only functional where cheap land is available and where neighbors will not be too upset by malodors.

Composting

Most wastewater sludges, even well-digested varieties, are not readily amenable for direct use on land. They are not, by the previous definition, stabilized, because they are either offensive to the senses or (such as digested/ sand bed-dried sludge) still contain substantial quantities of pathogenic microorganisms and/or chemical toxins. Composting, an aerobic biological method of sludge stabilization, can produce a product which has high utility, is reasonably safe, and is aesthetically acceptable. Why this technique of sludge stabilization has not found wider acceptance in the United States is discussed later.

Composting is, as noted above, an aerobic biological process. The basic aerobic decay equation holds:

$$\text{(Complex Organics)} + O_2 \xrightarrow[\text{Microorganisms}]{\text{Aerobic}} CO_2 + H_2O + NO_3 + SO_4^{=}$$

$$+ \text{ Other Less Complex Organics}$$

$$+ \text{ Heat}$$

During this decomposition the temperature increases to about 70°C (160°F) in most well-operated composting operations. The rise in temperature in a compost pile is shown in Figure 3-7. As the reaction develops the early decomposers are mesophilic bacteria followed in about a week by thermophilic bacteria, actinomycetes and thermophilic fungi. Above 70°C, spore-forming bacteria predominate. As the decomposition slows, the temperature drops and mesophilic bacteria and fungi reappear. Protozoa, nematodes, millipedes and worms are also present during the later stages. The concentration of dead and living organisms in compost can be as high as 25%.

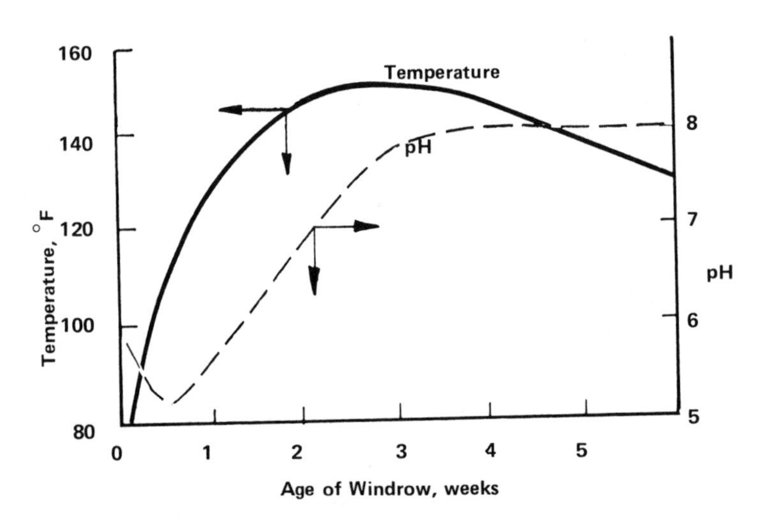

Figure 3-7. Temperature and pH variation in a compost pile.

The elevated temperatures destroy most of the pathogenic bacteria, eggs and cysts. Some of the more common pathogens and their survival at elevated temperatures are shown in Table 3-10. The product of thermophilic composting is essentially free of pathogens. Several extensive studies have been conducted on the fate of pathogens in composting, and in all cases the authors concluded that all potential pathogens, including resistant parasites such as *Ascaris* eggs and cysts of *Endamaeba histolytica* are destroyed (Wiley, 1962).

Table 3-10. Destructions of Some Common Pathogens and Parasites (Golueke, 1972)

Organism	Observations
Salmonella typhosa	No growth beyond 46°C; death within 30 min at 55°-60°C and within 20 min at 60°C; destroyed in a short time in compost environment
Salmonella sp.	Death within 1 hr at 55°C and within 15-20 min at 60°C
Shigella sp.	Death within 1 hr at 55°C
Escherichia coli	Most die within 1 hr at 55°C and within 15-20 min at 60°C
Entamoeba histolytica cysts	Death within a few minutes at 45°C and within a few seconds at 55°C
Taenia saginata	Death within a few minutes at 55°C
Trichinella spiralis larvae	Quickly killed at 55°C; instantly killed at 60°C
Brucella abortus or *Br. suis*	Death within 3 min at 62°-63°C and within 1 hr at 55°C
Micrococcus pyogenes var. *aureus*	Death within 10 min at 50°C
Streptococcus pyogenes	Death within 10 min at 54°C
Mycobacterium tuberculosis var. *hominis*	Death within 15-20 min at 66°C or after momentary heating at 67°C
Corynebacterium diphtheriae	Death within 45 min at 55°C
Necator americanus	Death within 50 min at 45°C
Ascaris lumbricoides eggs	Death in less than 1 hr at temperatures over 50°C

Because of the high rate of microbial activity, a large supply of nitrogen is required by the bacteria. If the reaction were slower, the nitrogen could be recycled, but since many reactions are occurring concurrently, a sufficient nitrogen supply is necessary. The requirement for nitrogen can be expressed as the C:N ratio.

Most experts agree that a C:N of 20:1 is the point below which nitrogen is not limiting the rate of decomposition. Above C:N of 80:1, thermophilic composting cannot occur since the nitrogen is severely limiting the rate of decomposition. Between these extremes, nitrogen is progressively limiting as the C:N ration increases.

The C:N ratio for activated sludge is about 6.3:1 and for mixed digested sludge the C:N is 15.7:1 (Golueke, 1972). Wastewater sludge, therefore, contains more than enough nitrogen. In fact, some of the nitrogen escapes as ammonium hydroxide above a pH of 7.0. Most sludge composting operates around a pH of 8.0 and thus cannot retain all of the available nitrogen.

The pH of the compost pile varies with time, showing an initial drop and then a leveling off at about 8.0, as shown in Figure 3-7. If the compost heap becomes anaerobic, the pH continues to drop. There is sufficient buffering within the compost to allow the pH to stabilize at an alkaline level provided the reaction stays aerobic. For educational purposes, the progression of pH and temperature in a compost pile can be readily demonstrated in a laboratory-scale apparatus (Vesilind, 1973).

The time required for a compost pile to mature depends on such factors as the putrescibility of the feed, the insulation and aeration provided, the C:N ratio, the particle size, and other conditions. Usually, 2 weeks is considered the minimum time for the adequate composting of municipal sludges. Mechanical composting plants, using inoculation of previously composted materials, can accomplish decomposition in 2 or 3 days. This material is, however, still quite active and usually requires further stabilization.

The completion of composting is judged mostly on the basis of a slight drop in temperature and a dark brown color. A more accurate measure is the determination of starch concentration in the compost. Starch is readily decomposable, and thus its disappearance is a good indicator of mature compost. A simple laboratory method for measuring starch in compost is available, although the technique yields only qualitative information (Lossin, 1971). This technique can also be applied to a composting demonstration project for the classroom (Vesilind, 1978).

It should be obvious that the composting process can work only in nonfluid situations where the microorganisms obtain sufficient oxygen. Mechanically dewatered sludge, which may have 70 to 80% moisture, compacts under its own weight and does not have the porous structure necessary to allow such input of oxygen. Further, the moisture is too high and would result in low composting temperatures even if the oxygen problem is solved.

Thus, one cannot simply pile up sludge and hope for composting to occur. The process can be achieved in one of three basic ways

- windrows with bulking agents
- aerated windrow with bulking
- mechanical composting.

The windrow technique consists of adding some bulking agent to the sludge, which allows oxygen penetration into the pile, and laying out the mixture in long piles or windrows, about 2.5 m (8 ft) wide, 1 to 2 m (4 to 6 ft) high.

The windrows are turned periodically to provide aeration. Bulking agents used can be previously composted dry sludge (Haug and Haug, 1977), wood chips (Mosher and Anderson, 1977; Epstein and Willson, 1974) or other available materials such as bark, rice hulls, etc. The bulking agent, as in the case of wood chips, can be recovered and reused. The ratio of sludge to wood chips is commonly between 1:2 and 1:3 by volume.

A logical extension of the windrow method is to use aerated windrows which would increase the rate of decomposition. Figure 3-8 shows one type of aeration employed. This reduces the time necessary to achieve decomposition and eliminates the need to turn the windrows.

The third composting technique is mechanical and includes a wide variety of proprietary and commercially available systems. The main difference among the different systems is the method of aeration, which ranges from rotary kilns that provide constant tumbling to auger screws in a digester. The major advantages cited by the mechanical systems are continuous (not cyclic, as with aerated windrows) operation and short retention time. Sometimes this short retention time is misleading, however, because the compost must be placed in windrows "to age," thus considerably extending the time and land requirements.

The operational control of composting is mainly in the control of moisture, volatile solids addition, and aeration.

Dewatered sludge ideally should be about 40% moisture before composting begins in order to assure proper porosity. With fairly dry bulking materials, however, a satisfactory 50% pile moisture can be readily obtained even for poorly dewatered sludge.

Figure 3-8, adapted from Haug and Haug (1977) facilitates such moisture calculations.

Example 3-3. Consider a sludge feed rate of 10 tons per day at 20% solids. Using dry wood chips as a bulking agent at a ratio of 1:2 sludge to chips, with 1 ton per day of new chips added and a requirement of 50% moisture in the pile, calculate the recycle ratio.

Materials balance by total wet solids is

$$T_o + T_b = T_c$$
$$10 + 1 = 11$$

Materials balance by dry solids

$$T_oX_o + T_bX_b = T_cX_c$$
$$(10)(0.2) + (1)(1.0) = (11)(X_c)$$
$$X_c = 0.273$$

The total moisture balance, assuming no net loss of moisture due to evaporation, is

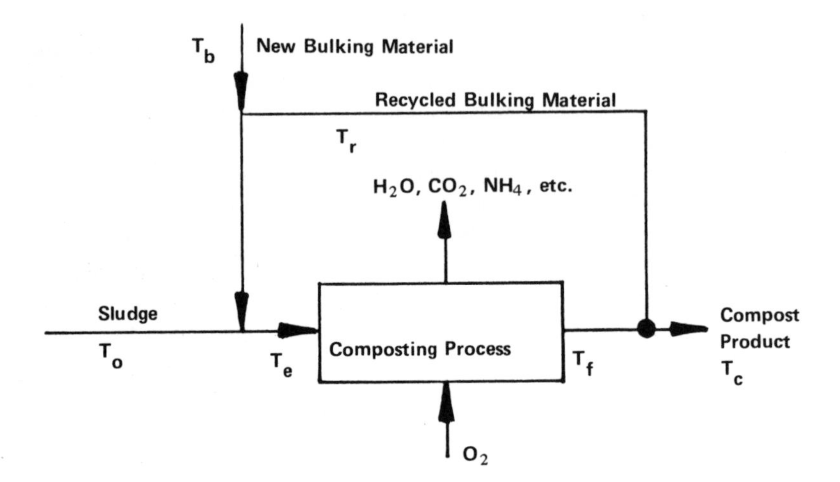

T = Wet weight of solids, mass units
V = Percent volatile solids
X = Fractional solids content
o = Incoming sludge
f = Final from composting process, before separation of bulking material
r = Recycle of bulking agent
b = Incoming bulking agent
c = Final compost product
e = Sludge and bulking agent mixture entering composting

Figure 3-8. Schematic of a composting operation. (after Haug and Haug, 1977)

$$T_f(1 - X_f) = T_o(1 - X_o) + T_b(1 - X_b) + T_r(1 - X_r)$$
$$T_f(0.5) = 10(0.8) + 0 + 0$$
$$T_f = 16 \text{ tons/day}$$

Since X_f is 0.5, the total solids input for the pile must be 8 tons/day

$$T_o(X_o) + T_b(X_b) + T_r(X_r) = 8 \text{ tons/day}$$
$$10(0.2) + 1(1) + T_r(1) = 8$$
$$T_r = 5 \text{ tons/day}$$

Hence 5 tons of dry bulking material must be recycled daily.

This example suggests that the total wood chips (bulking material) to be added to the pile is 6 tons per day for a sludge input of 10 tons per day at 20% solids. Data from the Beltsville USDA experiment station show that for a sludge of 22% solids, the mixing ratio should be 2:1, bulking material to sludge, by *volume* (Willson, 1977). The densities of the piles were not specified in the report.

A bulking material to sludge ratio of 3:1 by volume was reported for a project in Bangor, Maine. This mixture was stated to yield a moisture content of not more than 60% (Mosher and Anderson, 1977).

Volatile solids should be controlled in order to produce a sufficiently high rate of microbial degradation and hence temperature. At low temperature, destruction of pathogenic organisms is not achieved and the rate of composting is drastically reduced. The volatile solids concentration of the mixture of bulking materials and sludge can be calculated as in the example below.

Example 3-4. For the same conditions as in Example 3-3, assume that the incoming digested sludge is 50% volatiles, and the raw recycled bulking material volatiles are 98% and 80%, respectively. Calculate the percent volatiles in the pile. Also calculate the volatiles if the sludge is raw sludge with volatile solids of 70%.

With reference to Figure 3-8, the volatile solids balance is

$$T_oV_o + T_bV_b + T_rV_r = T_eV_e$$
$$(10)(0.5) + (1)(0.98) + (5)(0.8) = 16(V_e)$$
$$V_e = 0.62 \text{ or } 62\% \text{ volatiles}$$

If raw sludge is used,

$$V_e = 0.75 \text{ or } 75\% \text{ volatiles}$$

The final operational variable is aeration. The uptake of oxygen in the composting operation is a function of the rate of microbial activity, which in turn depends on the many variables discussed above. Theoretically, it should be possible to model the oxygen uptake rate similar to the aerobic digester. Practically, however, this would be difficult due to the variable temperatures, moisture levels and, hence, oxygen uptake rates within a compost pile. Also, except for mechanical or aerated pile composting, aeration is not continuous.

Experience with aerated wood chip/sludge composting, shown in Figure 3-9, indicates that the aeration rate should be between about 6 and 40 m^3/hr/tonne (200 and 1,200 ft^3/hr/ton) of sludge solids. It does not seem necessary to provide for more than about 8 m^3/hr/tonne (250 ft^2/hr/ton) to solids, since oxygen availability is no longer limiting at that rate (Willson, 1977).

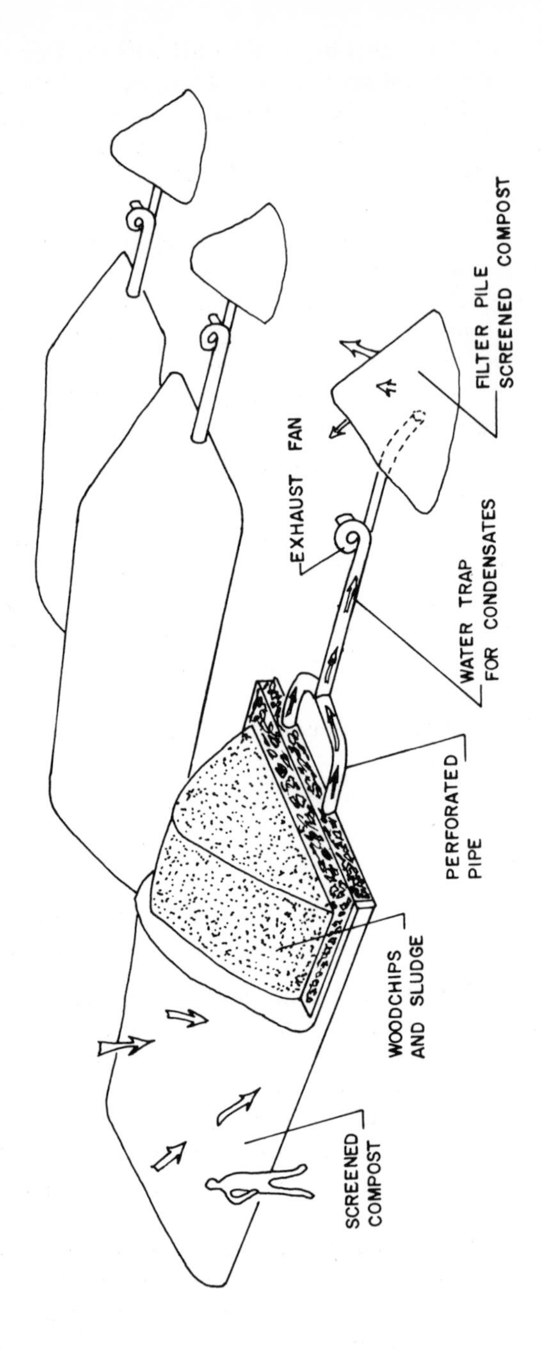

Figure 3-9. Aerated composting.

Composts produced from sewage sludges vary widely in quality and applicability to agricultural uses, dependent most on the type of community from which the sludge is derived. Table 3-11 shows some representative values of plant nutrients and sewage sludge compost.

Table 3-11. Typical Chemical Analysis of Compost from Raw Sludge (Epstein, 1977)

pH	6.8
Organic Carbon	23%
Water	35-58%
Potassium	0.2%
Phosphorus	1.0%
Calcium	1.4%
Magnesium	0.4%
Ammonia	235 mg/l
Nitrate	3 mg/l
Zinc	770 mg/l
Copper	300 mg/l
Nickel	55 mg/l
Cadmium	7.6 mg/l
Lead	290 mg/l

Recent data indicate that about 15 to 20% of the organic nitrogen in compost becomes available during a cropping period (Epstein *et al.*, 1977). Compost thus functions as a slow-release nitrogen fertilizer—a desirable attribute.

Potassium in sludge compost is low because it is water-soluble and remains with the liquid fraction in wastewater treatment.

Compost is usually considered a low-analysis fertilizer rich in organic matter. The addition of compost to soil improves the soil physical properties by (1) increasing the soil water content, (2) increasing the water retention capability, (3) enhancing aggregation, (4) increasing soil aeration by increasing permeability, and (5) decreasing surface crusting (Epstein and Parr, 1977).

Based on nitrogen requirements (See Chapter 11) the USDA recommendations for the application of aerated pile sewage sludge for different soils are shown in Table 3-12, and the suggested application rates for various uses are shown in Table 3-13.

One of the major criticisms of sludge composting over the years has been that although the compost is of demonstrated agricultural value, it cannot be sold because people just do not want it. Only a few cities have been able to market composted sludge, among them Los Angeles (County) and Chicago.

Table 3-12. Recommended Application Rates of Sewage Sludge Compost (Epstein and Parr, 1977)

Soil Texture	Soil Depth and Groundwater Conditions	Plants or Crops	Application Rate (tonne/ha)
Sand or Gravel	Shallow to groundwater, with no intervening soil	Grass or shrubs	60 to 110
	Deep to groundwater, with heavier material intervening	Grass or shrubs	110 to 220
Clay, Clay Loam, Silty Clay Loam	Shallow to groundwater	Grass	60 to 440
	Deep to groundwater	Grass or turf	220 to 440
Disturbed Soils	Deep to groundwater	Grass	220 to 440

Table 3-13. Recommended Application Rates of Sewage Sludge Compost (Epstein and Parr, 1977)

Plant or Crop	Application Rate (tonne/ha)	Method of Application
Sod, Turf, New Lawns	220 to 440	Till into surface prior to seeding
Sod, Turf, Established Lawns	50 to 80	Topdress
Agronomic Crops	Depending on nitrogen requirement	Till into soil prior to planting
Tree Nurseries	110 to 220	Till into soil prior to planting
Pastures	110 to 220	Till into soil prior to planting
Pastures	50 to 110	Broadcast periodically

A recent study conducted to determine why sludge composting has not been successful (Ettlich and Lewis, 1977) concluded that the main reason seems to be price. The material has been offered at a price which was too high in comparison with other commercial products. In addition, a successful operation requires favorable local publicity, delivery service, a catchy trade name, and the publication of guidelines and information on its use are necessary requisites for a successful program. Most composting operations seem to have failed because the municipality decided to make money on what should have been a public service.

CHEMICAL STABILIZATION

Lime Stabilization

Lime has long been used for sludge stabilization to reduce odor in latrines and two-holers all over the world. At sufficient high concentrations lime will increase pH until most biological life is destroyed, stopping odor production. Quicklime (CaO) also has the ability to dewater sludge by the reaction $CaO + H_2O \rightarrow Ca(OH)_2$. Lime is a good substance for sludge stabilization because nothing is oxidized and no dangerous substances are formed.

One function of all stabilization methods is the destruction of pathogenic organisms. It has been determined that a pH of 11-11.5, at 15°C and 4-hr detention time will destroy all *E. coli* and *Salmonella typhosa* (Riehl *et al.*, 1952). Some spore-forming bacteria may survive (Grabow *et al.*, 1969), but few pathogens form spores. Higher organisms such as hookworm and amoebic cysts may survive at least 24 hr at high pH (Farrell *et al.*, 1974). Smith (1974) suggests that a pH of 12.2 and a 30-min detention time with air mix is essential for the destruction of pathogens.

Lime also tends to eliminate odors and improve sludge dewaterability. Farrell *et al.* (1974) found that the addition of lime to raw primary sludge previously precipitated with alum produced a significant drop in specific resistance to filtration, a measure of how well a sludge dewaters (See Chapter 6). Some of his data are shown in Table 3-14.

Table 3-14. Lime Stabilization of Two Sludges (Farrell *et al.*, 1974)

	Alum Sludge		Iron III Sludge	
	Before	After	Before	After
pH	6.7	11.5	6.5	11.5
Specific Resistance (m/kg x 10^{12})	19.0	3.8	14.9	7.0

Sontheimer (1966) introduced the "lime bonding capacity." Lime reacts with sludge to form a calcium salt and other compounds containing calcium. Lime bonding capacity is the amount of lime required by the sludge to maintain a certain pH. If the pH is raised to 11.5, a steady pH drop to about 11.0 will occur in most sludges. Their lime bonding capacity is the amount of lime required to maintain a pH of 11. It is possible to neutralize this lime bonding capacity by bubbling in carbon dioxide, which results in the formation of calcium carbonate and thus further improves sludge dewaterability.

Paulsrud (1973) has found that even though lime bonding capacity might be calculated for short-term effects, long-term storage in lagoons, for example, will gradually decrease the pH and eventually produce nuisance conditions. In other words, lime-stabilized sludge is not chemically stable, and it is impossible to maintain a high pH even with very high lime dosages. Chemical and/or biological action eventually seems to bring the pH down (Sørensen, 1973).

In order to avoid rapid pH drop and the onset of nuisance conditions, it is necessary to keep the pH of the sludge above 11.0 for at least 14 days. The amount of lime required for this was determined by Paulsrud and Eikum (1974) and is shown in Table 3-15.

Table 3-15. Lime Dose Required to Hold the pH of a Sludge Above 11.0 for 14 Days (Paulsrud and Eikum, 1974)

Type of Sludge	Dose [lb $Ca(OH)_2$ per ton sludge solids]
Raw Primary	200-300
Septic Tank Sludge	200-600
Secondary Biological	600-1000
Alum (Secondary Precipitation)	800-1000
Iron (Sedondary Precipitation)	700-1200

Note: To obtain kg/tonne, multiply by 0.5.

Chlorine Stabilization

The BIF Corporation has introduced the *Purifax* system for sludge stabilization using heavy doses of chlorine (about 2,000 mg/l). This sludge dewaters well on drying beds and has excellent long-term stability.

Chlorine-stabilized sludges are somewhat difficult to dewater on filters, since the low pH (ca. 2) interferes with the action of the chemical conditioners. Test results indicate that the pH should be greater than 4 to allow for acceptable conditioning. The low pH and high chlorine levels promote the production of various chlorinated compounds of undetermined character, and caution should be exercised in the ultimate disposal of such sludges.

Oxygen Stabilization

Oxygen can be used to stablize sludge in either of two ways–biologically or chemically at elevated pressure and temperature.

Biological stabilization with oxygen resembles aerobic digestion except that pure oxygen is used instead of air. There are several oxygen wastewater treatment systems that have stabilized sludge up to 5% solids. Such thick sludge produces considerable biological energy, resulting in an elevated temperature and thus more rapid metabolic activity (Smith, 1974).

Chemically, oxygen can be used to stabilize sludge at a sufficiently high temperature, pressure and reaction time. The *Zimpro* process utilizes pressures ranging from 150-3,000 psi and steam to increase temperature to a self-sustaining oxidation reaction. Sludges thus produced are sterile and normally are easily dewatered.

Heat-stabilized sludges compact well, and 50% solids are common in vacuum filtration. The resulting filtrate, however, has a BOD of around 2,500 mg/l. Such high filtrate BOD values make it necessary to include this load in the design of the plant. The use of heat for sludge conditioning is discussed further in Chapter 7.

Heat Stabilization

Heat is the major cause for the destruction of pathogens in both anaerobic digestion and in composting, as noted above. In composting, the heat is derived from exothermic biological reactions within the sludge, and is sufficient to achieve almost complete sterilization. There seems to be an order of magnitude difference in the destruction rate between mesophilic digestion (35°), thermophilic digestion (50°) and composting (60°). Temperatures in composting systems can be uneven, however, and some pathogens can survive.

The next higher temperature system for the sterilization of sludge cannot be a biological system since obviously microorganisms must survive. Pasteurization at 70°C for 30 min to 1 hr, using an external heat source, has been successfully used to accomplish such sterilization. Direct steam injection is used to prevent organic fouling and inorganic scaling (Stern and Farrell, 1977).

Pasteurization has been used in Europe for some years with sludge spread on pastures during the summer growing season (Triebel, 1967). The total heat requirement is about 1,400 x 10^6 cal/tonne (5 x 10^6 Btu/ton) of wet sludge. This energy cost has been usually considered excessive in the United States.

Thermophilic digestion also has been shown to be effective in reducing the presence of pathogenic organisms. In one study (Ohara and Colbaugh, 1975) an anaerobic digester was run at 50°C in parallel with mesophilic digesters at 35°C. Thermophilic digestion consistently reduced the *Salmonella* densities to below detectable limits whereas in mesophilic digestion, although the reduction was greater than 99%, the sludge was not free of these pathogens. Similarly, viruses were measured using coliphages, and the average PFU per gram of wet sludge was 25.4, 2.1 and 0.03 for raw mesophilically digested and thermophilically digested sludges, respectively. Interestingly, there was no difference in the concentration of *Ascaris lumbricoides* between the two types of digestion. It is necessary to increase the temperature to about 60°C, as in composting, before this pathogen is destroyed.

Irradiation

High-energy irradiation of sludge for obtaining disinfection is a predictable and controllable operation that does not depend on high temperatures and the resulting corrosion and fouling associated with such systems. In addition, the costs are reputed to be about equal to alternative heat treatment systems. These systems are versatile and potentially beneficial in breaking down some refractory organic compounds such as PCBs. The disadvantages of high-energy irradiation are their high cost and technology and problems of safety (Stern and Farrell, 1977).

The sources of radiation used to disinfect sludges are (1) gamma radiation with cobalt-60 and cesium-137 as the principal radioisotopes, (2) waste products from nuclear reactors, and (3) high-energy electrons produced by accelerators.

The size of the microorganism affects its susceptibility to radiation, with larger microorganisms being most sensitive. For example, *Ascaris lumbricoides* is destroyed at 500 krad, 300 to 400 krad is needed to destroy bacteria, and 400 to 1,000 krad produces inactivation of virus. High temperature and oxygen both increase the effectiveness of irradiation (Sinskey *et al.*, 1976).

The destruction of pathogens by radiation is shown in Figure 3-10. It is of interest here that *E. coli* are more sensitive than *Salmonella* or bacteriophages. Obviously, *E. coli* are inadequate indicator organisms in this case.

In one U.S. research application, energized electrons are used as the source of ionizing energy (Trump *et al.*, 1976). As shown in Figure 3-11, the sludge is spread in a thin layer over the top of a rotating stainless steel drum and is ionized throughout its volume by a downward directed high-energy electron beam. The source of energy is a 750,000-V, 50-kW electron accelerator. The cost of disinfection of dewatered sludge is estimated at about $5.00 per ton of dry solids (Trump *et al.*, 1976).

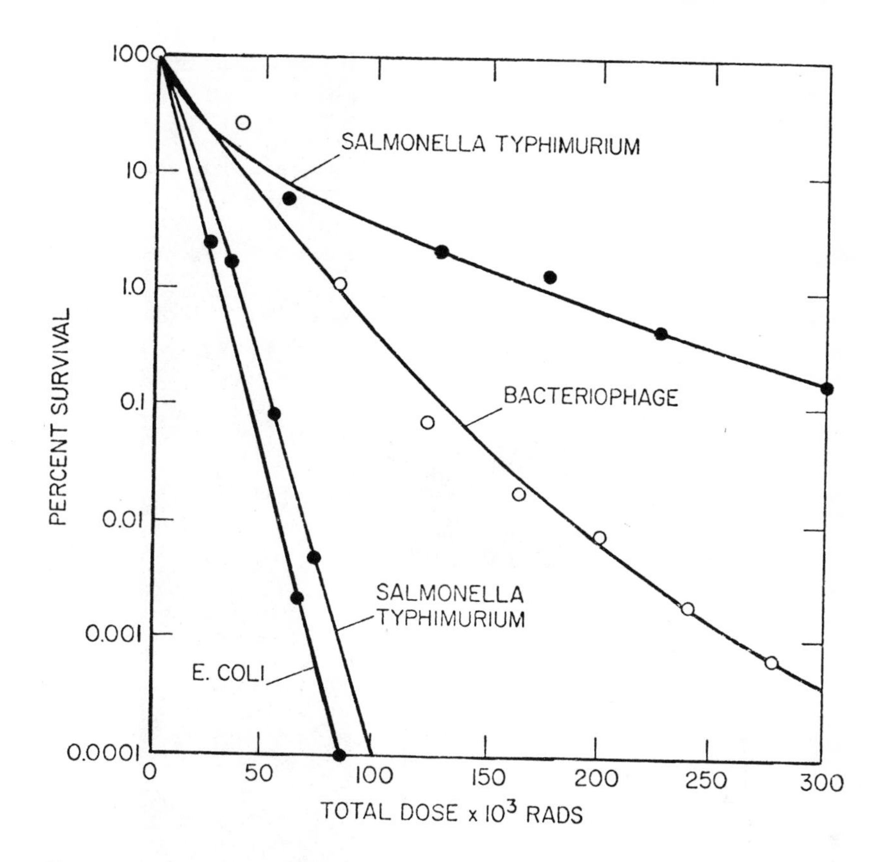

Figure 3-10. Survival of test organisms of differing sensitivity to radiation (Trump, ***et al.*** **(1976).**

CONCLUSIONS

There are many ways to measure sludge stability, depending on the desired criterion and the problem involved. The location and method for final disposal of the sludge must be specified before the required degree of sludge stability can be determined and analyzed rationally. It has been suggested that sludge stability be judged on the basis of both potential damage to the environment and production of nuisance conditions. If either condition is not met, the sludge is not stable, and further treatment is required. Such treatment obviously must be designed to solve the specific problem, and thus the selection of a method for measuring stability must depend on the proposed method of ultimate disposal.

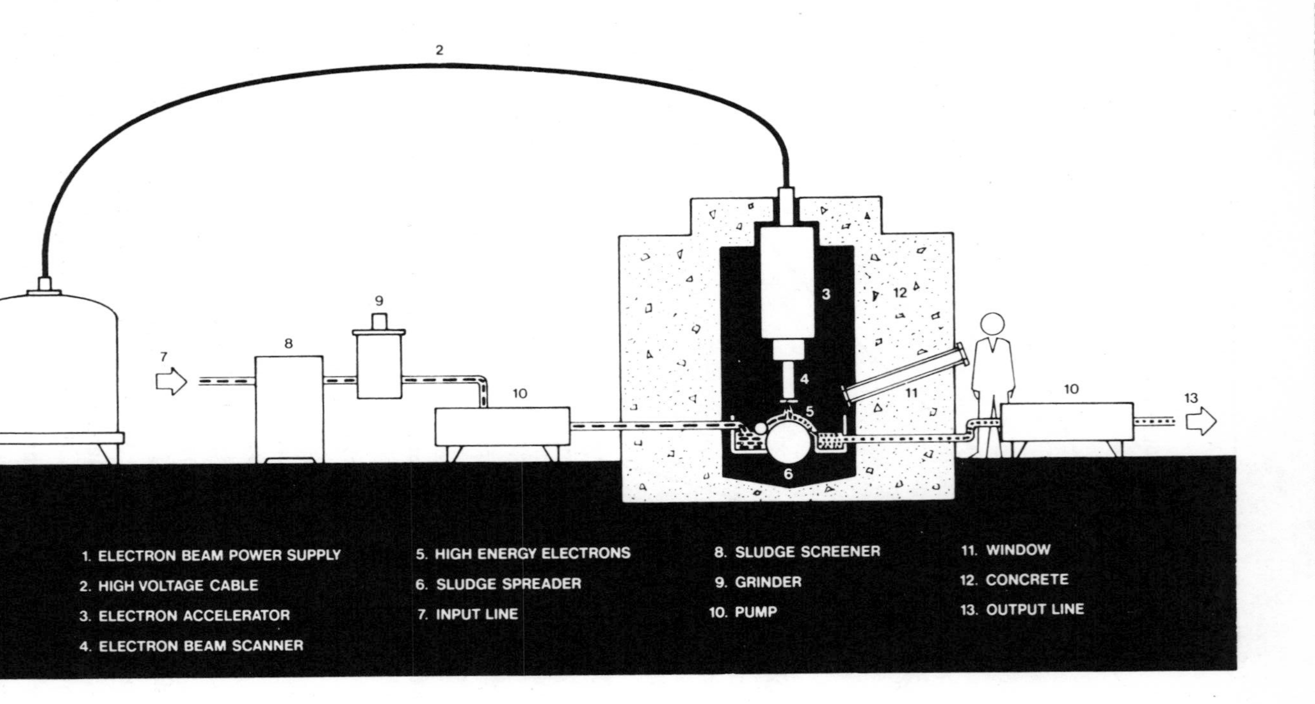

Figure 3-11. Irradiation of sludge (Trump *et al.*, 1976).

REFERENCES

Andrews, J. F. (1969). "Dynamic Model of the Anaerobic Digestion Process." *J. San. Eng. Div. ASCE* 95:SA1.

Andrews, J. F., and S. P. Graef (1970). "Dynamic Modeling and Simulation of the Anaerobic Digestion Process," *Advances in Chemistry*, ACS.

Banta, A. P., and R. Pomeroy (1934). "Hydrogen Ion Concentration and Bicarbonate Equilibrium in Digested Sludge," *Sewage Works J.* 6:234.

Barber, N. R. (1978). "Lime/Sodium Bicarbonate Treatment Increases Sludge Digester Efficiency," *J. Environ. Sci.* March/April, p. 28.

Bisogni, J. J., and W. L. Alonzo (1971). "Relationship between Biological Solids Retention Time and Settling Characteristics of Activated Sludge," *Water Res.* 5:753.

Burd, R. S. (1968). "A Study of Sludge Handling and Disposal," FWPCA (EPA) Pub. WP-20-4.

Buswell, A. M. (1939). "Anaerobic Fermentation," Ill. State Water Survey Bulletin, Vol. 32.

Dague, R. R. (1968). "Application of Digestion Theory to Digester Control," *J. Water Poll. Control Fed.* 40:12.

Eckenfelder, W. W., Jr. (1967). "Mechanism of Sludge Digestion," *Water Sew. Works* 114:6.

Eikum, A. S. (1973). "Aerobic Stabilization of Primary and Mixed Primary/Chemical (Alum) Sludge," PhD Thesis, Univ. Washington, Seattle.

Epstein, E., D. B. Keane, J. J. Meisinger and J. O. Legg (1977). "Mineralization of Nitrogen from Sewage Sludge and Sludge Compost," presented at Am. Soc. Agronomy meeting, Los Angeles, CA, November.

Epstein, E., and G. B. Willson (1974). "Composting Sewage Sludge," Proc. Nat. Conf. Mun. Sludge Management, Pittsburgh, PA.

Epstein, E., and J. F. Parr (1977). "Utilization of Composted Municipal Wastes," Proc. Nat. Conf. Composting of Municipal Residues and Sludge, Rockville, MD.

Ettlich, W. F., and A. E. Lewis (1977). "User Acceptance of Wastewater Sludge Compost," U.S. EPA 600/2-77-096, Washington, DC.

Farrell, J. B., J. E. Smith, Jr., S. W. Hathaway and R. B. Dean (1972). "Lime Stabilization of Primary Sludges," *J. Water Poll. Control Fed.* 46:1.

Golueke, C. G. (1972). *Composting* (Emmaus, PA: Rodale Press, Inc.).

Grabow, W. O. K., N. A. Grabow and J. S. Burger (1969). "The Bactericidal Effect of Lime Flocculation Flotation as a Primary Unit Process in a Multiple System for the Advanced Purification of Sewage Works Effluent," *Water Res.* 3:943.

Great Lakes-Upper Mississippi River Board of State Sanitary Engineers (1968). "Recommended Standards for Sewage Works," Health Education Service, Albany, NY.

Haug, R. T., and L. A. Haug (1977). "Sludge Composting: A Discussion of Engineering Principles," *Compost Sci.* November/December, p. 6.

Jaworsky, G. W., and G. A. Rohlich (1963). "Aerobic Sludge Digestion," *Advances in Biological Waste Treatment*, Eckenfelder and McCabe, Eds. (New York: Pergamon Press, Inc.).

Kempa, E. S. (1969). "Uber aerobe Schlammstabilisierung bei niedrigen Temperaturen," *Wasser Luft Betrieb* 13:336 (from Eikum, 1973).

Krauth, K. (1969). "Erfahrungen bei der Aeroben Schlammstabilisation," *Bev. Abwassertech. Vereining* 20:271 (from Eikum, 1973).

Kugelman, I. J., and K. K. Chen (1970). "Toxicity, Synergism and Antagonism in Anaerobic Waste Treatment Processes," presented at ACS meeting, Houston, TX.

Lamb, J. C., III, W. C. Westgarth, J. L. Rogers and A. P. Vernimmen (1964). "A Technique for Evaluating the Biological Treatability of Industrial Wastes," *J. Water Poll. Control Fed.* 36:10.

Lawrence, A. W. (1971). "Application of Process Kinetics to Design of Anaerobic Processes," *Anaerobic Biological Treatment Processes*, F. G. Pohland, Ed., ACS Adv. in Chem. Ser. 105, Washington, DC.

Lawrence, A. W., and P. L. McCarty (1969). "Kinetics of Methane Fermentation in Anaerobic Treatment," *J. Water Poll. Control Fed.* 41:2.

Lawrence, A. W., I. J. Kugelman and P. L. McCarty (1964). "Ion Effects in Anaerobic Digestion," Tech. Rep. No. 33, Department of Civil Engineering, Stanford Univ., Stanford, CA.

Lossin, R. D. (1971). "Compost Studies," *Compost Sci.*, March/April.

Lynam, B., G. McDonnell and M. Krup (1967). "Start-up and Operation of Two New High-Rate Digestion Systems," *J. Water Poll. Control Fed.* 39:4.

Middlebrooks, E. J., and C. F. Garland (1968). "Kinetics of Model and Field Extended Aeration Wastewater Treatment Units," *J. Water Poll. Control Fed.* 40:4.

Monod, J. (1942). "Recherches sur la Croissance des Cultures Bacteriennes," *Hermann et Cie*, France (from Lawrence and McCarty, 1969).

Mosher, D., and R. K. Anderson (1977). "Composting Sewage Sludge by High Rate Suction Aeration Techniques," U.S. EPA SW-614d, Washington, DC.

Mudrack, K. (1966). "Die Aerobe Schammstabilisierung," *Munchner Beit. Zwo Abw.-Fisch-und Fluzbiologie* 13:291 (from Eikum, 1973).

Ohara, G. T., and J. E. Colbaugh (1975). "A Summary of Observations in Thermophilic Digester Operations" in *Sludge Management and Disposal* (Rockville, MD: Information Transfer).

O'Rourke, J. T. (1960). "Kinetics of Anaerobic Treatment at Reduced Temperatures," PhD dissertation, Stanford Univ., Stanford, CA.

Patterson, J. W., P. L. Brezonik and H. D. Putnam (1970). "Measurement and Significance of Adenosine Triphosphate in Activated Sludge," *Environ. Sci. Technol.* 4:7.

Paulsrud, B. (1973). Research Engineer, Norwegian Institute for Water Research, Oslo, unpublished data.

Paulsrud, B., and A. Eikum (1974). Internal report, Norwegian Institute for Water Research, Oslo.

Pearson, E. A. (1968). "Kinetics of Biological Treatment," in *Advances in Water Quality Improvement*, E. F. Gloyna and W. W. Eckenfelder, Jr., Eds. (Austin, TX: University of Texas Press).

Randall, C. W., J. K. Turpin and P. H. King (1971). "Activated Sludge Dewatering: Factors Affecting Drainability," *J. Water Poll. Control Fed.* 43:1.

Reynolds, T. D. (1967). "Aerobic Digestion of Waste Activated Sludge," *Water Sew. Works* 114:2.

Riehl, M. L., H. H. Weiser and B. T. Rheins (1952). "Effect of Lime-Treated Water on Survival of Bacteria," *J. Am Water Works Assoc.* 44:5.

Ruffer, H. (1966). "Untersuchungen zur Characteriserung Aerob Biologisch Stabilisierter Schlamme," *Vom Wasser* 13:255 (from Eikum, 1973).

Schroepfer, G. J., and N. R. Ziemke (1959). "Development of the Anaerobic Contact Process," *Sew. Ind. Wastes* 31:2.

Sinskey, A. J., D. Shah and T. J. Metcalf (1976). "Biological Effects of Irradiation with High Energy Electrons," Proc. Nat. Conf. Sludge Management Disposal and Utilization, Miami Beach, FL.

Smith, J. E., Jr. (1974). Ultimate Disposal Research Program, EPA, personal communication.

Sontheimer, H. (1966). "Effects of Sludge Conditioning with Lime on Dewatering," Proc. Third Int. Conf., Munich, *Advances in Water Pollution Research*, Vol. 2.

Stern, G., and J. B. Farrell (1977). "Sludge Disinfection Techniques," Proc. Nat. Conf. Composting of Mun. Residues and Sludge, Rockville, MD.

Stern, G. (1974). "Pasteurization of Liquid Digested Sludge," Proc. Nat. Conf. on Mun. Sludge Management, Pittsburgh, PA.

Sørensen, P. E. (1973). Research Engineer, Water Quality Institute, Sørborg, Denmark, unpublished data.

Summer, W. (1971). "Odour Pollution of Air," (Cleveland, OH: CRC Press, Inc.).

Swanwick, J. D., W. J. Fisher and M. Foulkes (1972). "Some Aspects of Sludge Technology Including New Data on Centrifugation," *Water Pollution Manual* (Br.) HMSO.

Triekel, W. (1967). "Experiences with Sludge Pasteurization in Niersverband; Techniques and Economy," Inf. Bulletin, Int. Res. Group on Refuse Disposal (IRGRD) No. 21-31, p. 330.

Trump, J. C., *et al.* (1976). "Large-Scale Electron Treatment of MDC Boston Sludge, Physical and Chemical Aspects," Proc. Nat. Conf. Sludge Management Disposal and Utilization, Miami Beach.

Vesilind, P. A. (1973). "Laboratory Exercise in Composting," *Compost Sci.*, September.

Vesilind, P. A. (1978). "Solid Waste Engineering Laboratory Manual," (Durham, NC: Duke University Press).

Weston, Roy F., Inc. (1971). "Upgrading Existing Wastewater Treatment Plants," EPA Technology Transfer Contract 14-12-933.

Wiley, J. S. (1962). "Pathogen Survival in Composting Municipal Wastes," *J. Water Poll. Control Fed.* 34:80.

Willson, G. B. (1977). "Equipment for Composting Sewage Sludge in Windrows and in Piles," Proc. Nat. Conf. Composting of Municipal Residues and Sludges, Rockville, MD.

PROBLEMS

1. Using a plot similar to Figure 3-3, show and explain what would happen in a digester if
   (a) the temperature suddenly drops,
   (b) a toxic material is added which first affects the acid-forming bacteria,
   (c) the feed is discontinued.
2. List the analytical tests you would wish to perform to determine the "stability" of a sludge if it were to be
   (a) placed on the White House lawn,
   (b) dumped in a river,
   (c) sprayed on a golf course.
3. According to Buswell's formula, how many kg of methane should theoretically be produced from a kg of formaldehyde? Why would this be difficult to achieve in practice?
4. What digester volume would normally be required for a secondary (activated sludge) plant serving a city of 100,000?
5. Using the data in Table 3-6, calculate the percent reduction in total solids. Do you think the Chicago operation included supernatant withdrawal? Why or why not?
6. Using the data and information in Example 3-1, calculate the substrate removal efficiency, and the concentration of microorganisms in the digester if the feed (substrate) COD increased to 12,000 mg/l, and if the solids retention time remained the same.
7. Note that the yield (Y) in the examples in the text is positive. In Table 3-6, however, total solids in a digester are reduced. Does this make sense?
8. Calculate the COD removal efficiency in Example 3-1 if the digester were run at thermophilic range instead of mesophilic.
9. Suppose, in the case of Example 3-1, the maximum possible solids retention was 1.0 days. Would the digester still be operational?

10. What aerobic digester volume would be required for a secondary (activated sludge) plant serving a city of 100,000?
11. The method for calculating aerobic digester capacity from pilot plant data uses MLVSS as the "substrate" to be reduced. However, substrate is measured by COD in the anaerobic digester analysis. Does this make sense?
12. Design (including a plan of the plant layout) an aerated compost pile to serve a city of 100,000.
13. Suppose, in Example 3-3, it rains for 2 weeks, at a drizzle of 0.5 in./day, and that half of this penetrated the piles. What would this do to the composting operation? How can the operation be changed to allow for continued processing?
14. Using the data in Table 3-10, what is the C:N ratio for this sludge, and is it within reasonable limits?
15. If a sludge is sand bed dry (40% solids) before composting, would water have to be added to a wood chip/sludge composting process in order to attain required moisture content? Under what circumstances might sand drying sludge before composting be reasonable?

CHAPTER 4

SLUDGE PUMPING

Efficient and economical sludge disposal often involves pumping sludge in liquid form over fairly great distances. As a rule of thumb, the cost of pumping sludge for about 15 kilometers (10 miles) equals the cost of sludge dewatering and disposal as a solid.

As discussed in Chapter 2, concentrated sludge is not a Newtonian fluid, and thus the analysis of the fluid mechanics of sludge pipeline flow is rather difficult. The major design problem is the estimation of the energy (head) loss due to pipe friction.

DETERMINATION OF FRICTION FACTORS

The standard friction factor versus Reynolds number relationship used in the design of pipelines for Newtonian fluids includes the fluid viscosity. The fluid viscosity for Newtonian fluid is independent of the shear rate (by definition). For non-Newtonian fluids such as a sludge, viscosity depends on the shear rate and thus cannot be specified without also specifying the shear, (*e.g.*, as by liquid velocity in a pipe). Further, sludges are thixotropic fluids, and thus the viscosity is further dependent on the immediate past history. Determining the friction factor for sludge flow is therefore quite difficult.

One potentially useful method of non-Newtonian fluid transport was developed by Metzner and his co-workers. Recall that for laminar Newtonian flow, where the Reynolds number is less than 2,000,

$$f_F = \frac{16}{R} = \frac{16}{\frac{DV\rho}{\mu}} \qquad (4\text{-}1)$$

where f_F = Fanning friction factor

R = Reynolds number

D = pipe diameter, m (ft)
V = average velocity, m/sec (ft/sec)
ρ = fluid density, kg/m^3 ($lb\ sec^2/ft^4$)
μ = fluid viscosity kg/m sec ($lb_f\ sec/ft^2$)
(Note: 1×10^{-3} kg/m sec = 1 poise.)

Metzner and Reed (1955) were able to derive an expression applicable to all types of fluids. Their general relationship is

$$f_F = \frac{16\,(K8^{n-1})}{D^n\,V^{2-n}\rho} \tag{4-2}$$

where K = "consistency index," equal to viscosity (μ) for Newtonian fluids, poise or kg/m sec ($lb_f\ sec/ft^2$)
n = "flow behavior index," which indicates degree of departure from Newtonian fluids. For $n = 1$, the system is Newtonian, for $n < 1$, the system is pseudoplastic (as are most sludges)
(See p. 27 for definition of n.)

The modified Reynolds Number is thus

$$R' = \frac{D^n V^{2-n}}{K8^{n-1}}\rho \tag{4-3}$$

The head loss, h_L, is then calculated as

$$h_L = 4f_F\left(\frac{L}{D}\right)\left(\frac{V^2}{2g}\right) \tag{4-4}$$

where L is the pipe length.

To use the method, K and n must first be calculated from a rheogram by curve fitting. The generalized Reynolds number is then calculated and the friction factor found from $f_F = 16/R'$. The pressure loss is then determined from Equation 4-4.

Most sludge pipelines, however, exhibit turbulent flow, and the above analysis is no longer applicable. The first attempt to develop real-life equations for homogeneous slurries was by Caldwell and Babbitt (1941) who concluded that conventional pipe-flow analyses can be used if the density of the slurry and the viscosity are defined as

$$\mu = \frac{g\tau_w D}{8V}$$

where μ = viscosity of the slurry
D = pipe diameter
V = velocity of flow
g = acceleration due to gravity
τ_w = shear stress at the wall of the pipe

Dodge and Metzner (1959) have published a generalized plot of the Fanning friction factor vs the modified Reynolds number, shown as Figure 4-1.

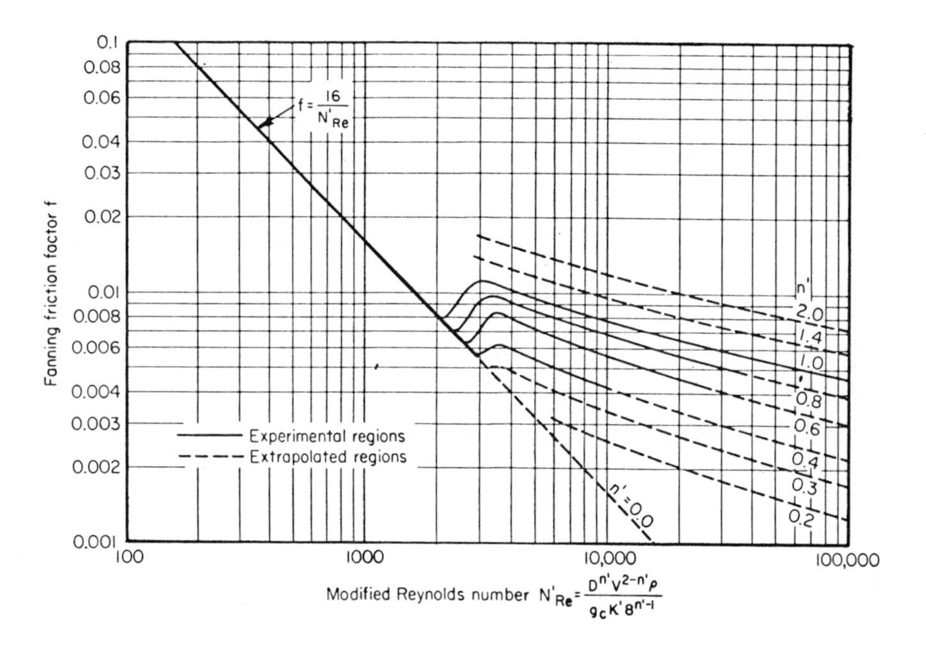

Figure 4-1. General pipe friction diagram (Dodge and Metzer, 1959).

The procedure for estimating head loss in a pipeline is first to calculate the modified Reynolds number from rheological data, enter Figure 4-1 to obtain f_F, and use Equation 4-4 to calculate the head loss.*

Example 4-1: Viscosity measurements on a sludge, using a rotational viscometer, yield a rheogram (τ vs du/dy), that can be described by the equation

$$\tau = K \left(\frac{du}{dy}\right)^n$$

with K = 0.03 poise and n = 0.9. If the sludge is pumped for 200 m in a 0.2-m pipe at a velocity of 1 m/sec, find the head loss due to friction.

*The procedure presented here differs slightly from the method of Dodge and Metzner. It is simplified, but for sludge pumping problems will yield results well within desired accuracy.

$$R' = \frac{D^n V^{2-n} \rho}{K 8^{n-1}}$$

$$= \frac{(0.2)^{0.9} (1)^{2-0.9} (1{,}000)}{(3 \times 10^{-3}) (8)^{0.9-1}}$$

(Note: 1,000 is the density of the sludge as kg/m³ assumed to be about that of water. The K value is converted from poise to kg/m sec by multiplying by 0.1.)

$$R' = 9.6 \times 10^{-4}$$

From Figure 4-1, f = 0.0055

$$h_L = 4f\left(\frac{L}{D}\right)\left(\frac{V^2}{2g}\right)$$

$$= 4(0.0055)\left(\frac{200}{0.2}\right)\left(\frac{1}{2 \times 9.8}\right) = 1.12 \text{ m}$$

The difficulty in using this procedure for wastewater sludges is that one does not have a *typical sludge*, thus the calculation of K and n for a particular sludge is of no value for any other. It is necessary to calculate these two constants for every sludge, and it cannot be assumed they will not change with concentration of solids or some other parameter. In addition, the thixotropic (time-dependent) nature of sludges makes the application of Metzner's method difficult.

Because of the problems involved in finding a rigorous approach to the design of sludge piping systems, several practical methods have been developed. The most commonly used approach is based on an adjustment of the Hazen-Williams equation. This equation reads:

$$V = 1.318\ Cr^{0.63} \cdot s^{0.54}$$

where V = the velocity in feet per second,
r = hydraulic radius in feet, (area/wetted perimeter)
s = the slope of hydraulic gradient
C = the Hazen-Williams friction coefficient

In metric units, the Hazen-Williams equation reads

$$V = 0.85\ Cr^{0.63}\ s^{0.54}$$

where V is in m/sec and r is in meters.

Assuming that the solids in a sludge will make its pumping more difficult, it is argued that this extra difficulty is compensated for by simply decreasing the value of the Hazen-Williams coefficient (C), thus requiring more head to create a certain flow in a pipeline. Table 4-1 is a tabulation of such a suggested modification for raw sludge (Brisbin, 1957). Similar correction factors are often used in design offices for other types of sludges. As activated sludge seldom is concentrated enough to behave any different than water, the equations applicable for water flow are used.

Table 4-1. Hazen-Williams Coefficients for Various Solids Concentrations of Raw Sludge (Brisbin, 1957)

Percent Total Solids	Apparent H-W Coefficient, C, Based on C = 100 for Water
0	100
2	81
4	61
6	45
8.5	32
10	25

Example 4-2: For the same conditions as given in Example 4-1, assuming the sludge is 4% solids (note 0.1 m = 0.328 ft, and 200 m = 656 ft.)

$$V = 1.318\, C\, r^{0.63}\, s^{0.54}$$

$$3.28 = 1.318(61)\left[\frac{\left(\frac{\pi(2 \times 0.328)^2}{4}\right)}{\pi(2 \times 0.328)}\right]^{0.63}\left(s^{0.54}\right)$$

$$s^{0.54} = 0.127$$

$$s = 0.022$$

$$h_L = sL = (0.022) \times 656 = 14.4 \text{ ft} = 4.4 \text{ m}$$

A third means of estimating head loss in sludge pipelines is to use an empirically derived graph such as Figure 4-2 (Stanley Consultants, 1972). This plot is derived mostly from actual data and can provide reasonable estimates of energy requirements, especially where positive displacement pumps are used and velocities are thus fairly well established.

Example 4-3: For the same conditions as given in Example 4-1, V = 1 m/sec = 3.28 ft/sec, for a 4% solids sludge, the friction head loss from Figure 4-2 is 3.2 ft/100 ft, or for a 656-ft pipe, h_L = 21 ft = 6.4 m.

It is interesting to note that the three examples, using the same data, gave head loss values of 1.12, 4.4 and 6.4 m, respectively. Sludge pumping, obviously, is not an exact science.

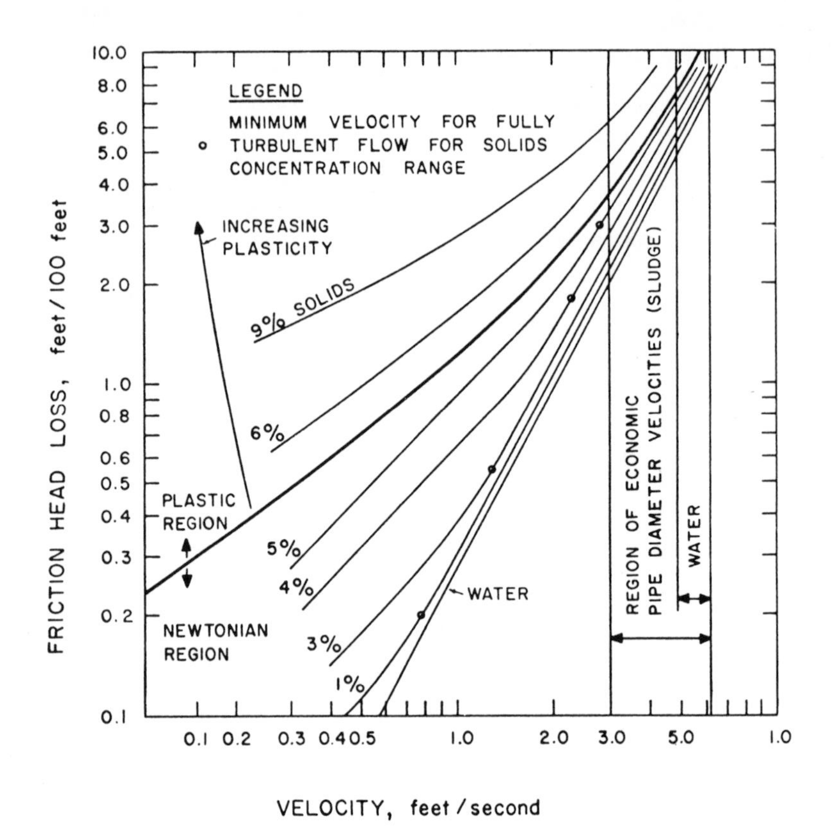

Sludge Pumping

The problem created by using the Hazen-Williams equation and decreasing the coefficients is that calculations for sludge pipelines often indicate the need for large-diameter pipes and low velocities that can result in clogging problems. Most design engineers try to maintain an average velocity of at least 5 ft/sec in order to avoid clogging.

This is also the critical velocity for maintaining turbulent flow. In general, velocities between 1.5 and 2.4 m/sec (5 and 8 ft/sec) have been found satisfactory (Sparr, 1971).

With some types of sludges, there is a possibility of the solids settling out while in the pipeline, especially for long lines. Newitt *et al.* (1955) have proposed that for slurries such as sewage sludge, homogeneity of the slurry can be assured if the velocity in the pipe exceeds a minimum value which is dependent on the settling characteristics of the solids. The proposed guideline is that

$$V > (1{,}800\ g\ D\ v_o)^{1/3}$$

where V = velocity in pipe, ft/sec
g = acceleration of gravity, ft/sec^2
D = pipe diameter, ft
v_o = free settling velocity of solid particles (Stokes velocity), ft/sec

Such velocities, however, may not solve the two types of clogging experienced in some plants: (1) large objects such as cans, broom handles and plastic toys (Fisichelli, 1970), and (2) grease accumulation.

The solution to the first problem is, of course, a better control of screening and comminution. The second problem demands more ingenuity.

One plant operator in Sweden fills the pipelines with a commercial solvent during the weekend. The solvent is reused several times before disposal. Another ingenious solution to grease accumulation has been to circulate periodically the hot liquor from the heat exchanger through the raw sludge pipelines, thus melting the accumulated grease. Ammonia and fuel oil solvents have also proved effective in dissolving grease (Fisichelli, 1970).

SLUDGE PUMPS

The last 50 years have seen tremendous improvement in the design of sludge pumps. As a result, the problems encountered in pumping sludge only a few years ago, when plugging of pumps was commonplace, are no longer of major concern.

Two types of pumps are used for moving sludge. The first is the well-known centrifugal pump, used almost exclusively for pumping filter humus and activated sludge which, as mentioned previously, are treated hydraulically as water.

It has been known for some time, however, that the centrifugal pump is inappropriate for pumping both raw primary and digested sludges and thick oxygen activated sludges. These sludges contain a great deal of large material as well as such things as rags and strings that can damage the pump and clog it with irritating frequency. It is necessary to use a type of pump that will have a positive displacement of the sludge and thus reduce clogging.

Two such pumps have become widely used in the field. One is the plunger pump that involves a positive displacement in two or more pistons. The second is the diaphragm pump that works by the movement of a diaphragm and the opening and closing of valves, thus again moving the sludge by positive displacement (Figure 4-3),

Both pumps produce a pulsating flow which might be unacceptable in some applications, such as feed to a centrifuge. As a result, the rotary screw

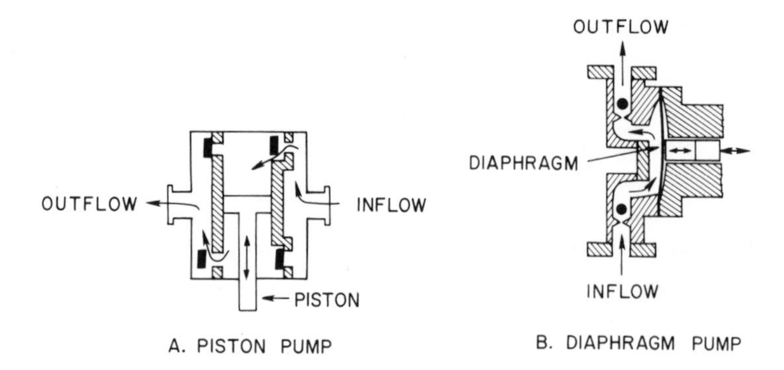

Figure 4-3. Plunger and diaphragm positive displacement pumps for heavy sludges.

pump, illustrated by Figure 4-4, has become widely used in pumping sludges. It is sometimes designated as semipositive displacement as some leakage between the rotor and the stator is possible.

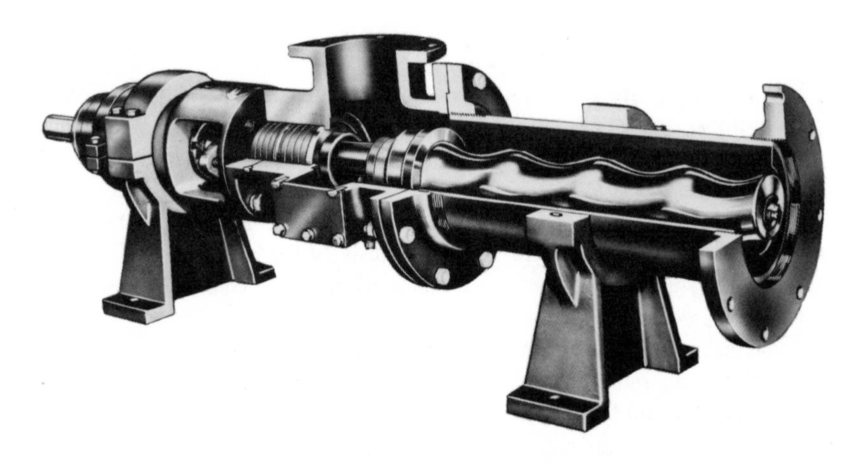

Figure 4-4. Rotary screw pump for heavy sludges where constant flow is required.

IN-LINE MEASUREMENT OF SLUDGE DENSITY

Good operation of clarifiers is greatly facilitated by obtaining an immediate and continuous record of sludge solids concentration. Significant digester volume can be saved, for example, by pumping only concentrated raw sludge from the primary clarifier.

One method of estimating sludge solids is to measure sludge density by the use of radioactive isotopes (Andrews, 1967). Several manufacturers now provide density meters especially constructed for raw sludge. They can be operated so that the sludge pumps start on a regular cycle. The density gauge indicates whether the sludge is dense enough and, if not, a feedback loop orders the pump to shut off. Such an operation is highly recommended in all larger treatment plants.

CONCLUSIONS

Sludge pumping is still much more an art than a science. However, the advances in sludge pumping equipment have to a great degree solved one of the largest problems in treatment plant operation. The design of pipelines can be a serious problem if the line is long and expensive, and thus good design criteria are necessary.

REFERENCES

Andrews, R. H. (1967). "The Radioactive Density Meter for Sewage Sludge Measurement and Control," Ont. Water Res. Comm., Paper No. 2008.

Brisbin, S. G. (1957). "Flow of Concentrated Raw Sewage Sludges in Pipes," *Proc., San. Eng. Div. ASCE* 83:SA3.

Caldwell, D. H., and H. E. Babbitt (1941). "The Flow of Muds, Sludges and Suspensions in Circular Pipes," *Trans. Am. Inst. Chem. Eng.* 37:237.

Dodge, D. W., and A. B. Metzner (1959). "Turbulent Flow of Non-Newtonian Systems," *J. Am. Inst. Chem. Eng.* 5:2.

Fisichelli, A. P. (1970). "Raw Sludge Pumping–Problems and Interdisciplinary Solutions," *J. Water Poll. Control Fed.* 42:11.

Metzner, A. F., and J. C. Reed (1955). "Flow of Non-Newtonian Fluids," *J. Am. Inst. Chem. Eng.* 1:4.

Newitt, D. M., H. F. Richardson, M. Abbot and R. B. Turtle (1955). "Hydraulic Conveying of Solids in Horizontal Pipes," *Trans. Inst. Chem. Eng.* (Br) 33.

Sparr, A. E. (1971). "Pumping Sludge Long Distances," *J. Water Poll. Control Fed.* 43:8.

Stanley Consultants (1972). "Sludge Handling and Disposal–Phase I; State-of-the-Art" prepared for the Metropolitan Sewer Board of the Twin Cities.

PROBLEMS

1. Waste activated sludge is to be pumped from the bottom of the final clarifier to the digester. The solids concentration is 2% and the desired flow rate is 3 m^3/min, with a static head of 16 m. Design a pipe to handle this flow and estimate the head required by the pump.

2. On Figure 4-1, plot the Hazen-Williams equation for a sludge of 6% solids, assuming $n = 0.9$ and $K = 0.03$ poise.
3. What diameter pipe is required to achieve a head loss not more than 1 ft per 100 ft at a velocity of 5 ft/sec?
4. A sand grain has a specific gravity of about 2.6 and a diameter of about 1 mm. What velocity would be required in a 0.25-m pipe in order to avoid settlement within the pipeline?
5. Raw primary sludge is to be pumped for 6 km against a static head of 26 m. Specify the required head and pipe diameter.

CHAPTER 5

SLUDGE THICKENING

Thickeners can be used in wastewater treatment plants where solids must be concentrated to increase the efficiency of further treatment. Thickening is economically attractive because considerable volume reduction is achieved with even modest increases in sludge solids concentration. The volume reduction is best illustrated as in Figure 5-1, showing that a 1% sludge with a volume of 100%, concentrated to a 2% sludge, will yield a volume reduction of 50%. This same sludge concentrated to 5% solids will be only 20% of the original volume.

Such volume reductions can result in considerable savings in plant costs. For example, anaerobic digesters are designed on the basis of solids detention, and a thicker sludge would allow for a reduction in the required digester volume—a very expensive and valuable commodity in wastewater treatment.

The proper location of the thickener in a wastewater treatment plant is important. If the sludge is to be digested, thickening a blend of raw primary and waste activated sludges has been advocated (Torpey, 1954). On the other hand, if raw primary and secondary sludges are to be dewatered and incinerated, the sludges should be thickened separately and blended immediately before dewatering (Harrison, 1972).

It is necessary to distinguish between sludge thickening and sludge dewatering. Although both processes result in solids concentration and volume reduction, the difference is in the degree. We can define thickening as the concentration of solids to less than 15% solids. Such a sludge is still pumpable by conventional means and has most of the characteristics of a liquid. Dewatering (discussed in the next chapter) is defined as the concentration of sludge solids to greater than 15%, and this sludge generally behaves as a solid.

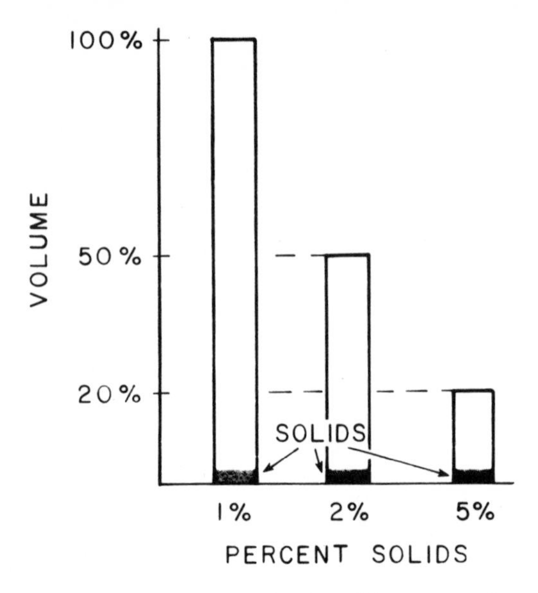

Figure 5-1. Volume reduction by solids concentration.

GRAVITATIONAL THICKENING

Operation

A conventional gravitational thickener is shown in Figure 5-2. The unit operates very much like a settling tank; the feed solids entering in the middle are distributed radially, and the sludge solids are collected as underflow in a sludge sump. The effluent exits over weirs.

The concentration of solids within a continuously operated thickener is also shown in Figure 5-2. There are three distinct zones in such a thickener. The clear zone on top is composed of liquid that eventually becomes the effluent escaping over the weirs. This liquid obviously has a low solids concentration. The next zone is called the feed zone although it is not necessarily at the same concentration as the feed solids. It is characterized by a uniform solids concentration. Below the feed zone is the compaction zone characterized by an increasing solids concentration to the point of sludge discharge.

The *sludge blanket* is defined as the top of the feed zone. Looking into an operating thickener, one can sometimes see the sludge blanket below the clear effluent. The height of this sludge blanket is the main operational control that a treatment plant operator has over a thickener. By increasing the underflow rate, the sludge blanket can be lowered, and this results in a lower solids residence time, a higher throughput of solids, and a lower solids

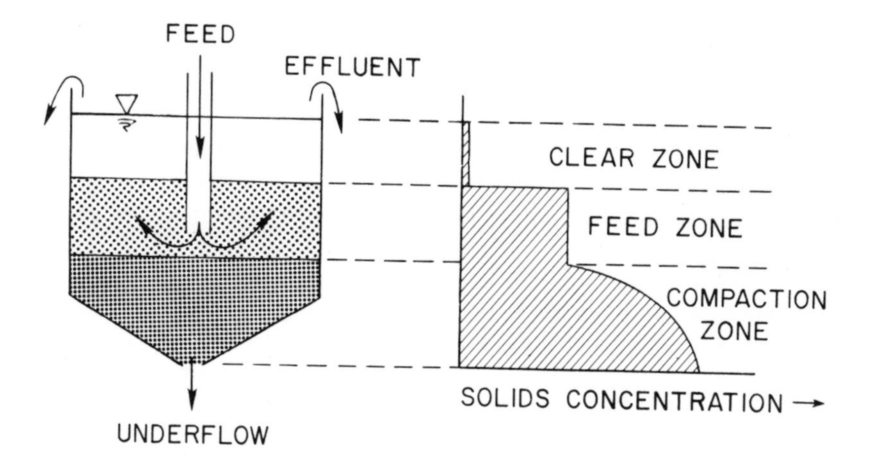

Figure 5-2. Typical continuous thickener.

concentration in the underflow. These shortcomings are countered, however, by the reserve volume the operator can maintain in the thickener to take care of unexpected heavy sludge loads (Munch and FitzPatrick, 1978).

Conversely, a high sludge blanket will yield a high solids retention time and a high concentration of the underflow (Kalbskopf, 1972; White and Lockyear, 1978). One danger in maintaining a high sludge blanket is that the solids residence time will be so high that gas formation will occur due to anaerobic activity. This gas formation will result in the flotation of solids and the creation of a dirty effluent. Chemicals such as chlorine may be necessary to inhibit biological activity and hence gas formation. A well-operated thickener will have a solids recovery of about 95%.

Design

Sludge thickening occurs in the bottom areas of any clarifier which is operated with a sufficiently long solids retention time so as to produce a sludge blanket. This situation is especially true with secondary clarifiers in activated sludge plants where the underflow solids concentration is of concern to the operator.

Since clarifiers such as the activated sludge final clarifier act as both a settling tank and a thickener, it makes sense to design for both of these processes. Dick (1970) has pointed out that it is likely that thickening in fact is often the most critical of the two processes and that both the settling and thickening requirements must therefore be considered.

The design procedures outlined below are thus applicable to tanks which are primarily gravity thickeners, as well as clarifiers in which sludge thickening occurs.

Two methods of thickener design are available to the engineer. Design can be based entirely on the experience of previous designers by copying their design criteria. This is necessary if no sludge is available for testing. Obviously, a better way is the utilization of laboratory data to development of design criteria.

Design Based on Experience. The settling of sludge is slow due to its low specific gravity, and hence solids throughput is an important criterion in the design of thickeners. The design basis of most settling tanks is expressed in terms of the solids loading or the solids flux, as kg solids/hr/m^2 (or in lb solids/day/ft^2).

Some typical solids flux values for domestic wastewater sludges are shown in Table 5-1. The design simply involves selecting a reasonable solids flux and calculating the required surface area by dividing the anticipated solids feed by the flux. Again, it must be cautioned that these design values are believed to be acceptable for existing thickeners, but cannot be used without considering the peculiarities and idiosyncrasies of each individual sludge and plant.

Table 5-1. Typical Solids Flux for Gravitational Thickeners

	Solids Flux		
Type of Sludge	(lb/ft^2/day)	kg/hr/m^2	Reference
Activated Sludge	4	0.8	Eckenfelder (1970)
	5	1.0	Newton (1964)
Trickling Filter Humus	9	1.8	Newton (1964)
Raw Primary Sludge	22	4.5	Eckenfelder (1970)
	25	5.1	Newton (1964)
Raw Primary and	19	3.8	Torpey (1954)
Activated Sludge	8	1.6	Newton (1964)
	12	2.4	Metcalf & Eddy (1972)
Pure Oxygen Activated Sludge	20	2.0	EPA (1974)

Design Based on Settling Tests. The common settling test used in the laboratory is conducted in a transparent cylinder filled with sludge and mixed to distribute the solids evenly. At time zero the mixing is stopped, and the solids are allowed to settle. If the solids concentration is sufficiently high, the solids will settle as a blanket in the *zone settling* regime. Zone settling means the solids will settle as a blanket without any interparticle movement and, therefore, all individual particles settle at the same velocity. This creates a distinct solid-liquid interface, the height of which can be measured with time. Figure 5-3 shows a typical test and the resulting time vs the sludge-liquid interface height curve.

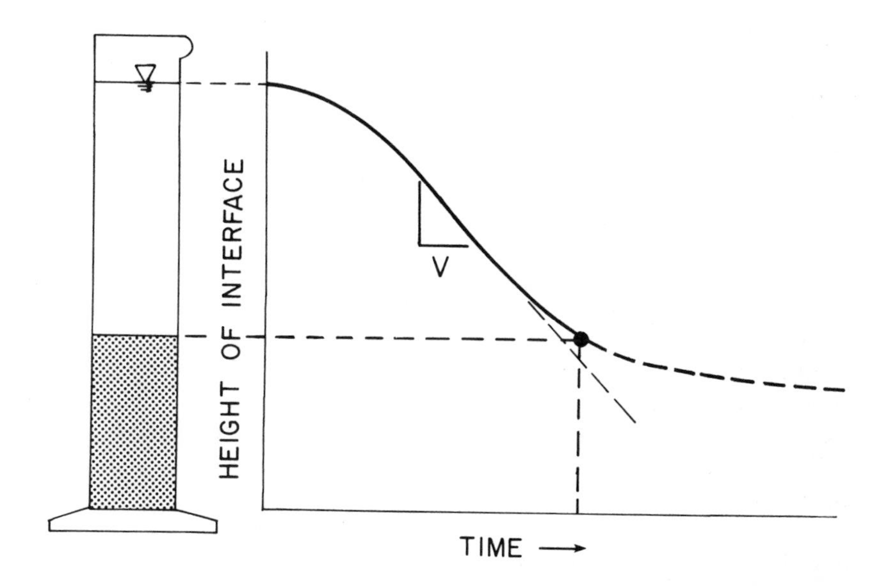

Figure 5-3. Thickening test in a cylinder with resulting interface height vs time curve.

The sludge will settle at a constant velocity which depends on its solids concentration (and, as noted below, other variables). It has been postulated that at time zero, concentration layers will begin to build up from the bottom and eventually intersect the sludge liquid interface. At that point, the constant velocity of the interface will decrease due to the higher concentration of solids at the interface. The settling velocity will gradually decrease until all settling ceases.

If we assume the straight-line portion of this curve is a settling velocity characteristic of the sludge at that solids concentration, it is possible to plot the velocity vs the concentration as in Figure 5-4. Further, by multiplying the velocity by the concentration, we can determine what is defined as the *solids flux.* (Note that velocity times concentration is m/hr x kg/m^3 = kg/hr/m^2, in other words, the same design dimension as used above.) A typical plot of the solids flux, G, versus concentration s also shown in Figure 5-4. Translation of the thickener design from such laboratory tests involves the selection of the solids flux that will limit the operation of the thickener–in other words the maximum solids loading.

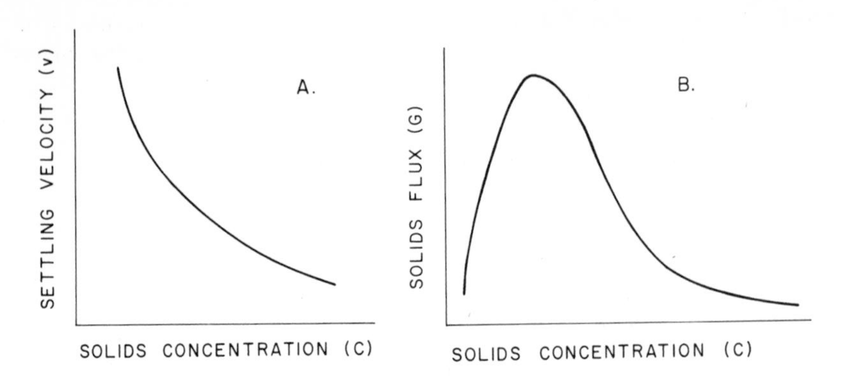

Figure 5-4. The results of several batch thickening tests plotted as (A) interface velocity vs initial solids concentration, and (B) solids flux vs initial solids concentration.

Figure 5-5 is a schematic of a continuously operating thickener. This thickener will operate as long as the application of solids (in terms of $kg/hr/m^2$) does not exceed the rate at which solids are transmitted to the bottom.

There are two ways that the solids move toward the bottom. The first is the bulk downward movement due to sludge removal of sludge as underflow. The continuous movement of solids (and liquids, of course) out the bottom, as sludge underflow, creates a velocity within the thickener that is independent of the solids settling rate. In other words, a thickener without solids concentration whatsoever (because there is no settling) will still have a solids flux, because the underflow creates a net downward movement. This movement has some velocity, u. The solids flux due to the underflow is thus $G_u = uC_i$ where C_i is the solids concentration at level i.

The second portion of the solids flux is due to the settling of solids within the thickener. If the solids did not settle, obviously no thickening would take place. This flux, called *batch settling flux* because it can be measured in a batch test, is $G_b = v_iC_i$ where v_i is the batch settling velocity at concentration C_i.

At some level, i, the total solids flux, G_i, is therefore equal to the sum of the flux due to the underflow and the batch settling flux, or $G_i = uC_i + v_iC_i$. Note that the solids flux is again expressed in terms of the solids loading, $kg/hr/m^2$.

This equation can be plotted as shown on Figure 5-6. The solids flux due to underflow, uC_i, is transcribed as a straight line or as a linear relationship of concentration with solids flux, whereas v_iC_i, the flux due to solids settlement as measured in a batch test, is simply a transposition of the flux curve from Figure 5-4. Notice that this flux curve has a minimum flux between the feed concentration, C_o, and the desired underflow concentration, C_u. This

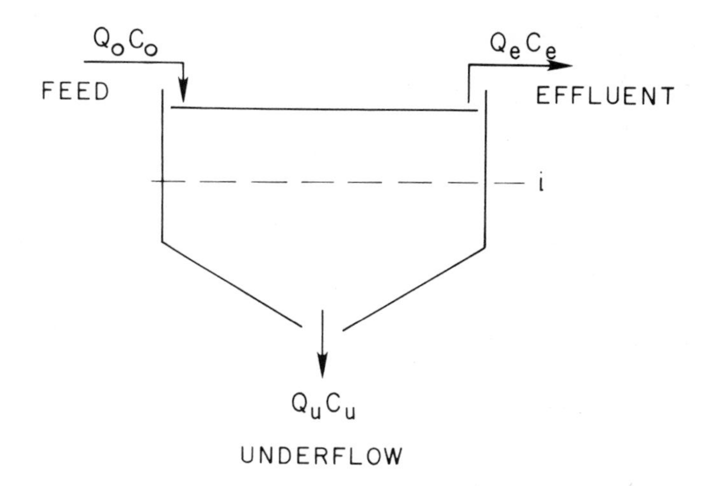

Figure 5-5. Schematic of a continuous thickener (Q is flow rate, C is solids concentration).

minimum flux is the maximum allowable solids loading if thickening is to be successful. If this limit is exceeded, the solids will overflow in the effluent. The design of a thickener is thus reduced to determining this minimum point. Several procedures are available for this, some of which are discussed below.

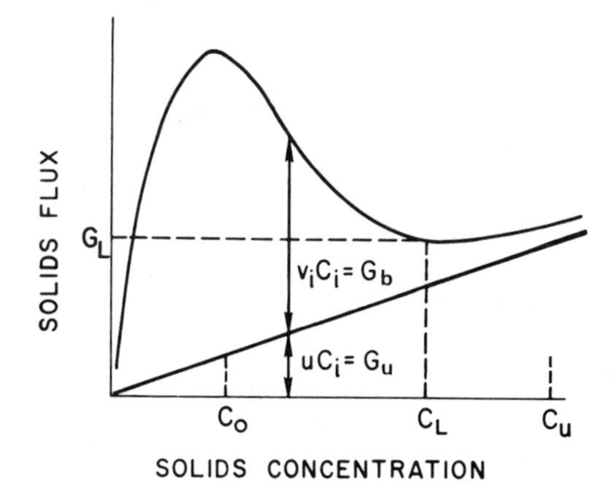

Figure 5-6. Total flux curve.

Coe and Clevenger Method (1916). Define the feed flow rate and solids concentration as Q_o and C_o, the effluent flow and solids as Q_e and C_e, and the underflow as Q_u with solids of C_u. If the thickener operates ideally, $C_e = O$.

From the solids balance:

$$Q_oC_o = Q_uC_u$$

and liquid balance:

$$Q_{o\ slurry} - Q_{o\ solids} = Q_e + Q_{u\ slurry} - Q_{u\ solids}$$

but

$$Q_{solids} = Q_{slurry} \times \frac{C}{\rho_s}$$

hence where ρ_s = solids density

$$Q_{o\ slurry} - Q_{o\ slurry}\left(\frac{C_o}{\rho_s}\right) = Q_e + Q_{u\ slurry} - Q_{u\ slurry}\left(\frac{C_u}{\rho_s}\right)$$

and

$$Q_o\left(1 - \frac{C_o}{\rho_s}\right) = VA + Q_u\left(1 - \frac{C_u}{\rho_s}\right)$$

where Q_e = VA, V = upflow velocity and A = area. But

$$Q_u = \frac{Q_oC_o}{C_u}$$

from the solids balance. Hence

$$Q_o\left(1 - \frac{C_o}{\rho_s}\right) = VA + \frac{Q_oC_o}{C_u}\left(1 - \frac{C_u}{\rho_s}\right)$$

$$Q_o - \frac{Q_oC_o}{\rho_s} = VA + \frac{Q_oC_o}{C_u} - \frac{Q_oC_o}{\rho_s}$$

$$Q_o - \frac{Q_oC_o}{C_u} = VA$$

$$Q_o\left(1 - \frac{C_o}{C_u}\right) = VA \text{ and } \frac{Q_oC_o}{A} = \frac{V}{\frac{1}{C_o} - \frac{1}{C_u}} = G$$

For thickening to be successful, V = v_i, the velocity of any concentration layer, C_i.

$$G_i = \frac{v_i}{\frac{1}{C_i} - \frac{1}{C_u}} = \frac{Q_oC_o}{A}$$

and

$$A = \frac{Q_oC_o}{v_i}\left(\frac{1}{C_i} - \frac{1}{C_u}\right)$$

It is necessary to find a number of values for v_i and C_i, and by knowing the feed solids and flow and the desired underflow concentration, the area can be calculated. A series of such calculations will define a maximum area, which will be the minimum area required for the thickener. If a smaller area is made available, the solids will not be able to move through this critical concentration layer and end up in the effluent.

Dick (1969) has shown that this relationship can also be obtained by recognizing that the underflow rate, Q_u, is

$$Q_u = \frac{GA}{C_u}$$

and that downward velocity, u, created by the underflow is

$$u = \frac{Q_u}{A} = \frac{GA}{C_u A} = \frac{G}{C_u}$$

Substituting this value in the general solids flux equation,

$$G_i = u\,C_i + v_i C_i$$

$$G_i = \frac{G}{C_u} u\,C_i + v_i C_i$$

and rearranging,

$$G_i = \frac{v_i}{\dfrac{1}{C_i} - \dfrac{1}{C_u}}$$

which is identical to the expression derived by Coe and Clevenger.

The Coe and Clevenger analysis has been confirmed experimentally by Dick and Young (1972). Laboratory experiments were used to predict underflow concentration, and these were compared to actual continuous thickener performance. The results, shown as Figure 5-7, are in close agreement (relative to the problems involved in conducting representative tests).

Yoshioka Construction (1957). It is possible to find the minimum flux using a graphical construction as shown in Figure 5-8. If a line is drawn from the desired underflow concentration, C_u, tangent to the underside of the flux curve, this line will intersect the limiting flux, G_L. The fact that this construction actually does provide the minimum flux can be seen by noting that by similar triangles

$$\frac{G_L}{v_L C_L} = \frac{C_u}{C_u - C_L}$$

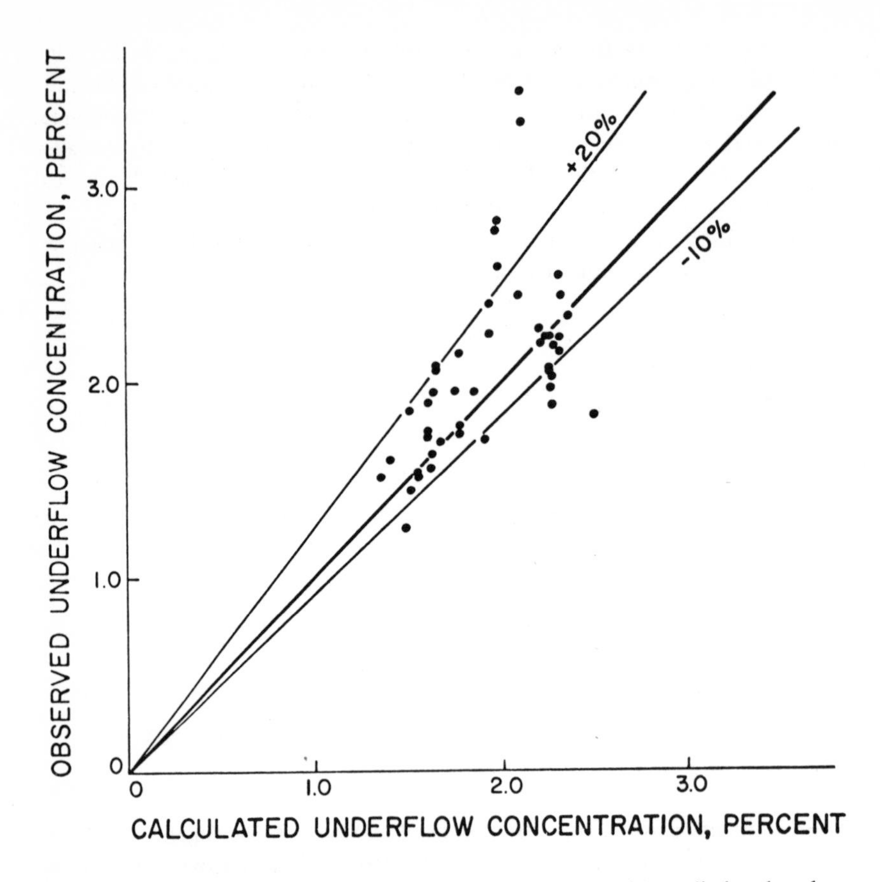

Figure 5-7. Comparison of continuous thickener performance with predictions based on batch settling tests (Dick and Young, 1972).

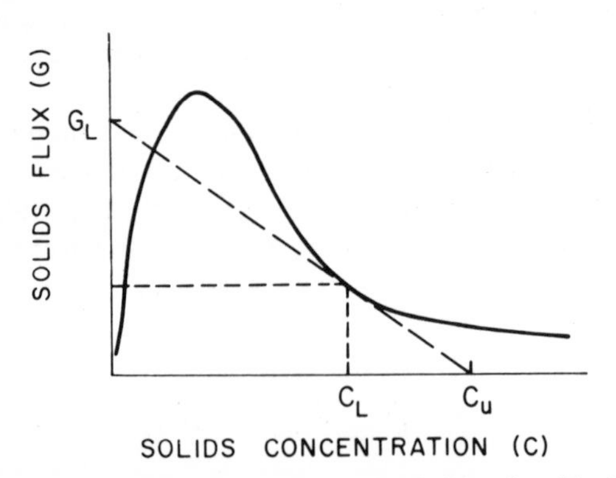

Figure 5-8. Yoshioka method of determining the limiting flux. For Example 5-1, assume C_u = 20,000 mg/l, and G_L = 13 lb/day/ft^2.

$$G_L = \frac{v_L C_L C_u}{C_u - C_L} = \frac{v_L}{\frac{1}{C_L} - \frac{1}{C_u}}$$

This is identical to the Coe and Clevenger solution at the limiting flux (i = L).

Example 5-1: If we are to find the minimum flux for a thickener which is to thicken 0.3 mgd of waste activated sludge from a concentration of 7,000 mg/l to a concentration of 20,000 mg/l, we can run a series of batch settling tests and attain a curve as in Figure 5-8. From this curve, starting at 20,000 mg/l as our desired underflow concentration, it is possible to draw a straight line and read the limiting flux as say 13 lb/ft²/day. The required area is then calculated as

$$A = \frac{Q_o C_o}{G} = \frac{[0.3 \text{ x factor}]\ [7 \text{ x factor}]}{13}$$

$$A = \frac{Q_o C_o}{G} = \left\{ \left[0.3 \text{ mgd x } 1.547 \frac{\text{cfs}}{\text{mgd}} \text{ x } 60 \frac{\text{sec}}{\text{min}} \text{ x } 60 \frac{\text{min}}{\text{hr}} \text{ x } 24 \frac{\text{hr}}{\text{day}} \right] \text{x} \right.$$

$$\left. \left[7 \frac{\text{g}}{\text{l}} \text{ x } 0.0022 \frac{\text{lb}}{\text{g}} \text{ x } 28.32 \frac{1}{\text{gal}} \text{ x } 7.48 \frac{\text{gal}}{\text{cu ft}} \right] \right\} \text{ x } \frac{1}{13 \text{ lb/ft}^2\text{/day}}$$

$$A = \frac{(0.3)\ (1.547)\ (60 \text{ x } 60 \text{ x } 24)\ (7)\ (0.0022)\ (28.32)\ (7.48)}{13} = 28{,}500 \text{ ft}^2$$

Since the thickener is usually included in a plan design for economic reasons (*e.g.*, reduce digester volume), it is sometimes useful to obtain continuous functions which can be used in optimization procedures. Martel and DiGiano (1978) have suggested developing a plot such as Figure 5-9 which clearly shows the effect of both feed and underflow solids concentration on required solids loading. These plots can be developed from batch thickening data and equations fit to the curves.

Differentiation. The relationship between the settling velocity and the suspended solids has not been described theoretically, but a number of investigators have found that this relationship seems to be an exponential one. It is thus possible to describe the settling velocity and solids concentration by the following equation:

$$v_i = ac_i^{-n}$$

where a and n are experimentally determined constants. If this relationship is substituted into the equation for the total flux, we obtain the following:

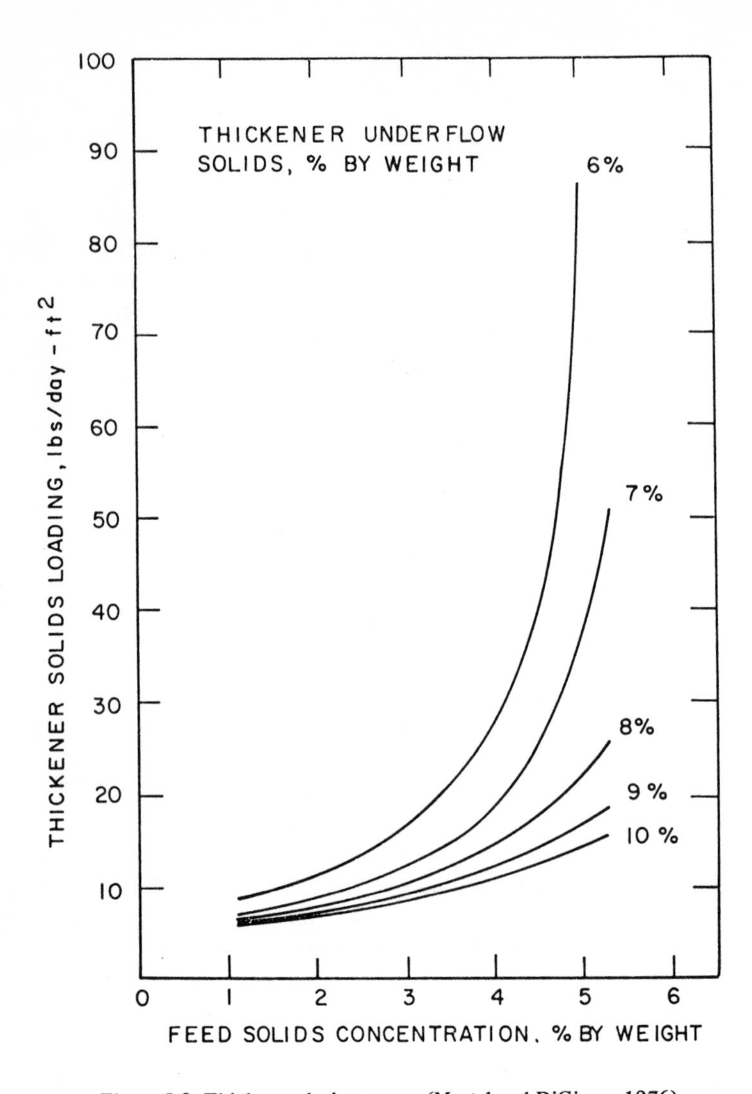

Figure 5-9. Thickener design curves (Martel and DiGiano, 1976).

$$G_i = C_i a\, C_i^{-n} + uC_i$$

$$\text{or } G_i = aC_i^{(1-n)} + uC_i$$

Then

$$\frac{\delta G_i}{\delta C_i} = a(1 - n)\, C_i^{-n} + u$$

The limiting flux is at

$$\frac{\delta G_i}{\delta C_i} = 0.$$

Solving for C_i and calling this the limiting concentration C_L at the limiting flux,

$$C_L = \left(\frac{a(n-1)}{u}\right)^{1/n}$$

For this to be a true minimum,

$$\frac{\delta^2 G_i}{\delta C_i^2}$$

must be greater than 1.

$$\frac{\delta^2 G_i}{\delta C^2} = \frac{an(n-1)}{C_i^{(1+n)}}$$

and the expression is positive if "a" is positive, which is the case from experimental observations. The limiting flux is therefore

$$G_L = [a(n-1)]^{1/n}\left(\frac{n}{n-1}\right)[u]^{n-1/n}$$

Dick and Suidan (1972) used this model to develop an on-line computer program for student use.

Much of the problem in developing general equations for thickening results from the difficulty of obtaining the correct relationship between velocity and solids concentration. A partial listing of the various proposed relationships between velocity and concentration are shown in Table 5-2. All are in part empirical and functions of the material being studied. It is thus dangerous to use any one of these equations without first checking its applicability to the slurry in question.

There seems to be considerable disagreement among these equations. As an example of such discrepancies, Lawler (1978) has plotted the reduced solids flux (C/v_o) vs the solids concentration (C), assuming K = 1, for several of the models. The results, shown in Figure 5-10, graphically illustrate the difference of opinion.

All of the velocity-concentration relationships listed in Table 5-2 are actually incomplete models of the thickening process, since they all ignore two important phenomena which occur during thickening.

The first of these is the recognition that the settling velocity is dependent on solids concentration only up to the point where there exists sufficient

Table 5-2. Relationships Between Solids Concentration and Settling Velocity

Equation	Source
$v = v_o(1 - KC)^{4.65}$	Richardson & Zaki (1954)
$v = v_o(1 - KC)^2 \cdot 10^{-1.82KC}$	Steinour (1944)
$v = v_o 10^{-aKC}$	Thomas (1964)
$v = ac^{-b}$	Cole (1968)
$v = v_o[1 - 2.78(KC)^{2/3}]$	Bond (1960)
$v = v_o e^{-ac}$	Vesilind (1969A)
$v = v_o \left[1 + \frac{3}{4} KC \left(1 - \frac{8}{KC - 3}\right)\right]$	Brinkman (1948)
$v = v_o(1 - KC)^a$	Maude & Whitmore (1958)
$v = v_o(1 - aKC)\left[1 - b(KC)^{\frac{1}{3}}\right]$	Oliver (1961)
$v = \dfrac{v_o\left(3 - \frac{9}{2}(KC)^{\frac{1}{3}} + \frac{9}{2}(KC)^{\frac{5}{3}} - 3(KC)^2\right)}{3 + 2(KC)^{\frac{5}{3}}}$	Happel (1958)

where
v = interface velocity of solids concentration, C
v_o = Stokes settling velocity for a single discrete particle
K = conversion factor, so that KC = volume fraction of solids in the slurry
a, b = constants (unique to each equation)

solid-to-solid contact to slow up the settling procedure. This area of settling is commonly referred to as "compression" or "compaction," and the solids settling is controlled not only by concentration, but also by the characteristics (size, shape, resilience, etc.) of the solids since further consolidation depends on the rearrangement and squeezing of the solid particles.

The fact that the weight of the overlying solids may influence the final solids concentration can be demonstrated by allowing sludges with different initial concentrations to settle to the final height. If the initial height is H_o and final height is H_∞, then

$$\frac{H_\infty}{H_o} = \text{constant}$$

only if the overlying solids have no effect on the compression of the underlying particles.

Brooks (1970) has shown that this is not true in most cases and generalized that

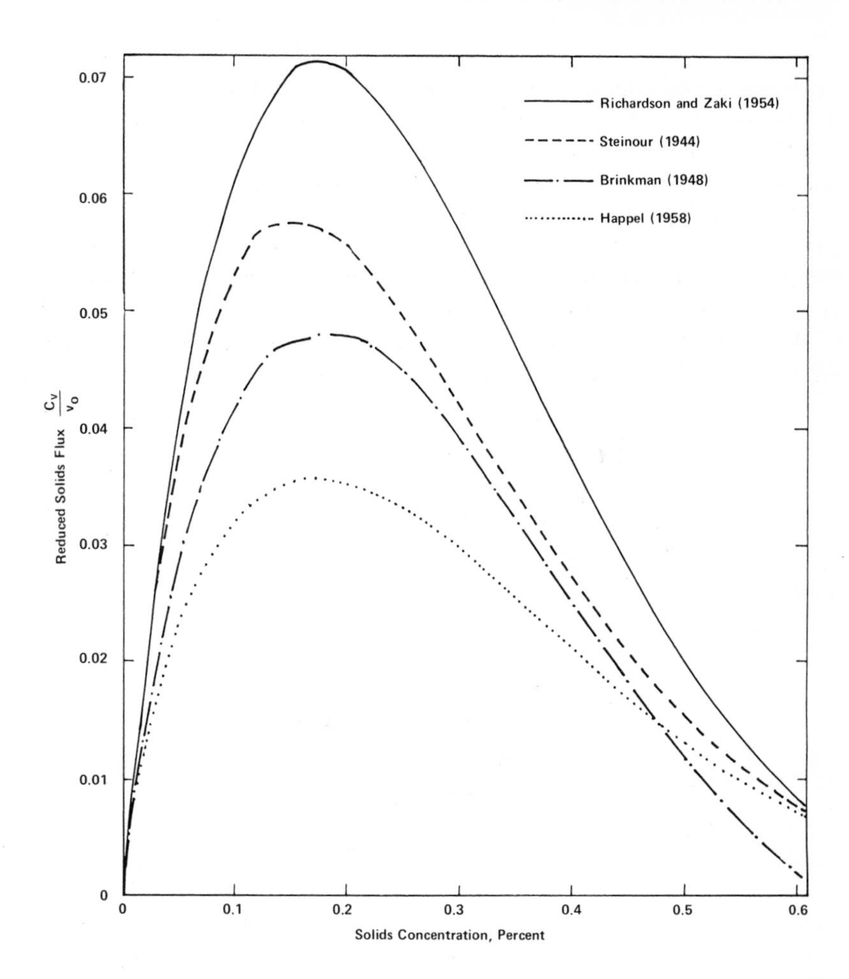

Figure 5-10. Comparison of velocity-concentration models when plotted as solids flux curves (Lawler, 1978).

$$\frac{H_\infty}{H_o} = a\,C^b$$

where C = initial solids concentration
a, b = constants

Taking the logs of both sides of this equation, we have

$$\log\left[\frac{H_\infty}{H_o}\right] = \log a + b \log C$$

Hence the data should be a straight line on log-log paper with b as the slope. If b is zero, then the sludge is noncompressible. Brooks has termed b the "coefficient of compressibility," which is a somewhat unfortunate name since a similar coefficient is defined in filtration (see page 156).

The second phenomenon is usually referred to as "channeling" (Fitch, 1966; Dell and Kaynan, 1968) and involves the formation of tubes or tubules within the sludge bed through which the water can escape. The existence of these channels obviously increases the settling velocity and produces a hump in the solids flux curve, as shown in Figure 5-11. As the concentration is increased, the channels disappear and normal thickening is resumed.

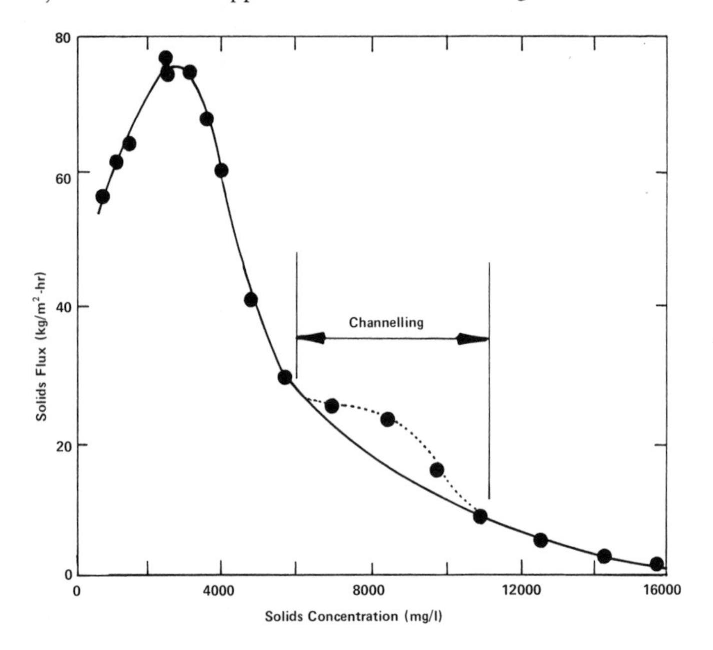

Figure 5-11. Solids flux curve for chromium hydroxide, showing channeling phase (Vesilind and Hayward, 1974).

Channeling has been visually observed and reported by a number of authors (Cole, 1968; Scott, 1966; Vesilind, 1969B). In fact, during the compression phase in batch settling, it is often possible to observe small volcanoes forming on the surface. Visible channels may also form along the cylinder wall, remain for some time, then disappear. Incidentally, such channelization is also supported by work done on sludge drainage by Randall *et al.* (1971).

A complete and rational model therefore must account for three modes of settling:

1. zone settling, where the velocity is a function of solids concentration
2. channeling, and
3. compaction, where the velocity is affected by solid-to-solid contact.

The difficulty of the problem is compounded by the inability, *a priori*, to establish the points at which these transitions occur.

Other Considerations in Thickener Design from Batch Tests. It is possible to use other methods to find the minimum flux, such as the development by Hassett (1964), but most are more tedious than the Yoshioka method and yield the same results.

Edde and Eckenfelder (1968) used a different approach to thickener design. Their method, although dependent on the determination of empirical constants, has been shown to give acceptable design values.

Hultman (1972) proposed the use of consolidation theory, borrowed from soil mechanics, as a means of continuous thickener analysis.

Perhaps the most rigorous (and controversial) analysis of thickening is the theory of settling proposed by Kynch (1952). Kynch assumes that during a settling test layers of increased concentration rise up from the bottom and eventually intersect the slurry liquid interface (Figure 5-9).

With reference to Figure 5-12, at time t_1, concentration layer C_1 has been propagated up from the bottom and is at the surface of the sludge blanket. At this point, all of the solids have passed through it, and their total weight is $C_o H_o A$ where C_o = initial concentration, H_o = initial height, and A = area of cylinder. The total solids passing through the layer is also $C_1 A t_1 (v'_1 + v_1)$ where v'_1 = upward velocity of concentration layer C_1 and v_1 = subsidence velocity

$$C_o H_o A = C_1 A t_1 (v'_1 + v_1)$$

$$v'_1 = \frac{H_1}{t_1}, \text{ then } C_o H_o = C_1 t_1 \left(\frac{H_1}{t_1} + v_1 \right)$$

and

$$C_1 = \frac{C_o H_o}{H_1 + v_1 t_1}$$

$$v_1 = \frac{H'_1 - H_1}{t_1} \text{ and } v_1 t_1 = H'_1 - H_1$$

$$C_1 = \frac{C_o H_o}{H_1 + (H'_1 - H_1)}$$

$$C_1 = \frac{C_o H_o}{H'_1}$$

This equation makes it possible to develop a complete velocity-concentration curve from only one settling test.

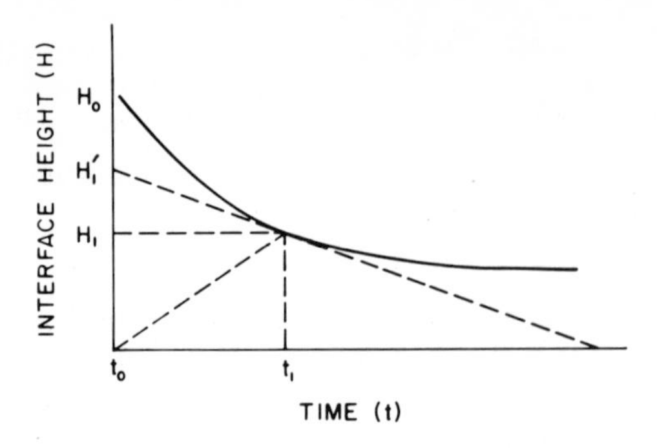

Figure 5-12. Kynch analysis of batch thickening.

Example 5-2. With reference to Figure 5-9, assume the initial solids concentration C_o as 4,000 mg/l and H_o as 1,000 cm. Drawing a line tangent to the settling curve at time equal to 20 min defines H'_1 as 800 cm. Using the above relationship,

$$C_1 = \frac{C_o H_o}{H'_1} = \frac{4{,}000 \times 1{,}000}{800} = 5{,}000 \text{ mg/l}$$

and the settling velocity at this concentration is

$$v_1 = \frac{800 - 400}{20} = 20 \text{ cm/min}$$

Drawing other tangents yields other corresponding concentrations and velocities that define a complete velocity-concentration, and only one experimental settling curve is necessary.

The introduction of Kynch's analysis was regarded as good news by design engineers who are always seeking to avoid laboratory work. But the theory had to be evaluated experimentally, and considerable effort was devoted to determining whether it was applicable to the design of thickeners such as those used in wastewater treatment.

Unfortunately, these experiments proved without doubt that Kynch's theory does not apply to highly compressible materials such as activated sludge and other wastewater sludges (Cole, 1968; Hultman, 1972). Methods for thickener design based on the Kynch analysis, such as the Talmage-Fitch procedure (1955), are thus not applicable in designing waste sludge thickeners. The Talmage-Fitch design method has been reproduced in many textbooks and articles and has been erroneously used by engineers for designing wastewater thickeners.

Most laboratory settling tests are conducted in small cylinders. It is assumed that the settling velocity thus measured is not influenced by container size and the experimental procedure. For many sludges, this is unfortunately a bad assumption, because laboratory artifacts have strongly influenced the results.

For example, Dick and Ewing (1967) showed conclusively that for poor settling sludges, the initial height has a very strong influence on settling velocity, whereas for sludges that settle very well the initial height had an influence only at very shallow initial depths. Some of their results are shown in Figure 5-13.

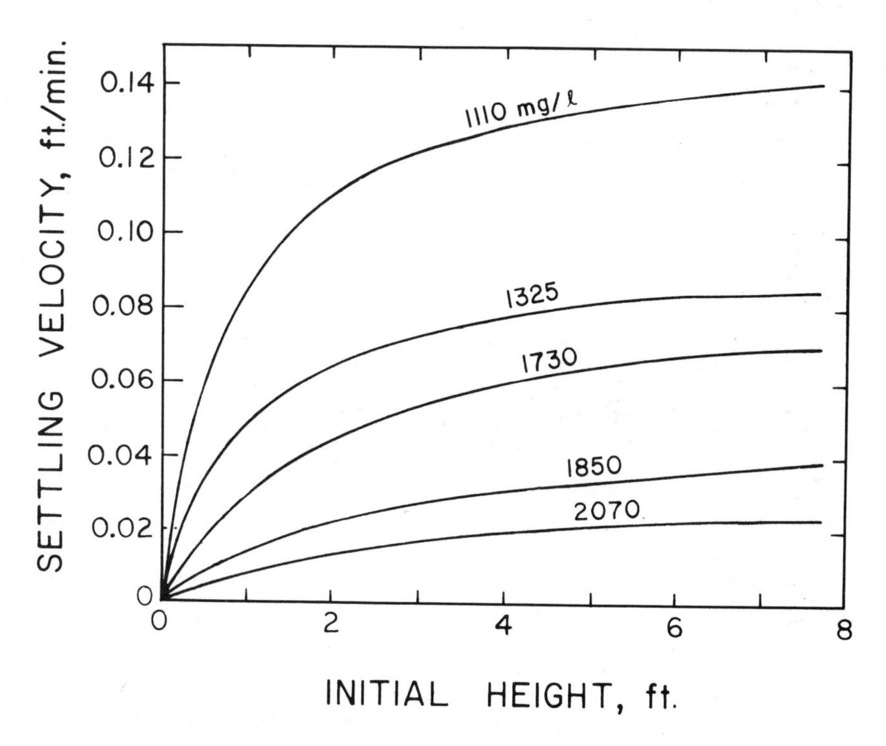

Figure 5-13. Effect of initial sludge height on settling velocity (Dick and Ewing, 1969).

Another effect generally not considered in such laboratory experiments is that of the cylinder diameter. Experiments with large cylinders and activated sludge have shown that cylinders seem to have two effects on sludge settling. First, the water trapped at the bottom of the cylinder selects the least tortuous route in attempting to find its way to the top, which happens to be along the smooth walls of the cylinder. Thus small cylinders with large wall surface area compared to the volume of sludge in the cylinder provide a convenient route by which this water can travel. As a result, smaller cylinders tend to increase the settling velocities.

In addition, there seems to be another laboratory artifact due to cylinder diameter. This can be called bridging, and is again most prominent in small cylinders. At high solids concentrations the sludge tends to bridge or arch across from wall to wall and, as a result, hinders the settling process. This results in slower settling for small cylinders and high solids concentrations.

These two processes were first suggested by the data showing the effect of cylinder diameter on an activated sludge as shown in Figure 5-14 (Vesilind, 1969A). It must be emphasized that these curves were obtained for one specific sludge using specific test conditions, and should not be translated to mean that all sludges behave in a similar fashion. In fact, it is entirely possible that many slurries will show very little effect due to diameter; others might have a much more pronounced effect.

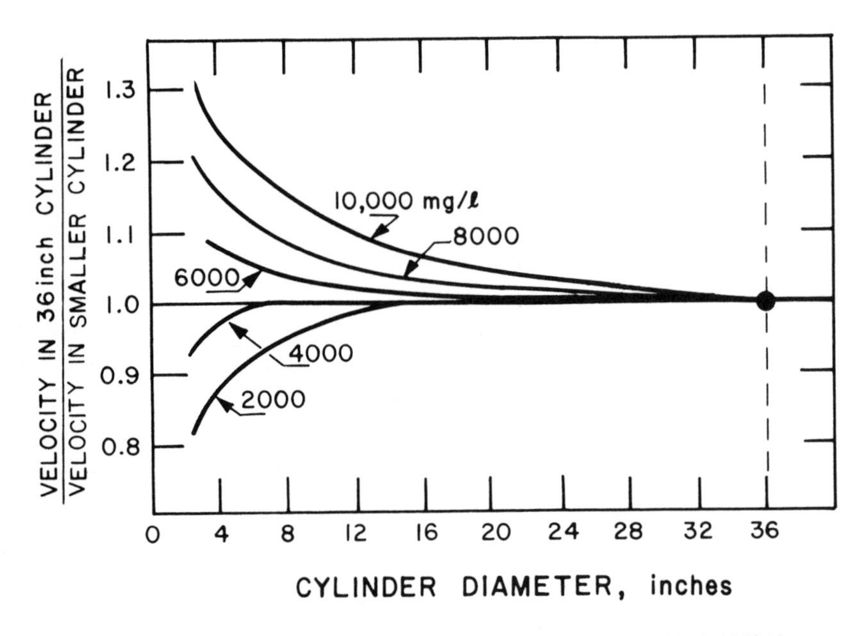

Figure 5-14. Effect of cylinder diameter on settling velocity (Vesilind, 1969A).

Incidentally, the data for Figure 5-14 were obtained by dropping bottomless cylinders of various diameters into a large transparent cylinder so that the interfaces in all cylinders could be read visually. This procedure guaranteed that the sludge had experienced similar flocculation.

Another variable not usually considered is the method of filling the test cylinder and dispersing the solids. Experiments have shown that the best results are obtained if the cylinders are filled from the bottom and only the turbulence created by the filling used to disperse the solids (Dick, 1967).

A fifth laboratory artifact generally ignored in settling tests is the effect of stirring. Sludge settling tests performed in small cylinders with and without stirring have shown that at slow speeds, stirring seems to have a beneficial influence on settling.

Large-scale tests, however, have shown that these effects can be attributed solely to the problem of sludge agglomeration in small cylinders. Agglomeration is the change in sludge texture from homogeneous to rough, much like the curdling of milk, as shown in Figure 2-1. The large agglomerated floc particles settle much faster due to ease at which the water can escape. Agglomeration seems to be hindered by the rapid dampening of turbulence in small cylinders, but is not depressed in large-scale settling tanks or large test cylinders. Plots of the percent of quiescent settling velocity vs stirring speed (such as Figure 5-15) for large cylinders show that stirring has no beneficial effect (Vesilind, 1969B). This was confirmed by Jordan and Scherer (1970) on a large continuous thickener and suggests that vertical picket fence rakes usually used in settling tanks might be of little value. Contradictory evidence has been published by some investigators. They reported installations where picket fence thickeners performed better than thickeners without pickets. This may have been because the pickets aided in the escape of gas bubbles thus promoting thickening.

Another practical conclusion is that, if the sludge behavior in large thickeners is to be duplicated in the laboratory, a slow stirring mechanism should be used to aid in sludge agglomeration.

The temperature variable is also often ignored, and yet the effect of especially cold climates can significantly influence the capacity of a thickener. The dependence of batch settling velocity on temperature is often conveniently expressed as

$$v = a\, e^{K_t T}$$

where v = batch settling velocity
a = constant
K_t = temperature coefficient
T = temperature

Table 5-3 is a listing of some coefficients for various sludges. Batch settling tests with dilute slurries seem to be affected more by temperature than concentrated slurries (Reed and Murphy, 1969).

In addition to the above artifacts, such variables as the age of the sludge and the angle of the cylinders can have substantial effects on the settling velocity.

It can thus be concluded that it is very difficult to duplicate the settling process in the laboratory. At the very least, these artifacts should be

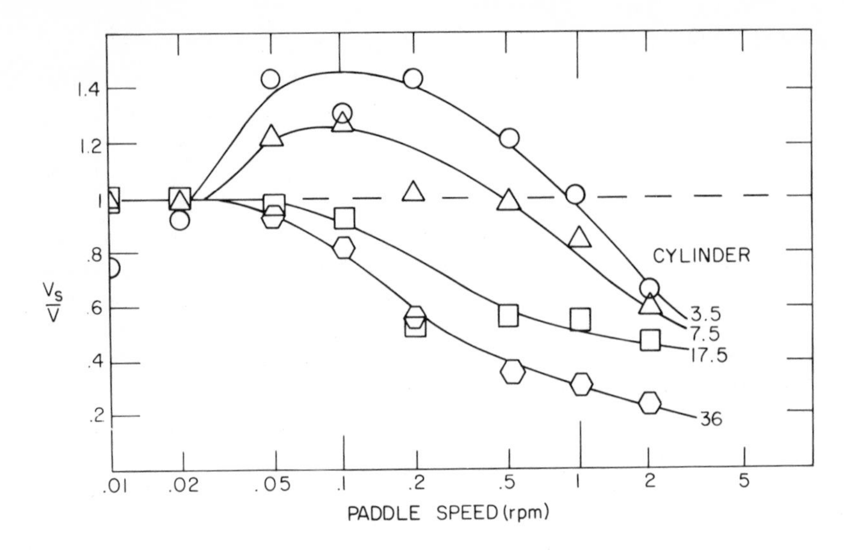

Figure 5-15. Effect of slow stirring on an activated sludge in four different cylinders (Vesilind, 1968B). V_s = settling velocity in a stirred cylinder, V = settling velocity in a quiescent cylinder. Cylinders are identified by the diameter in inches.

Table 5-3. Coefficients for Describing the Effect of Temperature on Batch Settling Velocity (Ronen, 1978)

Sludge	a	$K_t(^\circ C)^{-1}$
$CaCO_3$ Slurry	27	0.0114
Lime Sludge, pH 10.4	100	0.0138
Lime Sludge, pH 11.5	200	0.0114
Lime Sludge, pH 12.1	180	0.0060

Note: The settling velocity (v) is in cm/hr and temperature (t) is in degrees Celsius.

recognized and eliminated if possible. The following is a suggested procedure for conducting laboratory settling tests (Vesilind, 1968A).

1. The cylinder diameter should be as large as possible; 8 in. is a practical compromise.
2. The initial height should be the same as the prototype thickener depth. When this is not practical, 3 ft should be considered minimum.
3. The cylinder should be filled from the bottom, as shown in Figure 5-16.
4. The sample should be stirred throughout the test, but very slowly—0.5 rpm is a reasonable speed for an 8-in. cylinder.

Scale-Up from Continuous Thickeners. Conducting bench-scale continuous thickening experiments seems, on the surface, to be practical and even

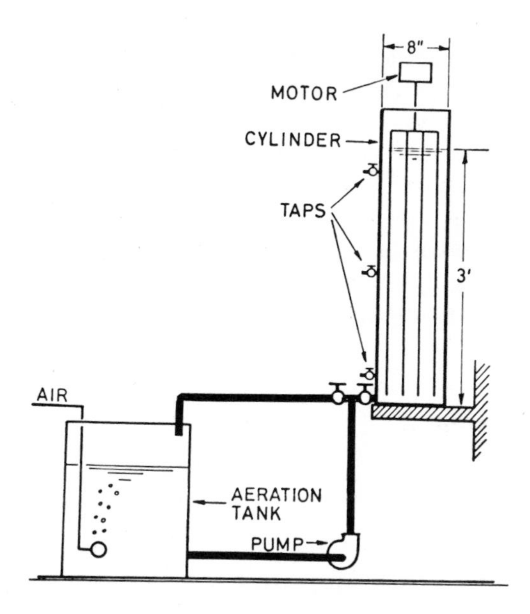

Figure 5-16. Suggested laboratory set-up for batch thickening tests (Vesilind, 1968A).

simple. Unfortunately, such experiments are very difficult to run and their results highly questionable. The most serious problem of continuous testing is maintaining equilibrium, or steady-state conditions. This is due for the most part to the necessity of removing underflow solids, that can be difficult to pump at the very low flow rates required. One means to enable continuous operation is to use a solenoid, a flexible hose and a pump. The solenoid is placed in the closed position between the thickener and the pump, and the pump is turned on. Periodically, the solenoid opens for a short time, sending a high rate of flow through the line. This high flow rate prevents clogging, and the pulse flow approximates continuous operation (Barth, 1968).

THE SLUDGE VOLUME INDEX

Perhaps the most widely used batch thickening test is the Sludge Volume Index (SVI). In light of the above considerations, it would be useful to consider the SVI a little more critically.

The SVI is defined as the volume in ml occupied by a gram of sludge after 30 min of settling. The test is conducted in a liter graduated cylinder, and the sludge level is read after 30 min settling. A convenient calculation is

$$\text{SVI} = \frac{\text{ml of sludge x 1,000}}{\text{suspended solids concentration, mg/l}}$$

The SVI was developed originally to aid the treatment plant operator in monitoring the quality of the activated sludge (Mohlman, 1934). Unfortunately, this has been used as a research tool to indicate sludge settleability, as well as to aid thickener design (Kalbskopf, 1972). As useful as this test is, its limitations must also be recognized.

Consider the settling curves for the two sludges shown in Figure 5-17. The settling characteristics of the two sludges differ greatly but the SVI is equal, demonstrating that it is difficult to describe settling by using only one point on the settling curve.

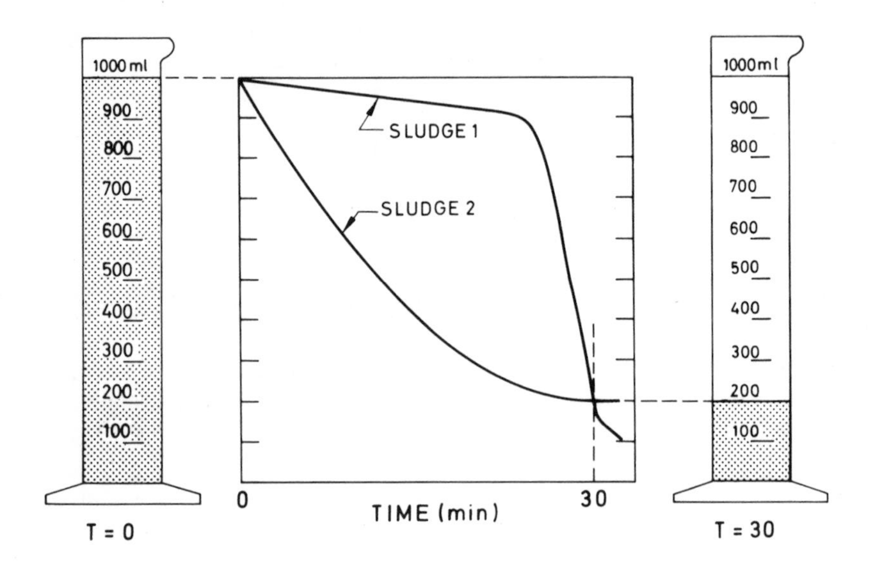

Figure 5-17. The SVI can be equal for two sludges having very different settling characteristics (Dick and Vesilind, 1970).

The SVI is not independent of the solids concentration, as is often assumed. For example, consider a sludge with an initial concentration of 10,000 mg/l and, after 30 min of settling, the sludge height is 1,000 ml (the sludge has not settled at all). The SVI is still calculated as 100, however, indicating a well-settling sludge. Figure 5-18 shows how the SVI varies with concentration for a typical sludge and also shows the maximum SVI possible for any concentration according to definition (Vesilind, 1969C). Thus it is misleading to speak of the SVI of a sludge at greater than 7,000 or 8,000 mg/l of solids. It should be standard practice to report the suspended solids concentration with SVI.

In cases where sludge settles extremely poorly, the SVI seems to be a totally inappropriate measure of settleability. Not only will it be difficult

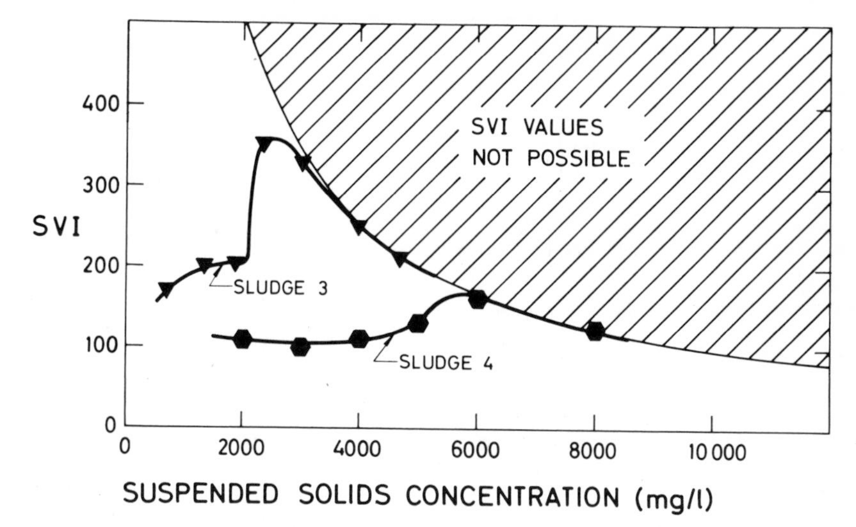

Figure 5-18. Effect of solids concentration on the SVI for two different sludges (Vesilind, 1968C).

to obtain high solids concentrations for running tests, but the data show poor precision. It is difficult to make many quantitative observations from results such as the data shown in Figure 5-19.

Dick and Vesilind (1970) have shown that the SVI is influenced by many of the laboratory artifacts discussed above. The SVI can have poor reproducibility and should not be used in research or design if other measures of sludge settleability are available. Eikum (1973) found not only that the SVI was a poor indicator of settling properties but in fact provided misleading information in the sedimentation properties of primary and mixed primary-chemical sludges. However, it must be emphasized that the SVI is a valuable tool for its intended purpose, *i.e.,* the day-to-day measurement of sludge settleability in the treatment plant. Comparisons of sludges using the SVI should be done with care.

Bulking Sludge

The activated sludge process in biological sludge treatment depends on the separation of sludge solids from the liquid. If this is not accomplished, the process will not work. The settleability of a sludge depends on its characteristics, and often activated sludge plants produce sludges which are difficult to settle by gravity. These are commonly known as *bulking sludges.*

There is no comprehensive definition for bulking sludge. Some regard any sludge with an SVI greater than 100 as a bulking sludge, while others prefer an SVI greater than 300 as the criterion of a severely bulking sludge.

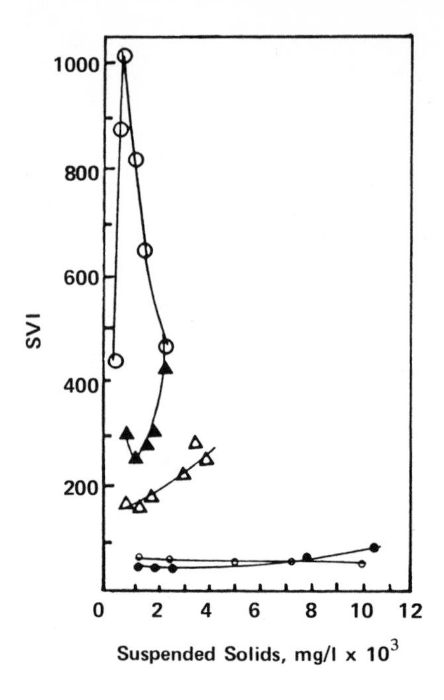

Figure 5-19. Some published SVI data for four different sludges.

Whatever the definition, bulking sludges are generally one of the greatest headaches of wastewater treatment.

The causes of bulking sludges are numerous and, to some degree, still unknown. It has been found that nutrient imbalances can often cause sludges to bulk (Forster, 1971). Waste high in carbohydrates generally produce bulking sludges although the SVI also tends to correlate well with the organic nitrogen content of the sludge (the SVI increasing with organic nitrogen). Sulfides have also been blamed for the growth of *thiothrix* bacteria which are filamentous and difficult to settle (Farquhar and Boyle, 1972). Where this growth is found, preaeration of the waste to drive off the sulfides seems the be an effective cure. Insufficient oxygen is often blamed for the bulking of sludge whereas some have blamed excessive aeration for poor settling.

Two types of bulking can occur. The first is known as "pin-floc" and consists of deflocculated solids particles which, by virtue of their small size and inability to form flocs, will settle poorly (Pipes and Weinke, 1969). These small particles are lost over the weir with the effluent.

The other, perhaps most prevalent in smaller industrial plants, is bulking due to the growth of filamentous organisms such as *thiothrix* and *sphaerotilus*. Once well established, it is difficult to get rid of these growths and sometimes the plant must be flushed out and started from scratch.

Such a process is drastic and sometimes impossible, so other solutions are needed. In 1940 Keefer listed a number of cures for bulking sludges. This list includes almost everything a treatment plant operator can do to change the activated sludge process. For example, one cure is decreasing air into the aeration tank. Should this not work, the operator can increase the amount of air into the aeration tank. A number of other suggestions are offered, only some of which are supported by scientific reason.

Because of the severe problem caused by bulking sludges, considerable research has been devoted to finding solutions. One cure seems to be to increase the aeration time of sludges (Ganczarczyk, 1970; Pipes and Meade, 1968). Generally for sludges with high SVI values, aeration will tend to decrease these values, that is, to increase the settleability of a sludge, whereas the opposite is true for sludges with initially low SVI values.

Anaerobic storage of activated sludge also tends to increase settleability. It has been suggested that extended periods of anaerobiosis will kill the filamentous organisms that are strictly aerobic and sensitive to low oxygen levels. Anaerobiosis also will result in the release of some precipated salts, such as calcium phosphate (Hartmann, 1963).

Chemicals also have been used to cure bulking problems, but for the most part these are not really cures but rather treatments that do not actually apply to the root of the problem. Chlorine has been used in many instances (Frenzel and Sarfert, 1971), whereas hydrogen peroxide has also been found to be a significant cure (Keller and Cole, 1971). The effect of H_2O_2 seems to linger a long time, producing good settleability, and this may be a very practical, if temporary, solution to a bulking problem.

Another cure successfully applied in some industrial plants is to add small quantities of heavy metals. Apparently a metal deficiency in wastewater promotes undesirable growths, and the addition of what are usually considered "toxic chemicals" results in better settling sludge.

In 1978 Sezgen, Jenkins and Parker attempted to bring some rationale to this situation by proposing a theory of sludge bulking based on the differential development of zoogleal (nonfilamentous) microorganisms and filamentous growths.

They suggest that the absence of filamentous organisms leads to the formation of pin-floc, a weak floc that has no "backbone" and shears into small aggregates which contribute to secondary effluent turbidity. On the other hand, an excess of filamentous microorganisms such as *Sphaerotilus natans* would produce an abundance of filaments extending from the flocs into the bulk solution, producing a bridging lattice which prevents the agglomeration of floc particles and hence thickening and compaction. The three different floc structures might be visualized as shown in Figure 5-20.

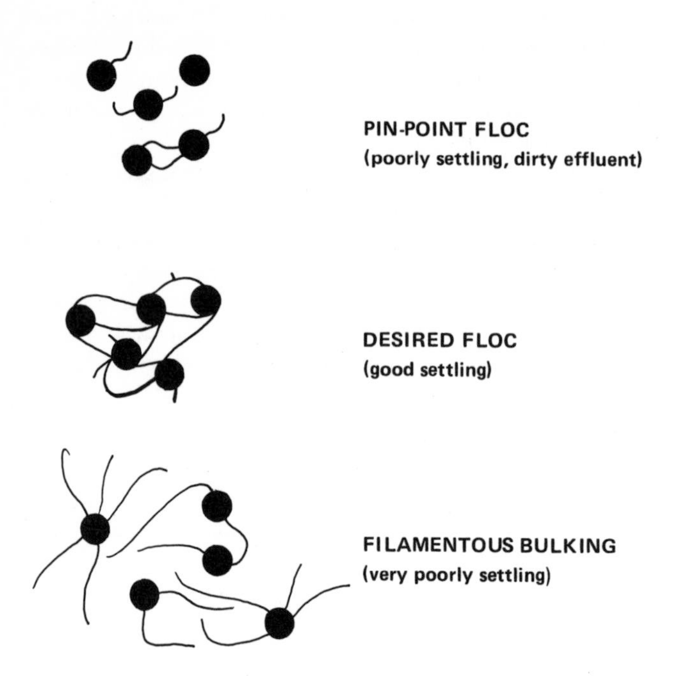

Figure 5-20. Pictorial representation of sludge flocs (Sezgin, Jenkins and Parker, 1978).

The theory was tested by growing cultures where the growth of filamentous organisms was promoted by maintaining dissolved oxygen levels within the flocs. The length of the resulting filaments was then related to high SVI values. It was concluded that in order to promote a balanced culture within an aeration tank, the bulk DO should never be lower than 2 mg/l at substrate removal rate of 0.3 g BOD/g VSS/day.

FLOTATION THICKENING

Obviously gravitational thickening works best with heavy sludges. However, many sludges encountered in wastewater treatment have specific gravities approaching 1.0 and thus settle poorly. Waste activated sludge, for example, is too light to compact appreciably in a gravitational thickener, and flotation thickening has been used successfully in concentrating hard-to-settle sludges.

Operation

Flotation thickening is simply gravity thickening upside down. Instead of waiting for the sludge particles to settle to the bottom of the tank, flotation utilizes tiny air bubbles that attach themselves to the sludge particles, make them lighter than the surrounding liquid, and thus buoy them to the surface where they are scraped off as thickened sludge. As shown in Figure 5-21, the air is introduced under pressure to recycled effluent, which is then mixed with the incoming sludge.

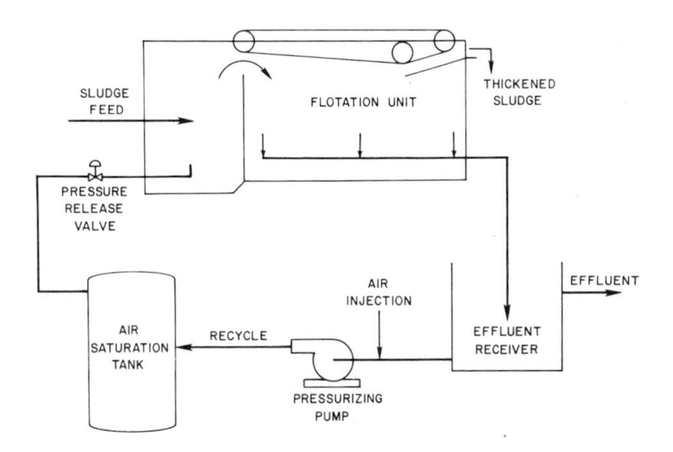

Figure 5-21. Schematic of a flotation thickener.

Flotation thickeners in wastewater treatment are often used to thicken activated sludge before digestion. Activated sludge, being a light, fluffy material, can be thickened by flotation quite readily. Other types of sludges, like raw sludge for example, are difficult to thicken by flotation because they are heavy and tend to settle.

Design

As with other types of devices, various design methods exist for flotation thickening, depending on the importance of the installation, the available sludge, and the skill of the engineer. The two most frequently used methods of sludge flotation design are, as with gravity settling, experience and laboratory tests.

Design Based on Experience. Design by experience simply involves the selection of a surface loading, detention time and a solids loading. Flotation equipment manufacturers, having installed numerous flotation devices in treatment plants, have a feeling for the kind of sludge to be produced, and are able to estimate quite closely the performance of a flotation thickener. As a result, they often design flotation units by simply selecting a surface loading, usually between 0.22 and 0.9 $m^3/hr/m^2$ (1 and 4 gpm/ft^2), a detention time of about 30 min, and a solids loading ranging from 2.5 to 25 kg/ m^2/hr (0.5 to 5 $lb/ft^2/hr$).

Such a design is reasonable only when it is impossible to run even a very crude laboratory test on the sludge. If sludge is available, it should be tested in the laboratory before a flotation thickener is designed.

Design Based on Flotation Tests. From laboratory tests, one should be able to estimate the following: (1) the air-to-solids ratio (A/S); (2) the surface area; (3) the detention time; and (4) the recycle rate. The depth is set by equipment limitations, so that for a given surface area the detention time is also fixed.

The air requirement can be analyzed as follows:

The concentration of air in the recycled flow is

$$C = fS_aP$$

where f = fraction of saturation achieved
S_a = saturation concentration of air at 1 atmosphere pressure
P = pressure in the retention tank in atmospheres

The total amount of air into the flotation tank is

$$A_{in} = (fS_aP)\,RQ + S_aQ$$

where Q = feed rate of influent
R = recycled rate as fraction of Q

The last term is air-dissolved in the feed, assuming full saturation. If the feed is aerobic sludge, the fraction contributed by oxygen may have to be subtracted.

After the pressure is released, the air at equilibrium is

$$A_{equil} = S_a\,(R + 1)\,Q$$

The amount of air available for flotation is the difference, or

$$A = A_{in} - A_{equil} = S_a\,(fP - 1)\,RQ$$

The solids feed is $S = QC_o$ where C_o is the concentration of solids. Hence

$$\frac{A}{S} = \frac{S_a\,(fP - 1)\,R}{C_o}$$

Commonly, S_a is in mg/l, P in atmospheres, and C_o in mg/l. At 20°C, S_a is about 24 mg/l. In practice, P is often about 3 atmospheres, f is equal to 0.5 to 0.8, and the recycle rate, R, is equal to 1.0 or more.

The laboratory procedure and equipment necessary to determine the design parameters are discussed by Wood and Dick (1973). The sludge and the water are pressurized by bubbling in air, and this mixture is bled into the flotation cell, thus simulating the operation of a thickener. By conducting a series of tests with various combinations of sludge and water, fed in under different pressures, it is possible to establish the optimum design criteria.

A more common laboratory set-up is illustrated in Figure 5-22, which includes only a flotation cell and a tank into which the sludge is placed and air is bubbled under pressure. In other words, the sludge is saturated with the air prior to the release of pressure.

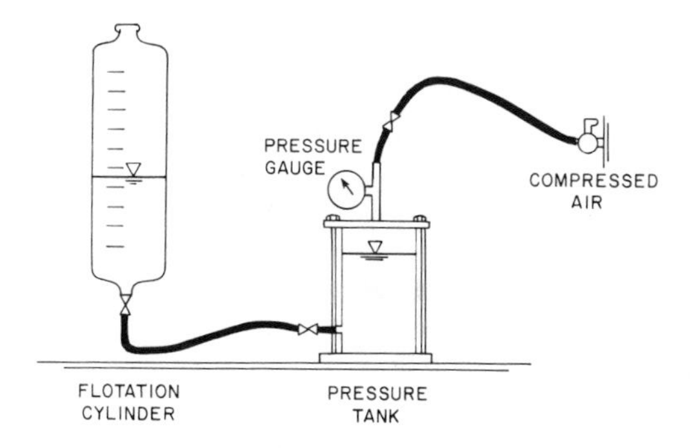

Figure 5-22. Simple laboratory flotation device.

The results of laboratory tests can be illustrated as shown in Figure 5-23. For various A/S ratios, the resulting curves for the solids in the effluent and the percent cake solids will be transcribed so that it becomes obvious that at some point any further increase in the A/S ratio would not greatly increase efficiency of operation. Based on such data, a design A/S ratio is selected.

The surface area for thickeners is often not designed at all but rather established from experience. This is partly due to the difficulties involved in establishing with any degree of confidence the required surface area from laboratory tests.

One way to design the surface area of a thickener is to consider the flotation thickener simply as an upside-down settling tank and to utilize the same procedure and methods outlined under gravity settling for flotation. It is quite reasonable to expect to obtain height vs time curves for flotation, the same way these curves are obtained for gravity thickeners except that for flotation these curves would be upside down. From these curves settling velocity can be obtained, plotted against the concentration, and the solids flux calculated. There is no reason to believe that a limiting flux would

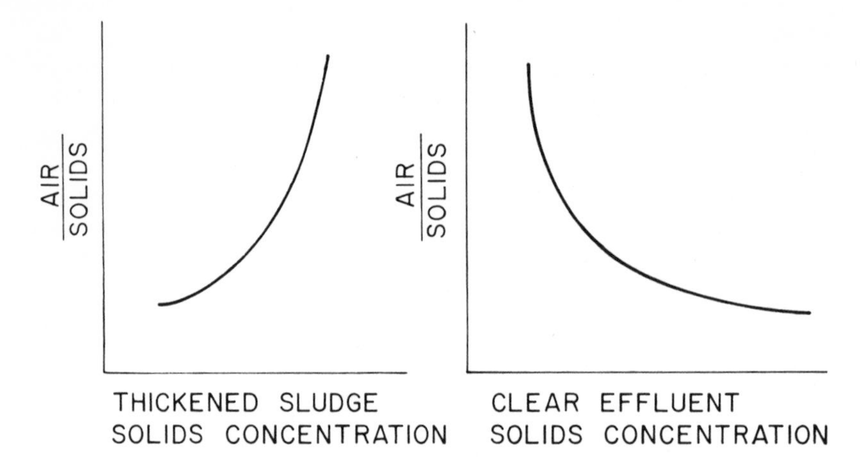

Figure 5-23. Typical results of laboratory flotation tests.

not also exist for flotation as it does for gravity thickening, and therefore it should be possible to design the area based on solids loading exactly as for gravity settling (Wood, 1970).

The recycle rate is calculated from the A/S equation. For a known A/S ratio and pressure, the recycle rate is the only unknown. Thus all required design parameters can be obtained from laboratory experiments.

Surprisingly, many of the problems associated with sludge settling by gravity also occur in flotation. A bulking sludge in gravity settling does not necessarily mean the sludge will be easy to thicken by flotation. On the contrary, bulking sludges also cause problems in flotation. Flotation units often operate well only after the addition of chemical conditioners such as organic polyelectrolytes (Katz and Geinopolos, 1967). The design engineer must consider the possible necessity of using polymers when flotation thickening is to be used in a wastewater treatment plant.

OTHER METHODS OF THICKENING

The centrifuge has been used as a thickener, especially for very light sludges that will not compact readily. Newer models of solid-bowl centrifuges (discussed in a later chapter), can effectively thicken light sludges, but have been known to experience problems in maintaining steady operation. The disc centrifuge (also discussed later), has operational problems, and a new screen-bowl device is advertised as being effective but is yet to be proven (*Chemical Engineering*, 1974). In short, although it is possible to thicken sludges by centrifugation, their application should be carefully considered.

Reverse osmosis, or ultrafiltration, has been used on a laboratory scale to thicken sludges. Reverse osmosis is a misnomer in that osmotic pressure is not involved in the process. Rather the sludge is placed under pressure, and water molecules are driven through a semipermeable membrane. By thus removing the water, it is possible to thicken sludges to very high concentrations, at least in the laboratory.

Activated sludge, when allowed to remain without aeration, will form nitrogen gas and CO_2, which is attached to the solid particles and causes them to float. This effect was used with good results in the Laboon process for thickening primary sludge (Laboon, 1952).

CONCLUSIONS

Sludge thickening is an intermediate process whose economic justification depends on the savings incurred in subsequent treatment. In most cases thickeners have readily paid for themselves in saving such things as digester space and in increased filter yields. They should not, however, be included automatically in every treatment plant. Each installation deserves a close study of just how much money can be saved by adding one more unit process to the treatment plant.

REFERENCES

Barth, E. F. (1968). "Device to Aid Pilot Final Settlement," *Environ. Sci. Technol.* 2:2.

Bond, A. W. (1960). "Behavior of Suspensions," *J. San. Eng. Div. ASCE* 86:SA3.

Brinkman, H. C. (1948). "On the Permeability of Media Consisting of Closely Packed Porous Particles," *Appl. Sci. Res.* A1:81.

Brooks, R. B. (1970). "Heat Treatment of Sewage Sludge," *Water Poll. Control (Br)* 69:(1).

Chemical Engineering (1974). "Centrifugal Thickener Cuts Costs of Dewatering Tasks," *Chem. Eng.* 81:1.

Chudoba, J., V. Ottova and V. Madera (1973). "Control of Activated Sludge Filamentous Bulking–I. Effect of the Hydraulic Regime or Degree of Mixing in an Aeration Tank," *Water Res.* 7:1163.

Coe, H. S., and G. H. Clevenger (1916). "Methods for Determining the Capacities of Slime Settling Tanks," *Trans. AIME* 55:356.

Cole, R. F. (1968). "Experimental Evaluation of the Kynch Theory," PhD Thesis, Univ. North Carolina, Chapel Hill.

Dell, C. C., and M. B. Kaynar (1968). "Channeling in Flocculated Suspensions," *Filtration & Separation (Br)* July/August, p. 323.

Dick, R. I., and B. B. Ewing (1967). "Evaluation of Activated Sludge Thickening Theories," *J. San. Eng. Div. ASCE* 93:SA4.

Dick, R. I., and B. B. Ewing (1969). Closure of "Evaluation of Activated Sludge Thickening Theories," *J. San. Eng. Div. ASCE* 95:SA2.

Dick, R. I. (1967). "Evaluation of Prevailing Activated Sludge Thickening Theories," PhD Thesis, Univ. Illinois, Urbana.

Dick, R. I. (1969). "Fundamental Aspects of Sedimentation," *Water Wastes Eng.* 7:2.

Dick, R. I., and M. T. Suidan (1972). "Modeling and Simulation of Clarification and Thickening Processes," Proc. 18th Annual Workshop, Assoc. Envir. Engr. Prof., Nassau.

Dick, R. I. (1970). "Role of Activated Sludge Final Settling Tanks," *J. San. Eng. Div. ASCE* 96(SA2):423.

Dick, R. I., and P. A. Vesilind (1970). "The Sludge Volume Index: What is it?," *J. Water Poll. Control Fed.* 41:7.

Dick, R. I., and K. W. Young (1972). "Analysis of Thickening Performance of Final Settling Tanks," *Proc. 27th Ind. Wastes Conf.* (Lafayette, IN: Purdue University).

EPA (1974). "Process Design Manual for Sludge Treatment and Disposal," U.S. EPA Technology Transfer 625/1-74-006, Washington, DC.

Eckenfelder, W. W. Jr. (1970). "Water Quality Engineering for Practicing Engineers," (New York: Barnes & Noble).

Edde, H. J., and W. W. Eckenfelder (1968). "Theoretical Concepts of Gravity Sludge Thickening: Scale-up Laboratory Units to Prototype Design," *J. Water Poll. Control Fed.* 40:8.

Eikum, A. S. (1973). "Aerobic Stabilization of Primary and Mixed Primary/Chemical (Alum) Sludge," PhD Thesis, Univ. Washington, Seattle.

Farquhar, G. J., and W. C. Boyle (1972). "Control of *Thiothrix* in Activated Sludge," *J. Water Poll. Control Fed.* 44:1.

Fitch, B. (1966). "A Mechanism of Sedimentation," *Ind. Eng. Chem., Fundamentals* 5:1.

Forster, C. F. (1971). "Activated Sludge Surfaces in Relation to the Sludge Volume Index," *Water Res.* 5:861.

Frenzel, H. J., and F. Sarfert (1971). "Experiences on the Prevention of Sludge Bulking by Chlorination of the Activated Sludge," *Gas- und Wassfach* 112:604.

Ganczarczyk, J. (1970). "Variation in the Activated Sludge Volume Index," *Water Res.* 4:69.

Happel, J. (1958). "Viscous Flow in Multiparticle Systems: Slow Motion of Fluids Relative to Beds of Spherical Particles," *J. Am. Inst. Chem. Eng.* 4(2):197.

Harrison, J. R. (1972). "Earn an A if you Cope with the C's and D's of Solids," *Water Wastes Eng.* 9:7.

Hartmann, L. (1963). "Activated Sludge Floc Composition," *Water Sew. Works* 110:7.

Hassett, N. J. (1964). "Concentrations in a Continuous Thickener," *Ind. Chemist* 40:29.

Hultman, B. (1972). "Gravitations–förtjockning," Proc. 8th Nordic Symposium on Water Research, Nordforsk Pub. No. 1972:3, Helsinki (in Swedish).

Jordan, V. J., Jr., and C. H. Scherer (1970). "Gravity Thickening Techniques at a Water Reclamation Plant," *J. Water Poll. Control Fed.* 42:1.

Kalbskopf, K. H. (1972). "Discussion of the Design Parameter for Secondary Sedimentation Tanks," *Water Res.* 6:429.

Katz, W. J., and A. Geinolosos (1967). "Sludge Thickening by Dissolved Air Flotation," *J. Water Poll. Control Fed.* 39:946.

Keefer, C. E. (1940). *Sewage Treatment Works* (New York: McGraw-Hill Book Company).

Keller, P. J., and C. A. Cole (1971). "Treatment of Filamentous Bulking with Hydrogen Peroxide," *Proc. 26th Ind. Waste Conf.* (Lafayette, IN: Purdue University).

Kynch, G. J. (1952). "A Theory of Sedimentation," *Trans. Faraday Soc.* 48:166.

Laboon, J. F. (1952). "Experimental Studies on the Concentration of Raw Sludge," *Sew. Ind. Wastes* 4:2.

Lawler, D. F. (1978). "A Practical Approach to the Thickening Process," PhD Dissertation, Univ. North Carolina, Chapel Hill (in preparation).

Lockyear, C. F. (1977). "Gravity Thickening of Biological Sludges," Water Research Centre, Tech. Rep. TR 39, Stevenage, UK.

Martel, C. J., and F. A. DiGiano (1976). "Production and Dewaterability of Sludge From Lime Clarified, Raw Wastewater," Dept. of Civil Engrg., Univ. of Mass., Amherst.

Maude, A. D., and R. L. Whitmore (1958). "A Generalized Theory of Sedimentation," *J. Appl. Phys. (Br.)* 9:477.

Metcalf and Eddy, Inc. (1972). *Wastewater Engineering* (New York: McGraw-Hill Book Company).

Michaels, A. S., and J. C. Bolger (1962). "Settling Rates and Sediment Volumes of Flocculated Kaolin Suspensions," *Ind. Eng. Chem.* 1:24.

Mohlman, F. W. (1934). "The Sludge Index," *Sew. Works J.* 6:1.

Newton, D. (1964). "Thickening by Gravity and Mechanical Means," *Sludge Concentration, Filtration and Incineration* Cont. Ed. Series, 113 (Ann Arbor, MI: University of Michigan).

Pipes, W. O., and F. S. Meade (1968). "Evaluation of the Kraus Model of Activated Sludge Bulking," *Proc. 23rd Ind. Waste Conf.* (Lafayette, IN: Purdue University).

Pipes, W. O., and C. L. Wienke (1969). "Differences Between Bulking and Deflocculation of Activated Sludge," presented at 42nd Annual Conf. WPCF, Dallas, TX.

Randall, C. W., J. K. Turpin and P. H. King (1971). "Activated Sludge Dewatering: Factors Affecting Drainability," *J. Water Poll. Control Fed.* 43:1.

Reed, S. C., and R. S. Murphy (1969). "Low Temperature Activated Sludge Settling," *J. San. Eng. Div. ASCE* 95:SA4.

Richardson, J. F., and W. N. Zaki (1954). "Sedimentation and Fluidization: Part I," *Trans. Inst. Chem. Eng. (Br.)* 32:1.

Ronen, M. (1978). "Characterization of Lime Sludges from Water Reclamation Plants," D.Sc. Dissertation, Univ. Pretoria, South Africa.

Scott, K. J. (1966). "Mathematical Models of Mechanism of Thickening," *Ind. Eng. Chem. Fundamentals* 5:1.

Sezgin, M., D. Jenkins and D. S. Parker (1978). "A Unified Theory of Filamentous Activated Sludge Bulking," *J. Water Poll. Control Fed.* 50(2).

Steinour, H. H. (1944). "Rate of Sedimentation," *Ind. Eng. Chem.* 36:840.

Talmage, W. P., and B. Fitch (1955). "Determining Thickener Unit Areas," *Ind. Eng. Chem.* 47:38.

Thomas, D. G. (1964). "Turbulent Disruption of Flocs in Small Particle Size Suspensions," *J. Am. Inst. Chem. Eng.* 10:517.

Torpey, W. N. (1954). "Concentration of Combined Primary and Activated Sludges in Separate Thickening Tanks," *Proc. ASCE* 80, September, No. 443.

Vesilind, P. A. (1968A). "Design of Thickeners from Batch Tests," *Water Sew. Works* 115:9.

Vesilind, P. A. (1968B). "The Effect of Stirring in the Thickening of Biological Sludge," PhD Thesis, Univ. North Carolina, Chapel Hill.

Vesilind, P. A. (1969A). Discussion of "Evaluation of Activated Sludge Thickening Theories," by R. I. Dick and B. B. Ewing, *J. San. Eng. Div. ASCE* SA1.

Vesilind, P. A. (1969B). "Quiescent Batch Thickening of Activated Sludge in Small Cylinders," *Vatten* 4:69 (Stockholm).

Vesilind, P. A. (1969C). "The Use and Misuse of the Sludge Volume Index," *Vann* 4:1 (Oslo).

Vesilind, P. A., and P. Hayward (1974). "Thickening of Metal Hydroxides," paper presented to ACS annual meeting, Chicago, IL.

White, M. J. D., and C. F. Lockyear (1978). "A Novel Method for Assessing the Thickening Ability of Sludge," Water Research Centre, Stevenage, UK, (in preparation), reported by Lockyear, 1977.

Wood, R. F., and R. I. Dick (1973). "Factors Influencing Batch Flotation Tests," *J. Water Poll. Control Fed.* 4:2.

Wood, R. F. (1970). "The Effect of Sludge Characteristics Upon the Flotation of Bulked Activated Sludge," PhD Thesis, Univ Illinois, Urbana.

Yoshioka, N. *et al.* (1957). "Continuous Thickening of Homogenous Flocculated Slurries," *Chem. Eng. (Tokyo)* 21 (in Japanese, with English abstract).

PROBLEMS

1. A sludge is thickened from 2,000 mg/l to 15,000 mg/l. What is the reduction in volume?
2. An activated sludge plant treating 100,000 m^3/day of influent is to have a gravitational thickener for its waste activated sludge. Estimate the required thickener area.
3. A laboratory batch thickening test resulted in the following data:

Solids Concentration (% solids)	Settling Velocity (cm/min)
0.6	0.83
0.8	0.35
1.0	0.25
1.2	0.17
1.4	0.11
1.6	0.088
1.8	0.067
2.0	0.050
2.2	0.041
2.4	0.033

   Calculate the required thickener area if the desired underflow concentration is 3% solids and the feed flow rate is 1 m^3/min at a solids concentration of 2,000 mg/l.
4. Using the data above, draw a curve showing thickener loading vs feed solids concentration (such as Figure 5-9) for the 3% underflow concentration.
5. Calculate the "coefficient of compressibility" for the following batch settling data:

Initial Height of Sludge (cm)	Final Compacted Height of Sludge (cm)
100	50
80	45
60	38
40	29

6. Construct a solids flux curve based on the batch test results below, using the Kynch method, and establish the limiting flux for a desired underflow concentration of 5% solids. The solids concentration of the sludge in the batch test is 2,000 mg/l.

Time (min)	Height (cm)	Time (min)	Height (cm)
0	100	12	71
2	95	14	67
4	90	16	64
6	85	18	61
8	80	20	59
10	75	22	57
		24	56
		26	55

7. Using Figures 5-13 and 5-14, estimate the possible error in settling velocity if the batch settling test with 8,000 mg/l sludge was conducted in a 20-cm-tall 3-cm-diameter cylinder.
8. Show by calculations how the maximum possible SVI vs solids concentration curve is derived.

CHAPTER 6

SLUDGE DEWATERING

The ultimate disposal of sludge is often facilitated by removing enough of the liquid portion so that the sludge behaves as a solid. This operation, called sludge dewatering, can be accomplished by many processes.

DRYING BEDS

Drying beds were historically the first method of sludge dewatering. They consist simply of shallow ponds with sand bottoms and tile drains (Figure 6-1). Sludge is pumped to these beds at a depth of 15-30 cm (6-12 in.). The time required for the sludge to dewater to a liftable consistency ranges from several weeks to several months and is obviously a function of climate and other conditions. In northern climates sand beds can be covered with a greenhouse roof to facilitate evaporation. Land costs generally limit the use of drying beds to small communities.

Operation

The removal of water from sludge in drainage beds is in two steps. Initially, the water is drained through the sludge into the sand and out tile drains. This process might last a few days until the sand is clogged with the fine particles and/or all the free water has drained away. Further dewatering occurs by evaporation.

The initial drainage is also thought to be a two-step process. A considerable fraction of the water is first drained by settling of the solids and general compaction. Secondary drainage occurs by the formation of channels that facilitate the movement of water (Randall *et al.*, 1971).

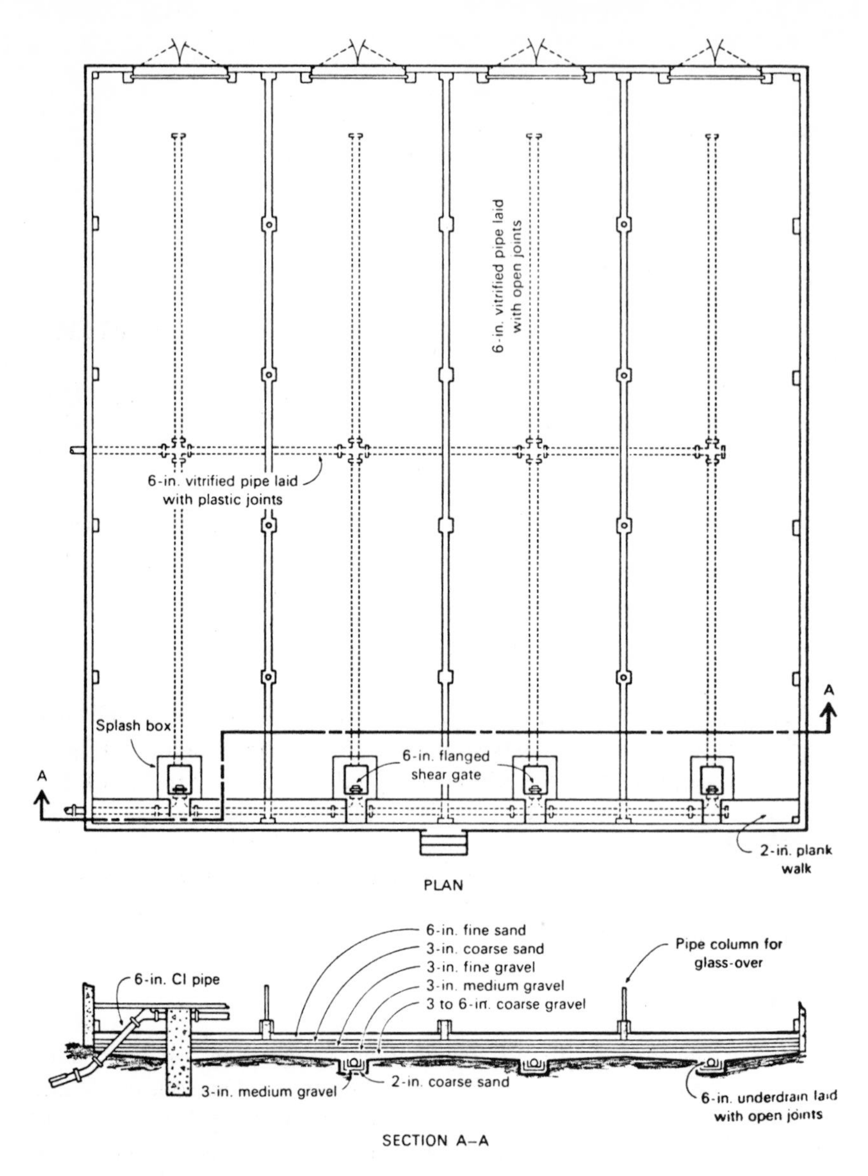

Figure 6-1. Typical sand drying bed (Metcalf and Eddy, 1973).

Penman and Van Es (1974) found that multiple applications on an impervious clay bed, draining the supernatant, allowing the sludge to freeze and taking it up while still frozen resulted in considerable savings in sludge disposal.

As with other methods of sludge handling, the design of sludge drying beds can be based on experience or, if the sludge is available, by scale-up from laboratory tests. Most often it is necessary to rely on experience because sludges are not available in sufficient quantities for testing.

Design manuals often include a table showing the required sludge bed areas in terms of population or quantity of wastewater expected at the treatment plant. Typical requirements for the northern United States are shown in Table 6-1. Solids loadings can vary from 50-125 $kg/m^2/yr$ (10-25 $lb/ft^2/yr$) for open beds to 60-200 $kg/m^2/yr$ (12-40 $lb/ft^2/yr$) for covered beds.

Table 6-1. Typical Standards for Drying Beds (WPCF, 1959).

Sludge	Square Feet per Capita	
	Open Beds	Covered Beds
Primary Digested	1.0-1.5	0.75-1.0
Primary and Humus Digested	1.25-1.75	1.0-1.25
Primary and Activated Digested	1.75-2.5	1.25-1.5
Primary and Chemically Precipitated Digested	2.0-2.5	1.25-1.5

By contrast, British experience dictates drying bed area between 3.5 and 5.5 ft^2 per capita, an increase from 1.5 ft^2 per capita considered adequate a few years ago. The increase is attributed to changes in sludge characteristics (Swanwick, 1972). It is possible that the standards in Table 6-1 are also obsolete.

When a sufficient quantity of sludge is available, it is far better to base design of drying beds on test results, either laboratory tests or by small-scale prototype drying beds.

According to Swanwick *et al.* (1962), it is possible to relate the specific resistance* of a sludge to the yield of dry sludge from a drying bed. Laboratory tests can thus provide some estimate of the required bed area. Nebiker *et al.* (1968) proposed a design equation based on a filterability test, but this is not readily usable due to a need for empirical constants.

A better design method is to place model beds in the field in actual climatic conditions. Such beds, with various types and depths of sand and with various depths of sludge applied, can yield the best possible information for the design of drying beds.

A device for estimating sand bed area requirements was developed by the Water Pollution Control Laboratory in Stevenage, England (Swanwick, 1972).

---

*Specific resistance is a measure of sludge filterability. It is discussed fully in the next section of this chapter.

Vertical tubes 4 cm in diameter are arranged with a layer of sand on the bottom to simulate drainage beds. Sludge is placed in these cylinders and then allowed to drain. When drainage ceases, the cake is lifted intact, its moisture content measured, and the moisture evaporated to a desired solids concentration. It has been observed that the rate of moisture loss in sand beds is approximately 75% of the loss from a clean water surface. Basic hydrological data yield the clean water evaporation rates, and the time requirements for sand drying can thus be estimated.

For open beds rainfall must be taken into account. Again experience has shown that about 57% of the rain is absorbed by the sludge and must later be evaporated, the remainder draining away (Swanwick, 1972). On the average, therefore, 57% of the rainfall must be added to the amount of water requiring evaporation.

A similar device was used by Randall *et al.* (1971) to study the effect of various actions on the filterability of aerobically digested sludge. It was reconfirmed that shearing of the sludge particles has a strong detrimental effect on filtration. Likewise, prolonged anaerobiosis decreases filterability.

Example 6-1. A mixed digested sludge of 6% solids is to be dewatered in open sand drying beds. The average rainfall is fairly evenly distributed over the year at 40 in., and the clear water evaporation rate is 60 in./yr. Laboratory tests indicate that the desired sludge moisture concentration is 50% and that drainage during the first few days increases the solids to 18%. Find the required bed loading.

Assume 10 in. of sludge is put on the beds. At 6% solids this is equivalent to 0.94 x 10 = 9.4 in. of water.

Sludge at 18% solids has (0.06/0.18) x 10 = 3.3 in. of sludge of which (1-0.18) x 3.3 = 2.7 in. is water.

Loss due to drainage = 9.4 - 2.7 in. of water.

Desired solids concentration is 50% solids, or (0.06/0.5) x 10 = 1.2 in. of sludge of which 0.5 x 1.2 = 0.6 in. is water.

Required evaporative loss = 2.7 - 0.6 = 2.1 in.

Evaporative rate = 0.75 x 60 = 45 in./yr.

Rainfall absorbed by sludge = 0.57 x 40 = 23 in./yr.

Net evaporation = 45 - 23 = 22 in./yr.

Since the required evaporation for each application is 2.1 in., the sand bed could theoretically be filled and emptied ten times a year. A safety factor would, of course, be appropriate.

Some years ago, it was suggested that sand drying beds should actually have flat, hard bottoms instead of sand and tile drains. The belief was the water would flow to the bottom and then flow underneath the sludge and out, thus not clogging the sand. A series of tests conducted by Randall (1969) show, however, that drying beds with sand bottoms perform in a far superior manner to the beds with impervious bottoms.

LAGOONS

Lagoons are often employed as an expeditious and low-cost means of sludge dewatering. Consisting of nothing more than a "hole-in-the-ground," sludge is pumped in at varying depths, usually 0.75-1.2 m (2.5-4 ft) and allowed to evaporate. Usually, there is no underdrain system for drainage. The process is obviously slow and requires large lagoon volume as well as odor-insensitive neighbors. The usual guidelines for drying lagoons are 35-39 $kg/yr/m^3$ (2.2-2.4 $lb/yr/ft^3$) of lagoon capacity. Other recommendations list lagoon requirements at 0.03 m^2 per capita (1 ft^2 per capita) for primary digested sludge to 0.1 m^2 per capita (4 ft^2 per capita) for activated sludge. The annual rainfall obviously must be taken into account.

WEDGE WIRE DRYING BEDS

Wedge wire drying beds, first developed in England, have seen some use in the United States. The wedge wire drying bed gets its name from the shape of the wires (triangular wedges) placed in parallel so as to create 0.25-mm openings between them. This design resists plugging and blinding. The filter, shown schematically in Figure 6-2, is operated by first filling the enclosure with water to about 1 in. above the wire bed. The water acts as a cushion for the incoming sludge and inhibits the movement of the solid particles through the screen. The water is next drawn off, and the sludge solids settle gently on the wire screen where further drainage occurs. If the sludge is allowed to remain on the screen for more than a few days, the combination of evaporation and drainage can produce quite dry cakes. Typical performance data are shown in Table 6-2.

The application of wedge wire screens is for small plants where the total amount of sludge is small and large capital investment is impractical. Considerable research is presently being conducted to develop devices to dewater small quantities of sludge—economically. One typical device under study at the Norweigian Institute for Water Research uses a roll of rough paper over a chicken wire support. As the sludge drains, the paper is simply pulled off with the sludge and a new piece is placed on the screen.

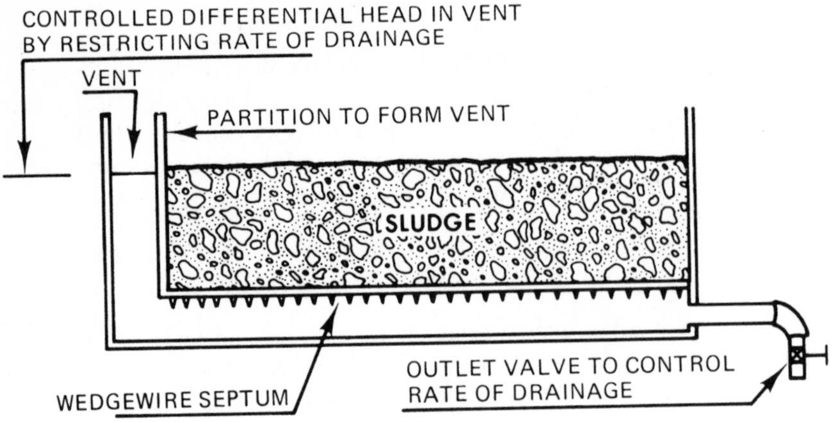

Figure 6-2. Cross section of a wedge wire drying bed.

Table 6-2. Typical Performance of Wedge Wire Drying Beds.

Sludge	Feed Solids (%)	Cake Solids (%)	Time of Drying	Solids Capture (%)
Primary	8.5	25	14 days	99
Trickling Filter Humus	2.9	8.8	20 hours	85
Mixed Digested	3.0	10.0	12 days	86
Waste Activated	0.7	6.2	12 hours	94
Waste Activated	1.1	9.9	8 days	87
Waste Activated	2.5	8.1	41 hours	100

VACUUM FILTERS

Vacuum filters have been used for 50 years in wastewater treatment for the dewatering of sludges. It is believed necessary to first digest sludge to increase filterability, although this is seldom true. Digestion does, however, reduce the odor problem, thus making the operation less objectionable than it might be otherwise.

Operation

The filter operates as shown in Figure 6-3. The chemicals used to condition the sludge and thus make it easier to filter are combined with the feed sludge and dumped into a trough, which sits underneath a large rotating drum. This drum is covered with a permeable fabric or other materials. A

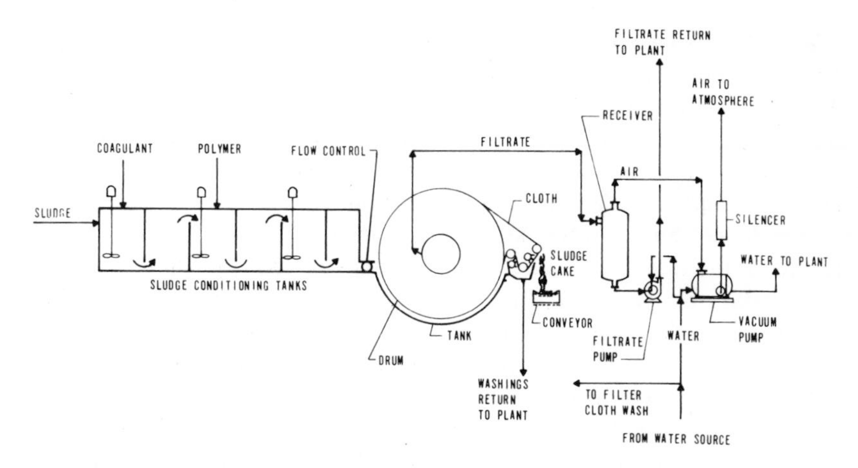

Figure 6-3. Operation of a vacuum filter (Weston,1971).

vacuum is drawn inside the drum, and water is sucked through the fabric into vacuum lines inside the drum and pumped out as the filtrate. The solids that cannot get through the fabric are caught on the surface of the drum and removed as the filter cake. Recent operational evidence indicates that belt filters (where the fabric is lifted off the drum during cake discharge, as in Figure 6-3) tend to require greater chemical doses than the older drum filters, where the sludge is scraped off with a doctor blade. The higher doses are required due to problems with cake release.

The objectives of vacuum filtration are to obtain:

1. a high solids concentration in the filter cake,
2. a clean filtrate, and
3. an acceptable filter yield.

The variables of importance in the operation of vacuum filters are listed in Table 6-3.

Table 6-3. Vacuum Filter Variables.

Machine Variables	Operational Variables
Vacuum Pressure	Type and Condition of Sludge
Drum Submergence	Chemical Conditioners
Drum Speed	Sludge Characteristics
Agitation	
Filter Medium	

The only real operational variable is the sludge to be filtered, because chemical conditioning can be considered a means of changing the sludge. The type of sludge filtered is obviously within the control of the operator, and it is also within his province to dictate what happens to the sludge before it gets to the vacuum filter. This might sound trite, but it is probably the most important variable in vacuum filtration.

The sludge solids concentration, for example, has a very significant influence on the filter yield (defined as the pounds of sludge obtained from the vacuum filter per square foot of filter area per hour). An increase in the feed sludge solids concentration usually results in a substantial increase in filter yield. Interestingly enough, the filter yield in terms of pounds per square foot per hour is often numerically equal to the percent of solids in the feed sludge. There is, of course, a practical limit of about 8-10% because, at greater solids concentrations, the sludge becomes difficult to pump. An increase in sludge solids concentration also tends to decrease the amount of chemical needed, hence making it doubly important for the treatment plant operator to try to introduce as concentrated a sludge as possible to the vacuum filter.

What he does with the sludge before it reaches the vacuum filter is also important. The amount of time that the sludge spends out of the treatment process, whatever that may be, always seems to decrease the filterability of a sludge. For example, if aerobically digested sludge is being filtered, removing the sludge from the aeration tank for any length of time tends to decrease sludge filterability. Similarly, digested sludge tends not to filter as well if the sludge is first taken out of the digester and allowed to cool. The action of chemical conditioners also deteriorates with time. Table 6-4 shows data for the Capillary Suction Time (CST), a measure of sludge filterability (see p. 157), for a digested sludge of various storage times. A high CST is indicative of poor filterability, as described on page 157. Prolonged storage increases the CST fourfold.

Table 6-4. Capillary Suction Time for a Conditioned Sludge After Various Storage Periods (*Water Pollution Research,* 1971).

Conditioner	Period of Storage (hr)	CST After Storage
Ferric Chloride	0	37
	0.5	46
	21	133
Primafloc C7	0	81
	0.5	173
	21	354

There are many other important variables that are not within the province of the treatment plant operator. For example, the viscosity of the fluid is difficult to control, as is the sludge temperature. However, both have a significant effect on sludge filterability.

The machine variables are also, to a limited degree, under the control of the operator. The vacuum imposed on the sludge can be adjusted, with the maximum practical limit being approximately 38 cm Hg (15 in. Hg). Unfortunately, many of the sludges filtered in wastewater treatment are very compressible, and higher vacuums tend to produce a nonpervious cake and hence can actually decrease the sludge yield from a vacuum filter.

It is possible to adjust the drum submergence, or the depth at which the drum is set inside the trough that contains the sludge. If the drum is dropped lower into the sludge, a greater percentage of the cycle time is devoted to the pickup of solids from the trough, and thus thicker but wetter filter cakes result. Decreasing the drum submergence tends to decrease the amount of time that the wet sludge is in contact with the filter, and thus produce thinner but drier cakes.

By increasing the speed of the drum, the length of time the sludge is in contact with the filter is decreased as is the time for drainage once the sludge has left the pool, thus obtaining a wetter cake. However, this tends to increase the sludge yield. Correspondingly, a slower drum speed tends to result in a drier cake and a lower filter yield.

The sludge in the trough must be agitated to prevent solids from settling and the extent of this agitation can be controlled. If the agitation is too violent, the flocs will be broken up and poor filtration will result. If the agitation is insufficient, settling will occur–resulting in operational problems.

Probably the most important machine variable is the selection of the proper medium for filtration, or the fabric that covers the drum. The available media can be categorized as open or tight. Open media have large pores while tight media have small openings. Although a tight filter will remove a higher percentage of the fines, it can also be so tight as to make filtration impossible. It can *blind,* or stop up completely, preventing further filtration.

The types of media in use today range from rayon acrylic, polyolefins and polyesters to wire screens and stainless steel coils. Historically, cotton has been the standard medium, but it is now very seldom used. Stainless steel coils have the advantage of never wearing out; however they have the disadvantage of being expensive compared with other types of media.

Another possible method of improving filter efficiency is to use compounds that can be mixed with the sludge to improve its filterability. The filter aids used in the past include paper pulp (Campbell *et al.,* 1978), coke, fly ash, clay and sawdust. Their disadvantage is that the quantity of sludge will obviously increase. The advisability of using filter aids is strongly dependent on the local cost, as well as the increase in filtration efficiency, so generalization is not possible.

Precoat filters are widely used in the chemical process industry but have not been found to be successful in the dewatering of sludges. In this operation, instead of mixing the filter aid with the sludge, the filter aid is placed on the filter fabric before the sludge is introduced. This prevents rapid premature clogging of the smaller pores and allows for a greater filter yield. Diatomaceous earth has been shown to be an effective but expensive filter aid for precoat filters.

Performance

In dewatering digested sludge, vacuum filters are typically able to form cakes of 20% solids, although 40% solids cakes have been obtained in some places. The filtrate quality can vary anywhere from 100 to 20,000 mg/l of solids, corresponding to solids recoveries of about 99 to 50%. The filtrate is almost always returned to the head of the plant for processing because it contains significant BOD and thus cannot be discharged with the plant effluent.

Waste activated sludge, unless mixed with primary sludges, is not normally dewatered on vacuum filters. An exception to this is pure oxygen sludges that have been filtered successfully to 14% solids with the use of ferric chloride conditioning in filter leaf and pilot filter tests.

Design

Vacuum filters can be designed on the basis of previous experience with a similar sludge, by the filter leaf test, or by using the specific resistance to filtration measured in the laboratory.

Design Based on Experience. When similar sludges are encountered, previous experience can be used for design. For example, if a filter operating on a typically domestic mixed digested sludge has a filter yield of 3.5 $lb/ft^2/hr$, it is quite likely that another filter operating on a similar sludge will perform in the same way. Table 6-5 lists some experiences in vacuum filtration.

Design Based on the Filter Leaf Test. A typical cycle of a vacuum filter is 30 sec of submerged operation, 60 sec of drying under vacuum but not submerged, and 30 sec off the filter. In other words, 25% of the drum circumference is submerged, and 25% is not covered by the fabric as the cake is discharged. Sludge drying occurs on the remaining 50%.

Such a cycle is simulated with a filter leaf. The filter leaf is in fact a small model of the prototype filter. As can be seen in Figure 6-4, the filter leaf

Table 6-5. Typical Filter Yields

Sludge	Chemical Conditioning	Filter Yield (lb/ft²/hr)	Cake Solids (%)	Reference
Digested Primary	Ferric chloride and lime	7.2	27	Burd, 1968
	Polymer	4-15	34-26	Burd, 1968
Mixed Digested	Ferric chloride and lime	4-8	21.5	Burd, 1968
	Polymer	4	32-24	Burd, 1968
Primary Lime	None	1.5-3.3	26-32	Martel and DiGiano, 1978

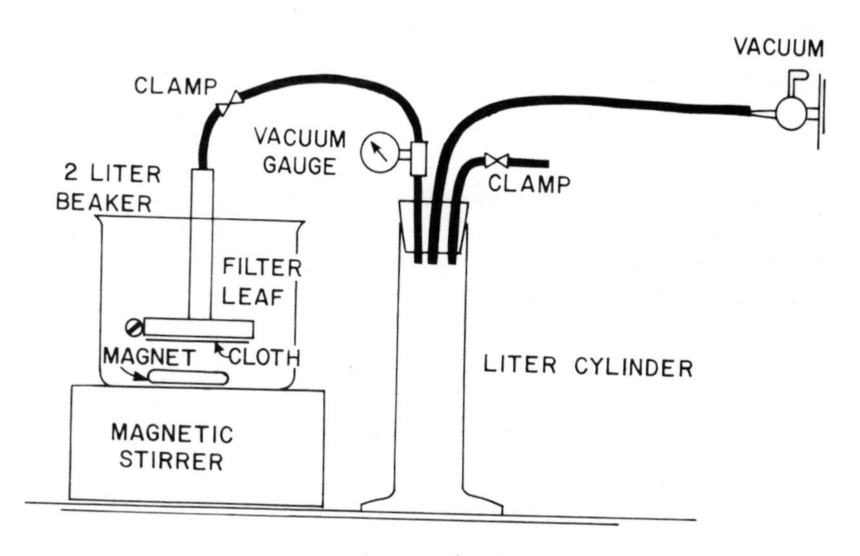

Figure 6-4. Filter leaf appartus.

consists of a round disc about 10 cm in diameter, over which the filter medium is stretched. This disc is connected to a vacuum source, through a graduated cylinder that is used to collect the filtrate. The filter leaf is placed into a beaker containing the chemically conditioned sludge and held there for 30 sec, or whatever the submerged time is on the prototype filter. The filter leaf is then removed from the sludge for as long as the filter would be in the drying cycle, and then the vacuum is turned off. The cake is scraped off the filter leaf and analyzed for solids content, and the total dry cake solids produced is calculated. The filtrate is analyzed for suspended solids, thus allowing for the calculation of the solids recovery. The diameter of the filter leaf is measured and used as the area in calculating filter yield.

Example 6-2. The following results were obtained from a filter leaf test:

Feed solids concentration = 4%
Cake solids concentration = 20%
Area of filter leaf = 0.1 ft^2
Total cake dry weight = 0.028 lb
Cycle: 30 sec submerged
60 sec drying
30 sec off the filter

2-min cycle
Filtrate solids concentration = 500 mg/l

The filter yield can be calculated as follows:

$$\text{Filter Yield} = \frac{(0.028\ \text{lb}/0.1\ \text{ft}^2)}{2\ \text{min} \times \dfrac{\text{hr}}{60\ \text{min}}} = 8.4\ \text{lb/ft}^2/\text{hr}$$

The solids recovery in any dewatering device is calculated from a solids balance. Define the feed flow rate and solids concentration as Q_o and C_o, the filter cake flow and solids as Q_k and C_k, and the filtrate flow and solids as Q_f and C_f. A liquid balance gives

$$Q_o = Q_f + Q_k$$

and solids balance

$$Q_oC_o = Q_fC_f + Q_kC_k$$

Substituting,

$$Q_oC_o = Q_fC_f + C_f\,(Q_f - Q_k)$$

$$Q_o\,(C_o - C_f) = Q_k\,(C_k - C_f)$$

$$Q_k = \frac{Q_o(C_o - C_f)}{C_k - C_f}$$

Percent recovery of dry solids (%R) is

$$\%R = \frac{\text{mass of dry solids as cake}}{\text{mass of dry feed solids}} \times 100$$

$$\%R = \frac{C_kQ_k}{C_oQ_o} \times 100$$

Substituting from above,

$$\%R = \frac{C_k \left[\frac{Q_o (C_o - C_f)}{C_k - C_f} \right]}{Q_o C_o} \times 100$$

and

$$\%R = \frac{C_k (C_o - C_f)}{C_o (C_k - C_f)} \times 100 \qquad (6\text{-}1)$$

Example 6-3. Using the data from Example 6-2, calculate the solids recovery.

$$\%R = \frac{200{,}000\ (40{,}000 - 500)}{40{,}000\ (200{,}000 - 500)} \times 100 \cong 99\%$$

A great advantage of the filter leaf test is that it is possible to use the same medium that would be used on the prototype. In other words, if a nylon cloth is to be used on the filter, a small square of the nylon can be used for the filter leaf. This affords a realistic approximation of the prototype operation.

It should also be realized that the filter cloth in the prototype is used repeatedly. As the cloth moves from the point where the cake is dropped off, it is often washed with water sprays, then dipped back into the trough containing the sludge. In the filter leaf test, the filter cloth should be washed between each run to determine the filter yield when the cloth is used repeatedly. A clean filter cloth will almost always provide a much higher yield than one that is used several times (Vesilind, 1973). Experiments at the British Water Pollution Laboratory suggest a 30-fold increase in the time required for filtration for polymer-conditioned sludge after five filtrations. Incidentally, ferrous sulfate and lime conditioning did not result in decreased filterability with repeated use of a filter cloth (*Water Pollution Research,* 1972).

The ability of the filter leaf to predict continuous vacuum filter performance has been well demonstrated. For example, data for a primary ferric chloride sludge are shown in Figure 6-5. Although there is some scatter, the overall trend is indicated, and it can be concluded that approximate filter yield predictions can be obtained from filter leaf data (Campbell *et al.,* 1975).

Data for a mixed sludge are shown in Table 6-6. Again it is obvious that the filter leaf provides useful design information (Christensen *et al.,* 1976).

Design Based on Specific Resistance. It is possible to analyze the filtration of sludge based on Darcy's equation describing the flow of fluid through a porous medium. First advanced over two hundred years ago, his well-known

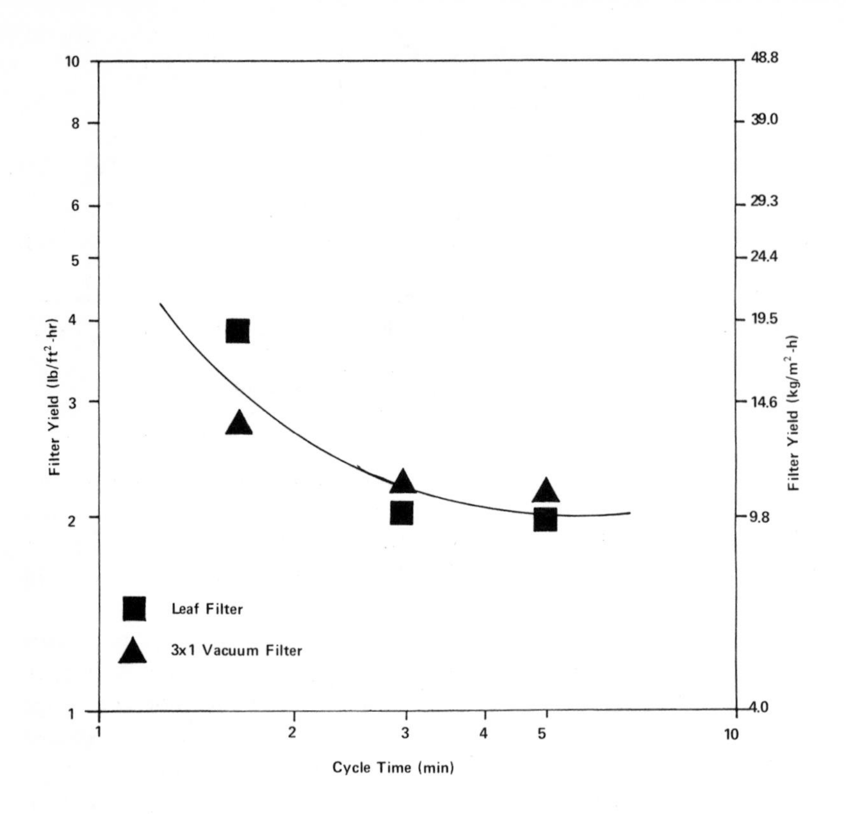

Figure 6-5. Prediction of filter yield with filter leaf (Campbell *et al.,* 1975).

formula is used in many fields, including hydraulics and soil mechanics. Carman (1933) adapted this equation to filtration, and Coackley and Jones (1956) adapted Carman's theoretical analysis to vacuum filter operation.

Using Darcy's law:

$$\frac{dV}{d\theta} = \frac{P}{\mu} \frac{AK}{L}$$

where $\frac{dV}{d\theta}$ = rate of flow, volume (V) per time (θ)

P = pressure difference

A = area

μ = viscosity

K = permeability

L = thickness

Table 6-6. Comparison of Filter Leaf and Vacuum Filter Performance (Christensen *et al.*, 1976)

	Filter Yield ($lb/ft^2/hr$)		Cake Solids (%)	
	Filter Leaf	Vacuum Filter	Filter Leaf	Vacuum Filter
	7.0	7.9	19.1	21.4
	6.3	7.6	19.4	21.8
	7.7	7.9	19.5	21.2
	5.9	6.8	20.0	21.2
	6.9	7.1	18.8	21.0
	7.6	7.1	18.7	21.0
	6.1	6.0	20.5	20.6
	6.0	5.8	19.5	21.7
Avg:	6.7	7.0	19.4	21.6

we can define resistance as $R = \frac{1}{K}$,

then

$$\frac{dV}{d\theta} = \frac{P}{\mu} \frac{A}{LR}$$

In a filter, resistance is contributed by both the filter medium and cake,

$$\frac{dV}{d\theta} = \frac{P}{\mu} \frac{A}{(LR + R_f)}$$

where R_f = resistance of filter medium.

The volume of the cake can be expressed as

$$LA = vV$$

where v = volume of cake deposited per unit volume of filtrate.

Substituting for L,

$$\frac{dV}{d\theta} = \frac{PA^2}{\mu(RvV + R_fA)}$$

It is more convenient to express cake as dry weight per volume instead of volume of cake per volume of filtrate. Also, R (resistance by a unit volume) is replaced by r (resistance by unit weight). Thus

$$\frac{dV}{d\theta} = \frac{PA^2}{\mu(wrV + R_fA)}$$

where w = weight of dry cake solids per unit volume of filtrate

r = *specific resistance*

Assuming constant pressure over time,

$$\int_0^\theta d\theta = \int_0^V \left(\frac{\mu w r V}{PA^2} + \frac{R_f \mu}{PA} \right) dV$$

$$\theta = \frac{\mu w r V^2}{2PA^2} + \frac{R_f \mu V}{PA} \quad \text{or} \quad \frac{\theta}{V} = \frac{\mu r w V}{2PA^2} + \frac{\mu R_f}{PA}$$

which is a straight line of type y = bx + a, where

$$b = \frac{\mu r w}{2PA^2} \quad \text{and} \quad a = \frac{\mu R_f}{PA}$$

It should thus be possible to measure the volume of the filtrate, V, at various times, θ; plot these as θ/V vs V, and obtain a straight line. The slope of this line is calculated, and since the slope b is equal to

$$\frac{\mu r w}{2PA^2}$$

it is possible to calculate specific resistance, r, the only unknown, as

$$r = \frac{2PA^2 b}{\mu w}$$

The experiments for obtaining such data are conducted with a Buchner funnel apparatus, shown in Figure 6-6.*

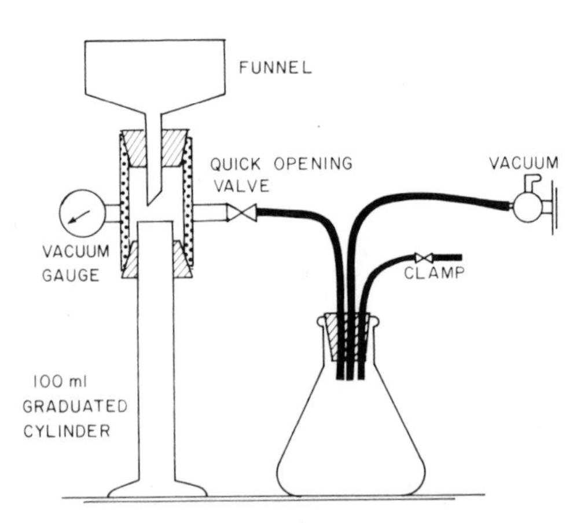

Figure 6-6. Buchner funnel apparatus for measuring specific resistance.

*It is also possible to use over pressure instead of vacuum (Tabasaran, 1971).

The test is conducted by pouring a reasonable (but constant) volume of sludge into the funnel (with filter paper) and imposing the vacuum at time zero. The amount of filtrate is then recorded with time.

The pressure P is measured with a vacuum gauge and converted to N/m^2. Area is calculated in m^2, and the viscosity μ in N sec/m^2.

The cake deposited per volume of filtrate, w, can be approximated by the feed solids concentration in kg/m^3. It can also be accurately calculated as

$$w = \frac{C_k C_o}{100\,(C_k - C_o)}$$

where C_k = cake solids concentration, %
C_o = feed solids concentration, %

This relationship can be derived from materials balance. A liquid balance gives

$$Q_o = Q_f + Q_k$$

and a solids balance yields

$$Q_o C_o = Q_f C_f + Q_k C_k$$

where Q is flow rate and C is solids concentration, and the subscripts o, f and k denote feed, filtrate and cake, respectively.

The weight of dry solids deposited as cake per volume of filtrate, defined as w, is

$$w = \frac{Q_k C_k}{Q_f}$$

Substituting from the liquid balance above,

$$w = \frac{(Q_o - Q_f)\,C_k}{Q_f} = \frac{Q_o C_k - Q_f C_k}{Q_f}$$

Rearranging the liquid balance and substituting,

$$Q_f C_f = Q_o C_o - Q_k C_k$$

$$Q_f C_f = Q_o C_o - (Q_o - Q_f)\,C_k$$

$$Q_f C_f - Q_f C_k = Q_o C_o - Q_o C_k$$

$$Q_f = \frac{Q_o\,(C_o - C_k)}{C_f - C_k}$$

Substituting,

$$w = \frac{Q_o C_k - \left[\dfrac{Q_o\,(C_o - C_k)}{C_f - C_k}\right] C_k}{\dfrac{Q_o\,(C_o - C_k)}{C_f - C_k}}$$

and rearranging,

$$w = \frac{C_k (C_f - C_o)}{C_o - C_k}$$

If the filtrate suspended solids are assumed to be negligible ($C_f = 0$),

$$w = \frac{C_k C_o}{C_k - C_o}$$

If the solids concentration is expressed in percent,

$$w = \frac{C_k C_o}{100 (C_k - C_o)}$$

Recently, there has been some confusion about the units of specific resistance. If the pressure P is measured in kg/cm, the specific resistance is in sec^2/g. This has been the common designation in the literature.

Unfortunately, pressure should not be measured in mass units, but in dynes/cm^2 instead. This makes r as cm/g. Since the adoption of the SI system, pressure is measured in Newtons/m^2, making r in terms of m/kg.

Table 6-7 is an attempt to sort out the dimensions found in the literature. It should be mentioned that most laboratories and publishers have adopted the SI system, and future work will be reported with r in m/kg.

Table 6-7. Conversion Factors for Filtration (based on Gale, 1971)

Term	To Convert from	to	Multiply by
P	g/cm^2	N/m^2	9.81 x 10
	mm Hg	N/m^2	1.33 x 10^2
	in. Hg	N/m^2	3.39 x 10^3
	psi	N/m^2	9.6 x 10^3
μ	poise	N s/m^2	1.00 x 10^{-1}
w	g/cm^3	kg/m^3	1 x 10^3
	mg/l	kg/m^3	1 x 10^{-3}
r	s^2/g	m/kg	9.81 x 10^3
L	kg/m^2/sec	kg/m^2/hr	3.60 x 10^3
	kg/m^2/hr	lb/ft^2/hr	2.05 x 10^{-1}
	lb/ft^2/hr	kg/m^2/hr	4.88

Example 6-4. The following data were obtained with a Buchner funnel apparatus:

Time (min)	θ (sec)	V (ml)	Corrected* V (ml)	$\frac{\theta}{V}\left(\frac{sec}{ml}\right)$
-2	–	0	–	–
0	0	1.5	0	–
1	60	2.8	1.3	46.3
2	120	3.8	2.3	52.3
3	180	4.6	3.1	58.0
4	240	5.5	4.0	60.0
5	300	6.1	4.6	65.2

*First 2 min are ignored; this is due in part to storage of water in the filter and funnel (Zingler, 1970). Gale (1967) argues that this must be neglected because the resistance due to the filter must be allowed to become negligible.

Other test variables:

$$P = \text{pressure} = 10 \text{ psi} = 703 \frac{g}{cm^2}$$
$$= 6.90 \times 10^4 \ N/m^2$$
$$\mu = \text{viscosity} = 0.11 \text{ poise} = 0.011 \ N/m^2$$
$$w = 0.075 \ g/ml = 75 \ kg/m^3$$
$$A = 44.2 \ cm^2 = 0.00442 \ m^2$$

The data are plotted as Figure 6.7. The slope of the line, b from the curve, is 5.73 sec/cm^6 = $5.73 \times \left(\frac{100 \ cm}{m}\right)^6 = 5.73 \times 10^{12}$

$$r = \frac{2 \times 6.9 \times 10^4 \times (0.00442)^2 \times 5.73 \times 10^{12}}{0.011 \times 75} = 1.86 \times 10^{13} \frac{m}{kg}$$

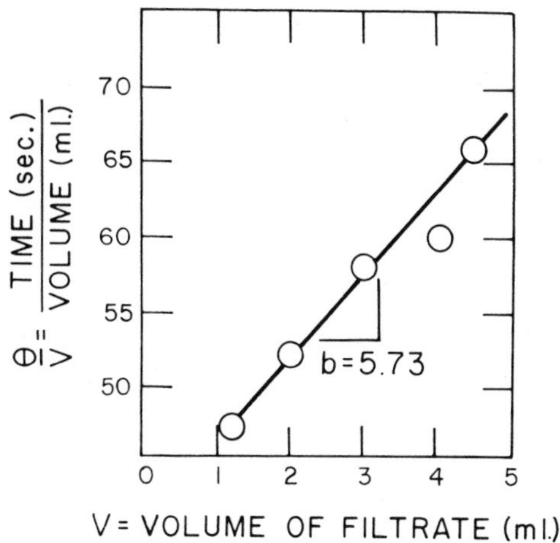

Figure 6-7. Typical laboratory filtration data, as used in Example 6-4.

Note:

$$r = \frac{\frac{N}{m^2} \times m^4 \times \frac{s}{m^6}}{\frac{Ns}{m^2} \times \frac{kg}{m^3}} = \frac{m}{kg}$$

Wuhrmann (1977) has developed a device for increasing the precision and speed of specific resistance determination. Basically, this device measures the time difference between two filtrate volumes in the graduated cylinder. If it is assumed that these two points accurately describe the straight line with a slope b, then

$$b = \frac{1}{V_2 - V_1}\left(\frac{\theta_2}{V_2} - \frac{\theta_1}{V_1}\right)$$

where V_1 and V_2 are the filtrate volumes at times θ_1 and θ_2. If we set both V_1 and V_2 at some constant volume, the only unknowns are the measured times, θ_2 and θ_1. It must, however, be repeated that this technique uses only two points on an often unpredictable curve, and seems to be of value only where the shape of the curve is known and repetitive tests are to be conducted.

The use of specific resistance for design is a logical step. Recall that from the specific resistance derivation,

$$\theta = \frac{\mu w r V^2}{2PA^2} + \frac{R_f \mu V}{PA}$$

Assume resistance of filter medium, R_f, is negligible, thus

$$\theta = \frac{\mu w r V^2}{2PA^2}$$

$$\text{or } \frac{V}{A} = \left[\frac{2P\theta}{\mu w r}\right]^{1/2} = \frac{\text{volume of filtrate}}{\text{filter area}}$$

Recall that w = weight of cake/volume of filtrate

$$w\ \frac{V}{A} = \left[\frac{2P\theta w}{\mu r}\right]^{1/2} = \text{weight of cake/filter area}$$

A drum filter operates so that the time of cake formation, θ, is only some fraction, k, of the total cycle time (time for one drum rotation), or $\theta = k\theta_c$. This the filter yield is

$$\frac{w}{\theta_c}\left(\frac{V}{A}\right) = \left[\frac{2Pwk}{\mu r \theta_c}\right]^{1/2} = F$$

where F = weight of cake/filter area/time

Example 6-5. Using the data from Example 6-4, calculate the filter yield.

$P = 6.90 \times 10^4 \text{ N/m}^2 \quad w = 75 \text{ kg/m}^3 \quad \mu = 0.0011 \text{ N s/m}^2$

$r = 1.86 \times 10^{13}$ m/kg. Assume $\theta_c = 120$ sec, $k = 0.25$

$$F = \left[\frac{2 \times 6.90 \times 10^4 \times 75 \times 0.25}{0.0011 \times 1.86 \times 10^{13} \times 120}\right]^{1/2} = 1.02 \times 10^{-3} \text{ kg/m}^2\text{/sec}$$

$$= 0.75 \text{ lb/ft}^2\text{/hr}$$

(Note: this is a poor yield).

From extensive correlative testing, it has been found that the above analysis does not give adequately accurate predictions of filter yield (Gale, 1970). Specifically, the sludge solids concentration seems to be an independent variable not adequately accounted for in the specific resistance equation (Bennett *et al.*, 1973). Empirical observations (Shepman and Cornell, 1956) indicate that

$$Y = 0.88\ C_f - 1.0$$

where Y = filter yield, lb dry solids/hr/ft^2

C_f = concentration of feed solids

These calculations result in a filter yield that could be used for design purposes, but the filter leaf test is a more accurate design method and should be used in preference to the specific resistance method.

The specific resistance can, however, be used to evaluate the effect of chemicals on sludge filterability. By conducting a series of tests with different doses of a chemical, curves similar to Figure 6-8 can be obtained. Often, especially for organic polymers, an optimum chemical dose is found, after which greater doses decrease filterability.

The specific resistance varies with pressure, area, solids concentration and liquid viscosity. The latter two are easy to measure and further comment is unnecessary. It should be recognized that the area term in the specific resistance equation is squared, so it becomes very important to have an accurate measurement of the Buchner funnel filter area. This is usually the area of the filter paper, although some investigators use the area of the porous part of the funnel.

The vacuum pressure is a variable that can be controlled in the prototype, and thus it is advantageous to have some idea how specific resistance varies with pressures. Theoretical analysis here is lacking, but empirically, it has been found that

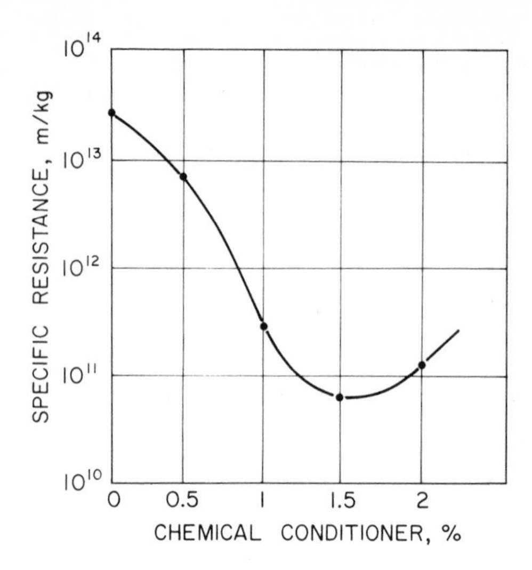

Figure 6-8. Typical specific resistance data with chemical conditioning.

$$r = r'P^{S_o}$$

where r' is a constant ($r' = r$ at $P = 1$) and S_o another constant which defines the slope on a log–log plot, such as Figure 6-9. S_o is called the *coefficient of compressibility,* since the slope of the line is an indication of how much the solids are compressed. Incompressible material like sand, for example, would have $S_o = 0$. For domestic sludges, S_o has been found to range between 0.4 to 0.85 (WPCF, 1969). Water treatment sludges have been found to have a

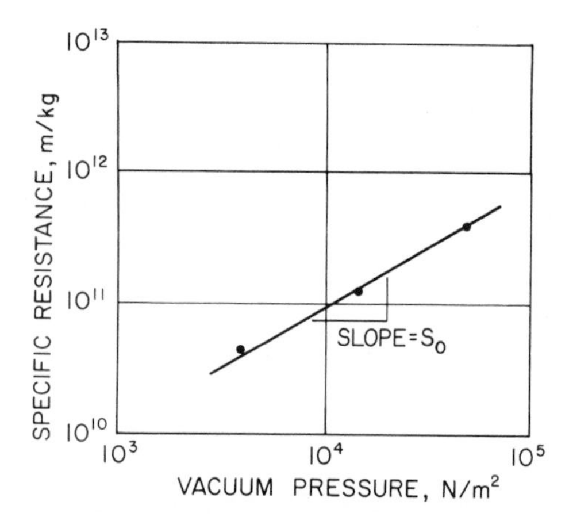

Figure 6-9. Compressibility of sludge as measured by the specific resistance test.

range of 0.8 to 1.3 in the coefficient of compressibility (Adrian *et al.*, 1968) while lime sludges have an S_0 of about 1.05 (Martel and DiGiano, 1978).

Obviously, the specific resistance test is not a very simple one to conduct because it requires some skill by the laboratory technician and considerable time. If it is necessary to run a series of tests with differently conditioned sludges, measurement of specific resistance becomes time-consuming. Accordingly, it has been shown that a test known as the capillary suction time (CST) can provide a reasonable approximation of the specific resistance for a particular sludge (Baskerville and Gale, 1968).

The CST is simply the time required for the liquid fraction of a sludge to travel one cm in a blotter paper. The device, pictured in Figure 6-10, consists of a timer which is activated when the liquid on the blotter touches one electrode and is stopped when the liquid reaches the second one. This time requirement for the filtrate to travel one cm is the capillary suction time in seconds.

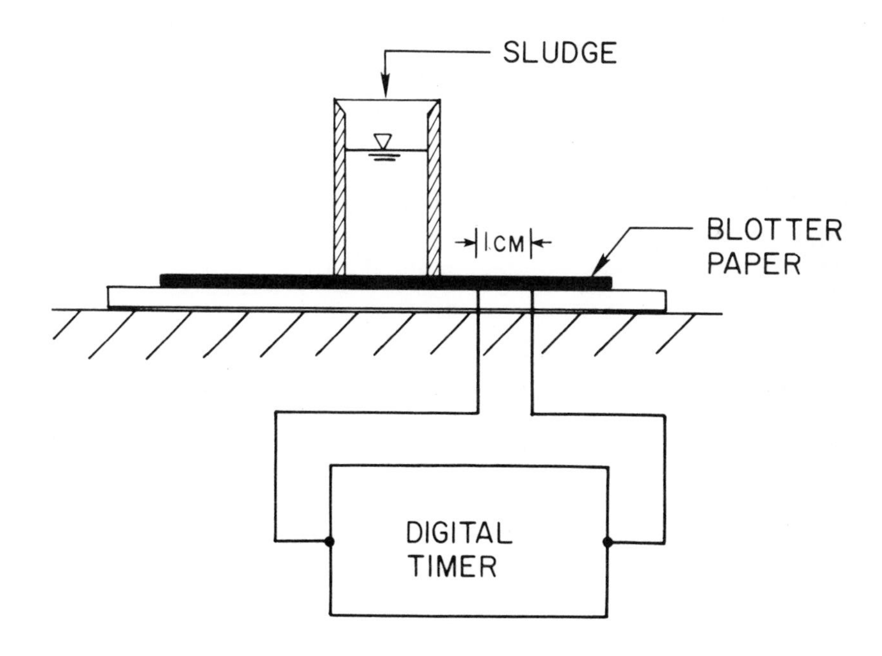

Figure 6-10. Capillary suction time (CST) apparatus.

The correlation between specific resistance and CST for a specific sludge must first be established by conducting a number of specific resistance tests with the Buchner funnel and simultaneously running the CST analyses.

Once this is obtained, it is a simple matter to run the CST and thus approximate the specific resistance from the correlation. The accuracy of this correlation is improved if the CST is plotted against the product of specific resistance and sludge solids concentration (Swanwick, 1972).

Summary

Several variations of the standard vacuum filter are available commercially. One filter uses rolled newspaper (or other paper) which continuously covers the drum. The operation is shown schematically in Figure 6-11. This filter seems to be especially applicable for sticky and difficult-to-dewater sludges. Additionally, the presence of the paper would not hinder subsequent incineration.

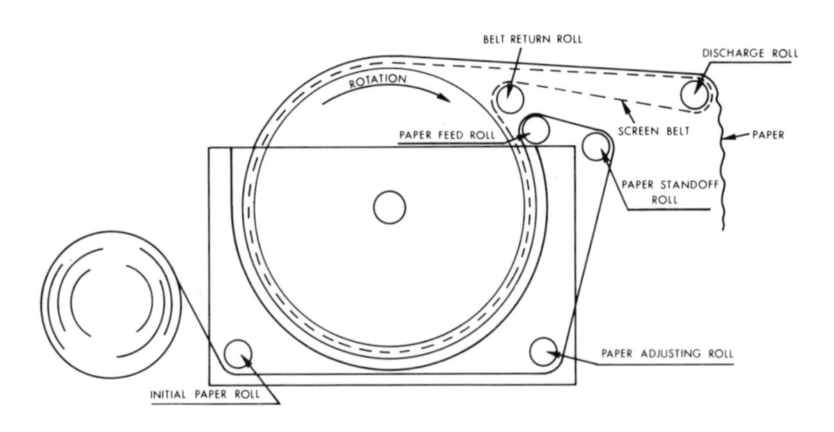

Figure 6-11. Vacuum filter using paper as filtering medium (Courtesy Technical Fabricators Inc., Piscataway, NJ).

Another variation of the standard rotary vacuum filter is the horizontal belt vacuum filter. The filter consists of an endless belt over a pan-type vacuum chamber.

All these variations attest to the basic soundness of the concept of dewatering by vacuum.

Vacuum filtration is an old established process. It has had many successes, but the landscape is also scattered with a number of monuments—unused and/or unusable vacuum filters. It is thus reasonable to assume that vacuum filtration also has problems.

Some of these are high operating costs, often resulting from the necessity for a skilled union-scale mechanic as an operator, and from the cost of chemical conditioners. Many treatment plants have had problems with blinding to the point where vacuum filtration was no longer possible, much less efficient. Another problem associated with vacuum filtration is that sludge with a strong odor can become a serious community concern. The sludge in vacuum filtration is always open to the atmosphere and often warm—two conditions producing the most severe odor problems. It is also quite clear that many vacuum filters no longer in operation are idle because of inexperienced operators. After the filter is installed in the treatment plant, neither the consulting engineer nor the manufacturer has responsibility for continued assistance, thus the treatment plant operator is left to his own designs to make the unit function. This sometimes ends in failure.

Nevertheless, vacuum filtration has been significantly successful in wastewater treatment and will remain in use in years to come.

PRESSURE FILTERS

Pressure filtration differs from vacuum filtration in that the liquid is forced through the filter medium by a positive pressure instead of a vacuum. A number of different kinds of pressure filters are on the market. The oldest and most widely used in the chemical process industry (and in Europe for wastewater treatment) is the *filter press*. This name is a misnomer since the sludge is not really pressed. As shown in Figure 6-12, the filter press operates

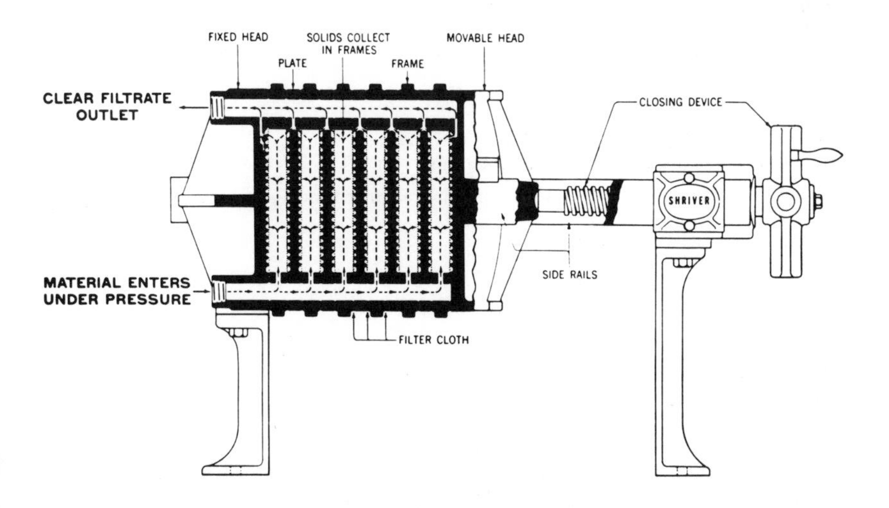

Figure 6-12. Filter press (Courtesy of T. Shriver Div., Envirotech Corp.).

by pumping the sludge between plates that are covered with a filter cloth. The liquid seeps through the filter cloth leaving the solids behind between the plates. When the spaces between the paltes are filled, the treatment plant operator separates the plates and removes the solids.

Filter pressing is a cyclic operation, a fact that has contributed significantly to its unpopularity in the United States. However, this cyclic operation is considered an advantage in Europe. In many smaller plants, the operator can start the filter press in the evening and empty it upon returning to work the next morning. The filter press operates all night without constant maintenance.

One of the more unpleasant aspects of the filter press is the removal of solids after the filtration is complete. This has prompted a number of different designs for automatic cake removal. One method is to blow the cake off by sending compressed air through the filtrate tubes. Another method is to make the filter cloth a moving belt so that after each operation the belt is shifted to allow the cake to drop off either side.

Although the specific resistance can also be used in the design of filter presses (Jones, 1956), the standard Buchner funnel apparatus operates on a vacuum, and thus cannot be a realistic model of pressure filtration.

Wilhelm (1978), however, reports excellent results in the use of specific resistance in scaling up pressure filters. He derives the cycle time, or time necessary to perform one dewatering operation, as

$$T = b\left(\frac{C_K \rho_s A l}{2C_F}\right)^2$$

where T = cycle time, sec

C_K = cake solids concentration, weight fraction

ρ_s = density of solids, g/cm^3

A = area of filtration, cm^2

l = cake thickness, cm

C_F = feed solids concentration, g/cm^3

b = slope of θ/V vs V plot using the Buchner funnel apparatus, sec/ml^2

Taking the log of both sides,

$$\log T = \log b\left(\frac{A^2 C_K{}^2}{C_F{}^2}\right) + \log(1\ \rho_s)^2$$

and plotting T vs $b[(AC_K)/(C_F)]^2$ on log-log paper was found by experiment to yield a straight line thus providing a means of estimating cycle time for various feed and cake solids.

An alternative method of scale-up is a laboratory-scale filter press used by several manufacturers (Carnes and Eller, 1972). Gerlich and Rockwell (1973) report that results from such a filter press correlate well with prototype yields.

In addition to the pressure filter as described above a number of *belt filter presses* have recently found wide use in sludge dewatering, especially at smaller installations (Austin, 1978).

The belt filter press has actually been used widely in Europe since the mid-1960s (Nyberg, 1970) but only recently have these low-capital-cost, low-speed and versatile devices become popular in the United States.

There are many different types of belt filter presses in use. One of the newer belt filter presses is shown schematically in Figure 6-13. In this press, the sludge is continuously squeezed between two belts, one a press belt and the other a filter belt. The rollers through which the belts move describe an S-shaped curve. This flexing of the belt is one reason why the belt filter press is so effective, as first noted by Klein in Germany (Villiers and Farrell, 1977). As the two belts with sludge in the middle are wrapped around a roll, shear is introduced and the differential movement allows the wet sludge to work its way toward the belt for subsequent dewatering (see Figure 6-14). Shearing the cake continually introduces new sludge to the belt surface, especially as the direction of shear is alternatively changed as the belt goes over rollers.

Filter presses have performed well in most cases. In one comparative study with a vacuum filter (Villiers and Farrell, 1977), a belt filter press achieved 33% cake solids with a combined primary and secondary sludge, while the vacuum filter only managed 17-19% solids.

Although accurate scale-up of filter presses requires a modeling of the shearing between the belts, a technique using a simple drainage and bench-scale hydraulic press has been shown to yield reasonable estimates of prototype belt filter performance (Baskerville *et al.*, 1978).

CENTRIFUGES

The first centrifugal dewatering of sludge was attempted in 1902 in Germany. Although the machine operated rather well, electrical current was too costly and the experiment was discontinued. In 1920, Milwaukee made a similar unsuccessful attempt. It was not until 1960 that a machine was placed in a treatment plant in San Mateo County in California and, with the aid of chemicals, performed admirably in dewatering digested sludge. The combination of increased skills in machining and especially in the production of organic polyelectrolytes for sludge conditioning finally made centrifugation a feasible and attractive alternative to vacuum filtration.

The solid-bowl centrifuge (or the decanter) is the type of machine used in wastewater treatment. This machine has the attribute of being able to dewater or at least separate out any solid from any liquid, as long as the solids are heavier. It is possible to use a centrifuge for many purposes in a wastewater treatment plant, for example, thickening activated sludge and, when

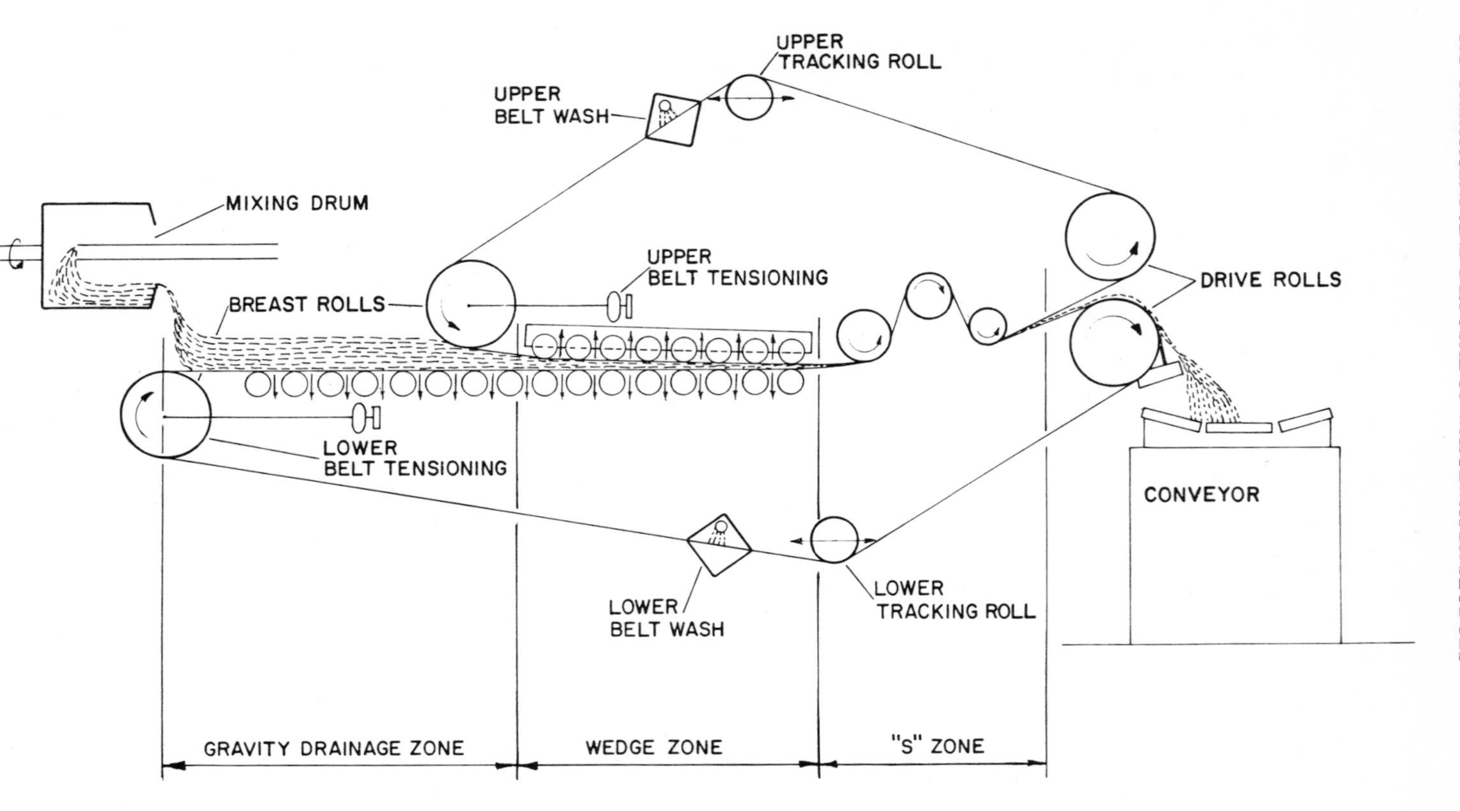

Figure 6-13. Belt press filter schematic (Courtesy Tait-Andritz).

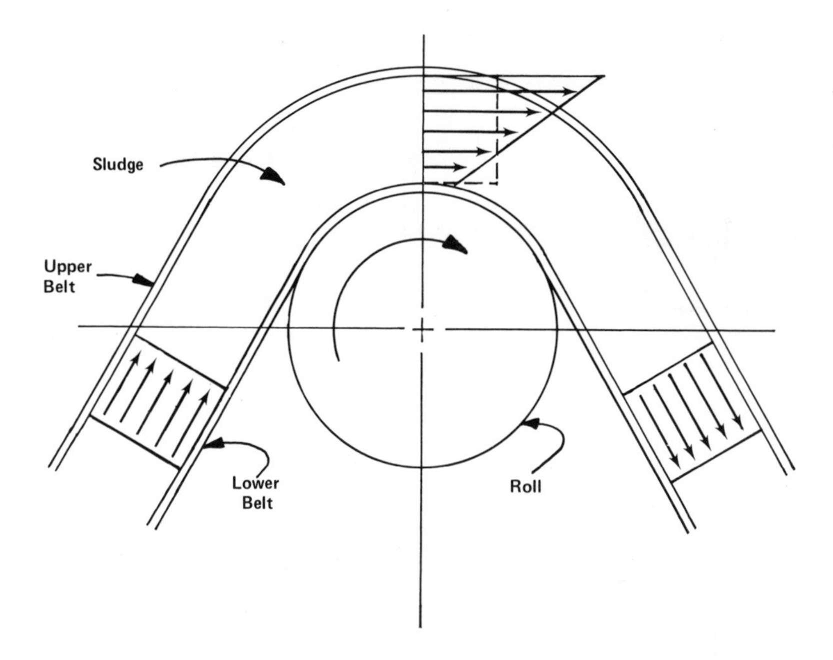

Figure 6-14. Schematic of sludge movement between two belts going over a belt press filter roll.

this is not needed, for the dewatering of digested sludge. This versatility increases the machine's value.

Operation

The conventional solid-bowl centrifuge is shown schematically in Figure 6-15A. It consists of an outer bullet-shaped bowl that rotates at high speed. The sludge is pumped through a central pipe into this rotating bowl and because of centrifugal force hugs the inside walls of the bowl. The heavier solids will sink to the bottom (that is, to the inner bowl wall) and the lighter liquid will remain pooled on top.(The bowl acts as a highly effective settling tank and nothing more.) It is necessary to remove the sludge to make the operation successful and in the solid-bowl centrifuge, this process is accomplished by the scroll, or screw conveyor, that is placed inside the machine and rotates only slightly slower than the bowl. This screw action tends to convey the solids up onto the inclined beach and out the open end. The centrate, or clear liquid, flows out the holes on the other end of the bowl.

Also shown in Figure 6-15B is a sketch of the Kruger centrifuge manufactured in Denmark which has been shown to have excellent operational and performance characteristics.

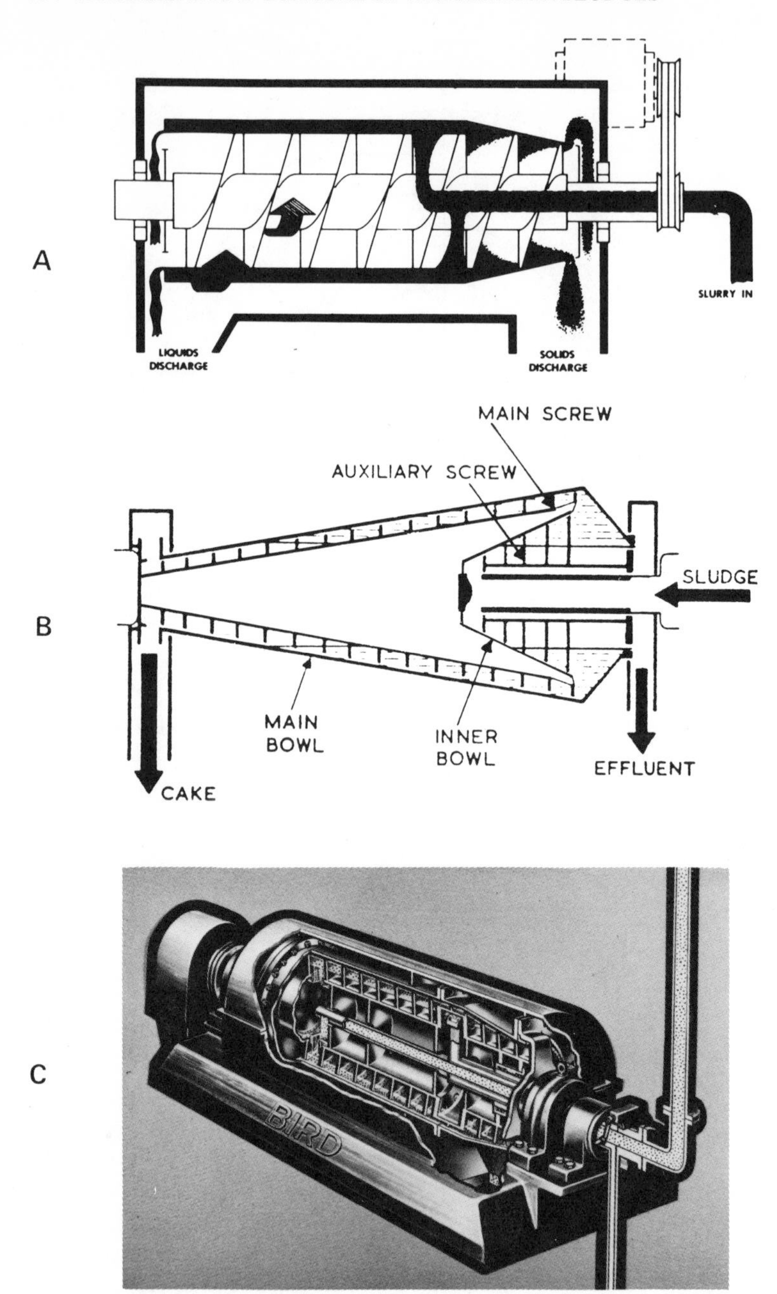

Figure 6-15. Three solid-bowl centrifuges (A. Courtesy of Penwalt, Sharples-Stokes Div.; B. Courtesy of I. Kruger A/S; C. Courtesy of Bird Machine Co.).

A third variation of the solid-bowl design is the concurrent solid bowl, manufactured by Bird Machine Co. As shown in Figure 6-15C, the solids and liquid move in the same direction; the liquid is scooped up by an internal skimmer mechanism and the solids continue on to the beach. This machine performs well with light sludges due to a reduction of internal turbulences from conventional solid bowl designs.

The objectives of centrifugation are similar to those of other dewatering devices. It is necessary to obtain a dry cake, a clear centrate and a reasonable throughput or, in the language of filtration, a centrifuge yield.

The variables involved in centrifugation are listed in Table 6-8. These may be classified as machine variables or operational variables, as before.

Table 6-8. Variables Affecting Centrifuge Performance

Machine Variables	Operational Variables
Bowl diameter	Residence time
Bowl length	Sludge characteristics
Bowl rotational speed	(including sludge
Beach angle	conditioning)
Beach length	
Pool depth	
Scroll rotational speed	
Scroll pitch	
Feed point of sludge	
Feed point of chemicals	
Condition of scroll blades	

The bowl diameter is a machine variable controlled by the design engineer when the unit is purchased. Increasing the bowl diameter (and maintaining the same centrifugal force, that is, slowing down the machine) will result in a longer retention time within the machine. This results in a higher solids recovery, much as it would in a settling tank, but at the expense of cake dryness. As the solids recovery is increased, the smaller particles will also escape as cake, increasing the moisture content of the cake and decreasing the solids concentration.

Increasing the bowl length will also increase the residence time which will generally result in a high solids recovery, but the changes in cake dryness are not always predictable. It is also possible, under some circumstances, to increase the cake solids by increasing the bowl length. This is discussed later in the chapter.

Bowl speed is one variable that can be designed into the machine simply by selecting the correct gear ratio. Increasing the bowl speed increases the centrifugal force and the solids recovery, with possibly a concurrent increase in the cake solids as well. This will occur only if solids recovery already approximates 100%, that is, when almost all the solids are driven out the cake end. An increase in centrifugal force at that point will then increase the cake solids concentration, but this increase is often marginal, depending on the sludge. In fact, most centrifuge manufacturers have recognized that high speeds are seldom necessary for wastewater sludges and are now offering low-speed machines. This substantially increases the life of the machine and decreases operating costs.

The pool depth is a variable that in some machines can be varied while the machine is in operation. Other machines require removing plugs at the end of the bowl to increase or decrease the pool depth. An increase in pool depth will result in higher retention time, hence better solids recovery and a wetter cake.

Increasing the conveyor speed will force the solids out of the machine more quickly, thus leaving some of the wetter solids and increasing the solids concentration in the cake while decreasing solids recovery. Similarly, the conveyor pitch will influence solids recovery and cake dryness. If the scroll pitch is increased, the solids will be moved out faster but only the larger, heavier solids will be pushed out leaving the wetter solids in the centrate.

Increasing the number of leads in the conveyor will likewise increase cake dryness at the expense of solids recovery.

It is sometimes possible to change the feed point of the sludge within the machine. If the feed point is changed toward the beach, a wetter cake will often result while solids recovery will increase; this is because the sludge will have a longer travel time to the end of the machine where the centrate exists.

The chemical conditioners can be applied to the sludge before it goes to the centrifuge, or the chemicals can be fed directly into the machine. There does not seem to be any one good way of doing this because there are instances where placing chemicals into the machine will improve performance while other operators have found that mixing chemicals with sludge in the feed pipe will increase performance. Schwoyer and Lottinger (1973) have in fact suggested that the polymer-sludge contact time be considered a design parameter because of wide variation with different sludges and polymers.

It has also been found that the concentration of the polymers fed to the centrifuge influences performance. Figure 6-16, for example, shows that for a constant polymer dosage the performance deteriorates as the polymer feed is diluted (Campbell and LeClair, 1974). Although these data do not indicate an optimum, others have found that there is in fact an optimum at a polymer flow of about 10% of the sludge feed flow (volumetric) (Schultz, 1967). The performance of a centrifuge can therefore be enhanced simply by adjusting the dilution of the polymer.

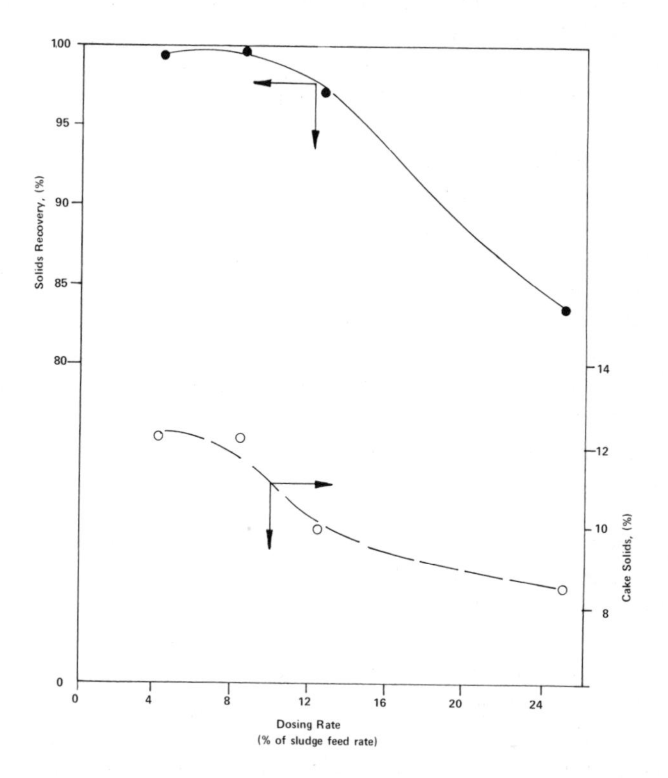

Figure 6-16. Concentration of polymer has an effect on sludge centrifuge performance (Campbell and LeClair, 1974).

The correct polymer (one that will result in best centrifuge performance) is often the one that produces strong flocs, not necessarily large flocs. The strength of the flocs can be judged visually (by an experienced eye) if the conditioned sludge is poured on a ceramic surface. A more accurate estimate can be obtained by calculating the floc strength (Petersen and Hansen, 1974) defined

$$F = \frac{S_1}{S_2}$$

where F = floc strength, as a fraction

S_1 = CST of conditioned sludge, sec

S_2 = CST of conditioned sludge after it has been stirred in a standard laboratory stirrer for a set amount of time.

Obviously, the concept of floc strength as a parameter by which different sludges and chemicals can be compared requires a standard method and a standard stirring apparatus.

The latter has been designed and constructed by the Water Research Centre in Stevenage, U.K., and has been accepted by most European water pollution control agencies and laboratories. The standard stirrer is shown in Figure 6-17 (Gale and Baskerville, 1970). The proposed European standard technique for evaluating the potential applicability of a chemical condition is to place about 100 ml of sludge into a 250-ml beaker and add 20 ml of conditioner. The mixture is then mixed for 10 sec and a sample run on the CST apparatus. The sludge is then mixed for progressively longer times and a CST run on each sample. The data are plotted as total mixing time versus CST. Typical data are shown in Figrue 6-18. A conditioned sludge for which the CST increases markedly with stirring will probably cause problems in centrifugation since the flocs will tend to be destroyed during centrifugation. It has also been found that resistance to shear is a function of polymer dose. It is thus reasonable that although a certain chemical dose yields adequate drainage on a filter (low specific resistance or CST) it may be highly susceptible to shear. With a little extra polymer, higher resistance to mechanical shear and better dewatering performance might be obtained (Swanwick, 1974).

In a series of interesting experiments (Swanwick, 1974), the CST of sludge before and after stirring (at various times) was compared to the CST of a feed sludge to a centrifuge and the CST of reconstituted sludge (cake plus centrate). It was discovered that the mechanical shear experienced by the sludge in the centrifuge can be simulated by a stirring time of between 5 and 20 *minutes* with the standard stirrer. This finding forcefully illustrates the violent shearing that occurs in the centrifuge, and suggests that the test might be used as a means of comparing machine designs.

It is difficult to develop generalized expressions for centrifuge performance since it is affected by the many variables discussed above. Researchers in Finland, however, have developed a multiple regression equation to describe centrifuge performance (Puolanne, 1977). The solids recovery was found to be

$$R = 48.4 - 9.36\,Q + 2.64\,P - 0.198 + 95.0\,T$$

where R = solids recovery, %

Q = feed rate to centrifuge, m^3/hr

P = polymer dosage, kg/ton of dry solids

S = volume of settled sludge in ml (1,000-ml sample, 30-min settling)

T = total solids in feed, %

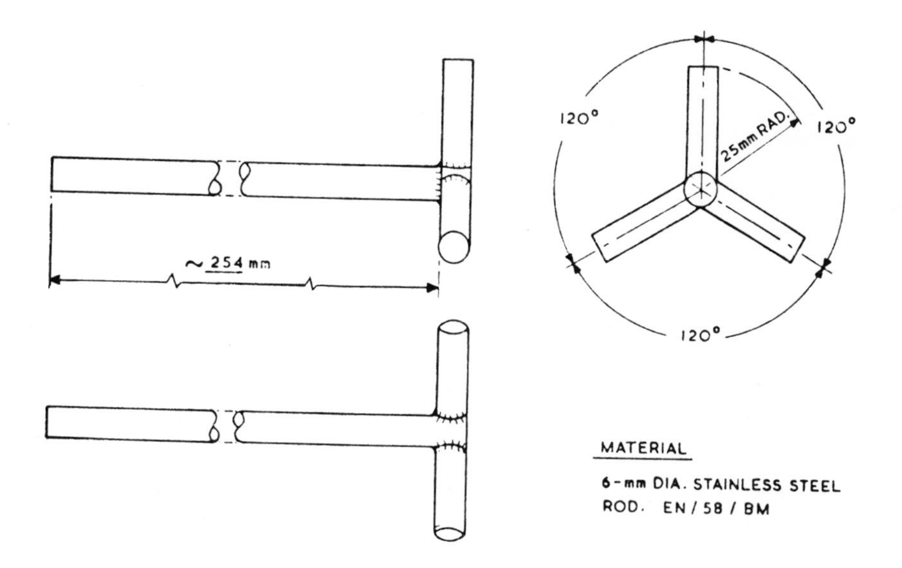

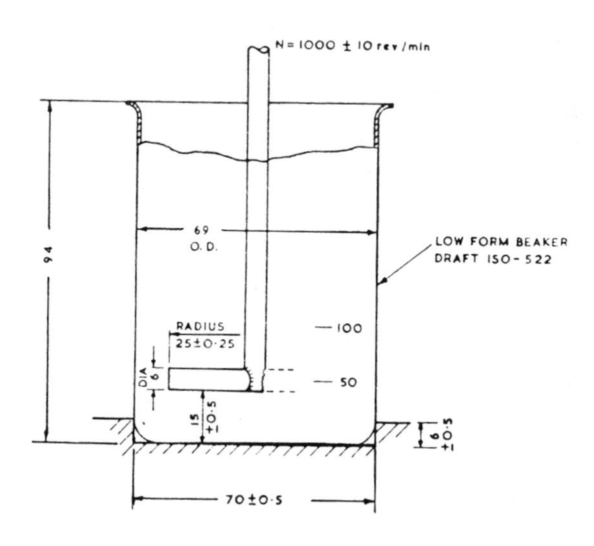

Figure 6-17. Standard stirrer (Swanwick, 1974).

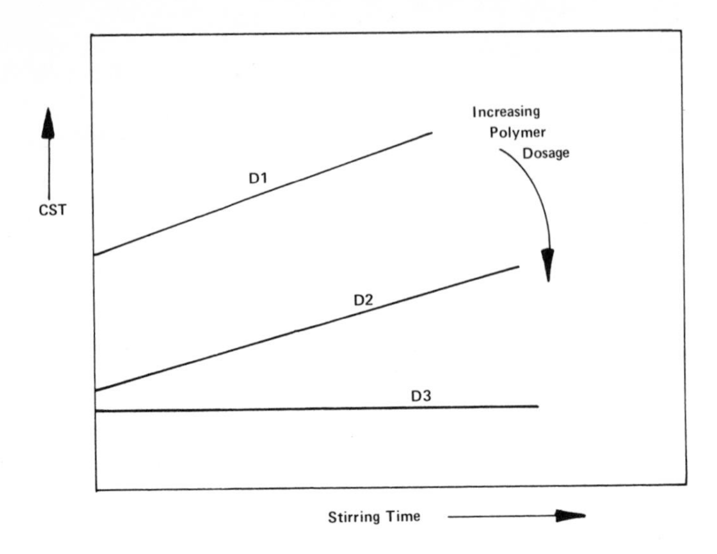

Figure 6-18. Effect of stirring on polymer-conditioned sludge. At doses D1 and D2 stirring produced a progressive deterioration in dewaterability. At a higher dose D3, stirring had only a minor effect on dewaterability (Swanwick, 1974).

The cake production was found to be

$$C = -5.96 + 26.3\,T - 0.0289\,S + 0.370\,P + 0.694\,Q$$

where C = cake production, kg dry solids/hr

The Finnish researchers specifically state that their equations are for only one sludge and machine, and should not be considered generalized expressions. Nevertheless, it is interesting to note the possibility of developing operational models for centrifuges.

One drawback of centrifugation is that the conveyor will be subjected to severe wear and tear because of its grinding against the sludge, especially if the sludge contains sand or other gritty materials. A worn conveyor will result in greater cake dryness since only the heavier particles are moved out, but this is, of course, at the expense of solids recovery.

The bowl angle, or the angle at which the conical section comes off the cylinder, also has a strong influence on both cake dryness and solids recovery if the sludge to be centrifuged comprises soft, fluffy materials. An increase in the angle will prevent some of the softer, fluffier solids from making their way up the beach, and these solids are hence discharged in the centrate, thus decreasing solids recovery and increasing the cake solids concentration.

The sludge on the inclined beach has, in addition to the force directed radially outward, a force component called slippage force that pushes the sludge back into the cylindrical section of the bowl, as shown on Figure 6-19 (Albertson and Guidi, 1967). The slippage force equals (sin α) x G where α is the bowl angle and G is the centrifugal force. If the solids still can flow, they will move under the conveyor blades, back to the cylinder and not be expelled as cake.

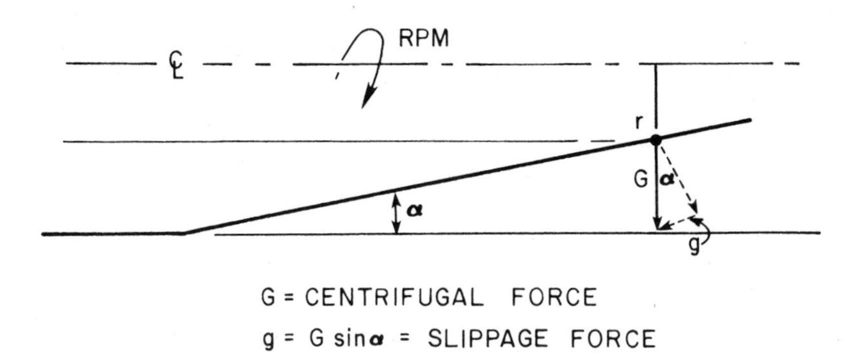

Figure 6-19. Effect of bowl angle on the movement of sludge.

The critical point of solids slip is reached when the solids emerge from the pool. It has been estimated that at that point the slippage force can increase by a factor of 10 (Albertson and Vaughn, 1971).

Because of this problem, modern machines built to handle soft sludges have small beach angles, or the machines operate with the pool level raised above the outlet. This method, called *super pool,* utilizes hydraulic force to help the solids out.

The amount of time sludge solids are allowed to drain on the dry beach (which is governed by the pool depth and conveyor differential speed plus, of course, machine variables) can influence cake dryness (DiGregorio and Shell, 1976). The residence on the beach can be calculated as

$$R_s = \frac{60\ L_b}{S\ \Delta\omega}$$

where R_s = residence time of the solids on the beach, sec
L_b = length of dry beach, cm
S = conveyor pitch, cm
$\Delta\omega$ = differential speed between conveyor and bowl, rpm

Another method used for very light materials is the addition of a false beach. Feeding a material such as gypsum and allowing the machine to centrifuge this material for a while results in a zero clearance between the conveyor blades and the gypsum-coated inner bowl, increasing the performance of a solid-bowl machine for materials such as activated sludge, alum and sludge.

The most important operational variable left to the operator is residence time within the machine as controlled by the flow rate of sludge to the centrifuge. By increasing residence time (or decreasing flow rate) it is possible to increase solids recovery at the expense of cake dryness.

The other important operational variable is the addition of chemicals. Both the kind of chemical and the way it is added can influence performance.

If a machine is operated at less than 100% recovery, which is usual, it is possible to obtain a cake solids concentration versus solids recovery curve similar to that shown in Figure 6-20 (Vesilind, 1973). This curve is a typical result of a series of tests in which many machine operational variables are changed. In other words, by changing one variable a higher solids recovery is traded for a lower cake solids concentration, or vice versa. Changing residence time (flow rate), for example, enables movement up and down this curve. The operator must select some operating point at which a sufficient solids recovery is obtained and a dry enough cake produced. The only means of moving off this curve is to increase the chemical feed or in some other way change the characteristic of the sludge (or, in some cases, change the centrifugal force by changing the bowl speed). Obviously, there are other machine variables that will also force the movement off this operating curve, but these are generally not within the control of a treatment plant operator. It is necessary to emphasize that the operator has only one option—that of moving up and down this curve, unless the sludge characteristics are changed by some means such as chemical conditioning.

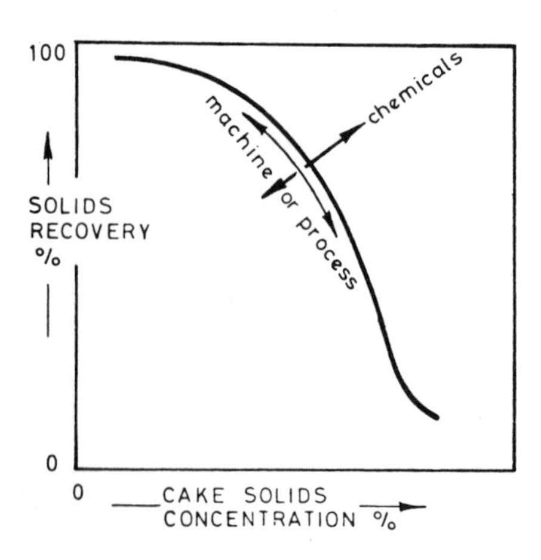

Figure 6-20. General solids recovery–cake solids curve for a centrifuge.

There has been some difference of opinion on how cake solids are related to solids recovery. Some published experimental data show both an increase and a decrease of cake dryness with an increase in solids recovery.

This apparent disparity may be rectified if Figure 6-21 is considered (Vesilind, 1969). At very low detention times (high flow rates) only the heavy, easy-to-dewater solids are removed as cake, yielding a dry cake and a poor solids recovery. As the detention time is increased (lower flow rates), the cake solids get wetter as recovery improves. At 100% recovery all solids are captured, and any decrease in flow rate (increased detention time) might actually result in a drier cake, due in part to decreased turbulence. Thus both a wetter and drier cake can result from increased residence time, depending on the solids recovery. Most centrifuges, it should be mentioned, operate at recoveries less than 100%, thus wetter cakes usually accompany increased recoveries.

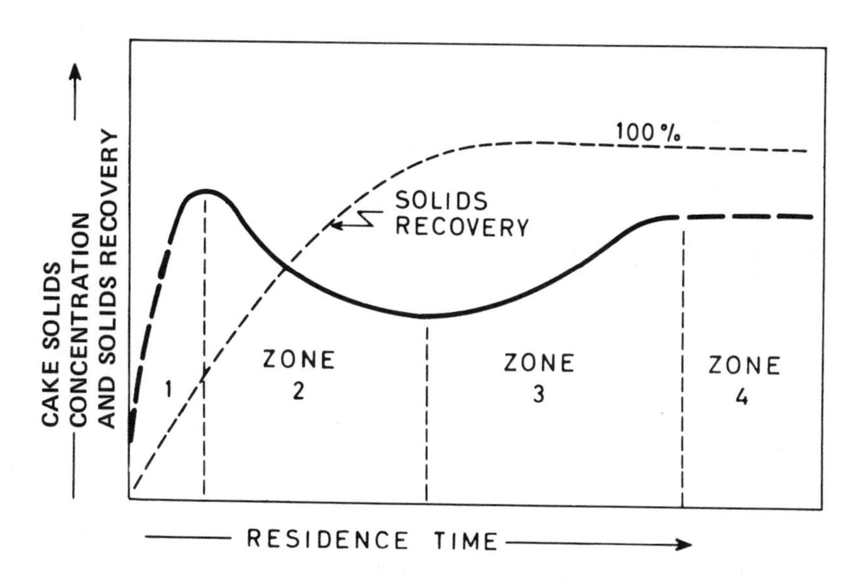

Figure 6-21. Idealized centrifuge performance.

The validity of this analysis was confirmed by Ronen (1978) using high lime sludges which contain various components. With reference to Ronen's Figure 6-22, a short residence time to the left of point A (or Zone 1 by the description in Figure 6-21), mainly the heavy granular particles such as grit and calcium carbonate are separated from the sludge. These particles do not retain much water and thus produce a dry cake. As the residence time is increased, wetter solids are recovered and the cake solids concentration deteriorates. Even longer residence times, however, produce an improvement in cake solids as well as recovery.

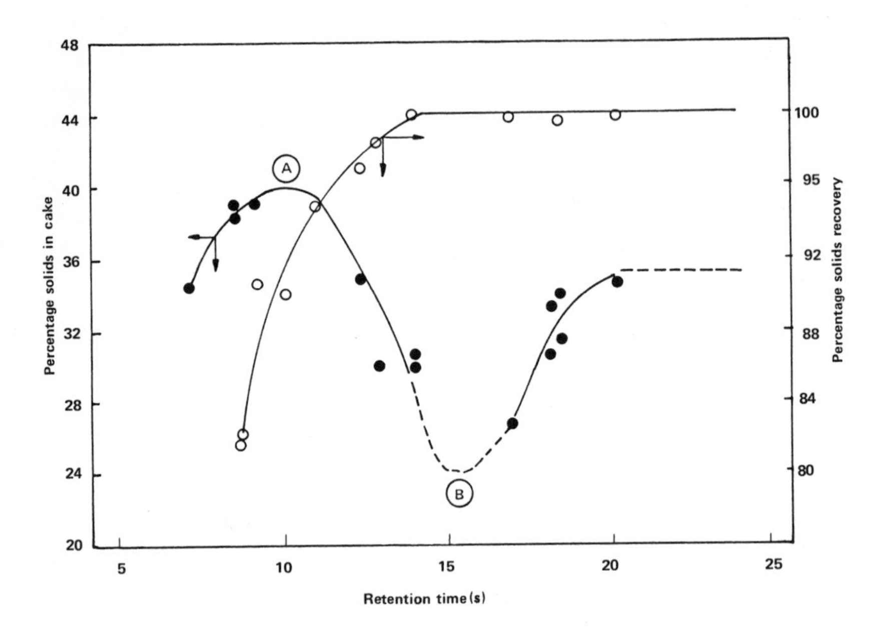

Figure 6-22. Centrifugation of a lime sludge (Ronen, 1978).

In addition, the treatment plant operator can control the history of the sludge before it is put into the centrifuge. As in vacuum filtration, almost anything the operator does to the sludge will decrease centrifuge performance. For example, the aging of sludge since its last treatment step can have a great influence, as older sludges dewater poorly. It has been suggested also that bruising sludge prior to dewatering has a detrimental effect (Parker *et al.*, 1972).

The dewatering characteristics of mixed digested sludges depend heavily on the type of secondary sludge produced. A plant with a mixed liquor with a high SVI will also have dewatering problems. Swanwick *et al.* (1972) concluded that "the dewatering characteristics of secondary sludges . . . is the major factor influencing the dewaterability of sewage sludges." It would thus behoove the operator, from the standpoint of dewatering, to operate his secondary aeration system to obtain the best possible settling sludge.

Performance

Because of the versatility of the solid-bowl machine, it is difficult to generalize on its performance. For example, in a large treatment plant in Los Angeles 66% recovery with a 30-35% cake solids concentration with digested primary sludge was, until recently, considered very reasonable. The centrate was pumped to the ocean, and it was the only necessary to remove larger material from the sludge. Now, stricter controls have forced the plant to achieve higher recoveries. Similar situations exist in West Chester near Philadelphia, where 50% of solids are recovered as a dry cake and then barged to the ocean, resulting in substantial savings in the cost of sludge disposal.

With most mixed digested sludges, a cake of 20% and a recovery of 70-95% can be expected. Some better machines and better chemicals can result in cakes as high as 35% and recoveries approaching 100%.

Activated sludge is notoriously difficult to dewater. Cake solids of 5% are common (chemicals might increase this to 10%) and recoveries of 85-90% can be expected with solid-bowl machines.

An exception is activated sludge produced in pure oxygen aeration systems. Although the data are skimpy, indications are that with no chemical conditioning, cake solids of 10-12% are attainable at 80-90% recovery.

It is interesting that the centrifugability of activated sludge can be related to the SVI. As shown in Figure 6-23 (Vesilind and Loer, 1970), a sludge that does not settle well in a liter cylinder will also have poor centrifugation properties. Incidentally, pure oxygen sludges commonly have SVIs around 35.

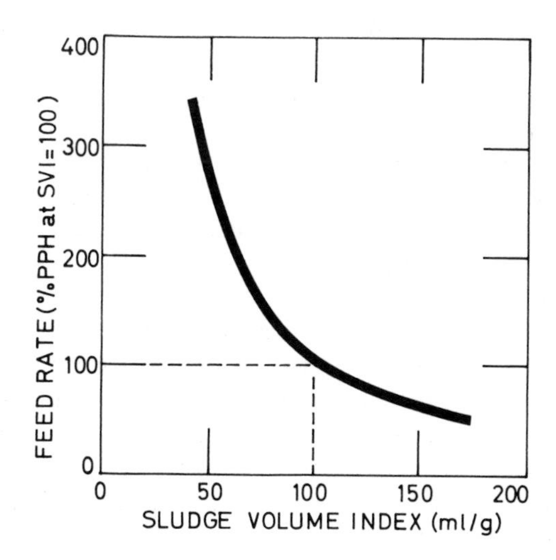

Figure 6-23. Effect of SVI on centrifugal dewatering of activated sludge.

Design

It is not possible at present to design a centrifuge for a specific application using fundamental principles. In fact, other than the work by Schnittger (1970), no attempt has been made to analyze the basic process of centrifugation. It is possible, however, to estimate the performance of a machine for a specific sludge by any one of three methods—experience, laboratory tests or scale-up from similar machines, the latter being the preferred method.

One serious drawback to the use of similar continuous machines for scale-up is that either the machine or the sludge must be transported. Moving the former is expensive (although several manufacturers use truck-mounted models) and moving the latter is dangerous, especially if the sludge produces gas. Nissen and Vesilind (1973) found that formaldehyde, at concentrations of 1% for activated sludge and 0.1% for digested sludge, will arrest biological activity while not appreciably changing the dewatering and thickening characteristics.

Design Based on Experience. As in vacuum filtration, this is based simply on the application of operating data from an existing plant to a new one. This is often necessary in estimating centrifuge performance for unconstructed treatment plants, and manufacturers are more often than not forced to design on the basis of experience. The loading is expressed in terms of gallons per minute (gpm) of sludge flow to the machine or pounds per hour (lb/hr, or as one manufacturer prefers, PPH). The proper use of such criteria is discussed later in the chapter.

All manufacturers have sales literature which lists their machines in terms of "capacity" for different types of sludges. Too often this "capacity" is ill-defined and intended only for sales purposes.

Design Based on Laboratory Test. The list of variables that influences centrifuge performance include many with only marginal effect. The two of greatest influence are the solids retention time (or the flow rate to the machine) and the centrifugal force. For any laboratory method to be successful in predicting prototype centrifuge operation, it is necessary to duplicate the effect of retention time and centrifugal acceleration.

One method of estimating how well a prototype continuous centrifuge will perform is to settle the sludge in test-tube laboratory machines (Vesilind, 1970). A common type holds 15-ml tubes at right angles to the shaft as it rotates. The sludge in question can be spun in such a machine, often as high as 1,500 gravities. Both detention time and centrifugal acceleration can thus be duplicated in the laboratory.

The centrifuge operates because two basic processes occur within its bowl. One is clarification or the settling of solids from the mother liquid. The second is the successful movement of these solids out of the bowl. If either process is not performed successfully, the centrifuge will not work. It is necessary then to measure in the laboratory test how a sludge will clarify as well as to estimate how well a sludge will be moved out of the bowl.

It is possible to denote the efficiency of centrifugation as*

$$\text{Percent Recovery} = \frac{(C_o - C_c)}{C_o} \times 100$$

where C_c = the solids concentration of the centrate
C_o = the feed solids concentration

This relationship does not really measure the amount of solids deposited as cake and therefore is not a proper measure of centrifuge performance. This relationship will, however, yield results very similar to those obtained using the correct recovery relationship.

If the centrate solids concentration obtained in the laboratory with a bottle centrifuge is designated by the symbol C'_c, this expression (applicable to laboratory results) is $(C_o - C'_c)/C_o$, and proposes to quantitatively describe sludge-settling characteristics in the laboratory. The feed solids concentration, C_o, is the same as for the prototype centrifuge. Term C'_c is determined by spinning a sample of sludge in the bottle centrifuge at the prototype centrifugal acceleration and residence time. The centrifugal acceleration is calculated as

$$\text{C.A. (x gravity)} = \left(\frac{r_1 + r_2}{2}\right) \omega^2 \frac{1}{g}$$

where r_1 = the radial distance from the center of the centrifuge to the top of the sludge
r_2 = the radial distance to the bottom of the tube, or to inner wall of the bowl
ω = the radial velocity, in radians per second
g = the gravitational constant

The residence time for the laboratory machine can be estimated by the power on time. By manipulation of the speed control, the starting and stopping times for most laboratory centrifuges can be made equal. It can then be assumed that the slight loss in centrifugation during the speed-up is compensated for during the braking.

The centrate solids concentration, C'_c, can be determined by placing four 10-ml samples in graduated tubes and spinning them at the centrifugal speed for the appropriate time. The tubes are then removed, and the centrate

*This is not an accurate definition, but is used only for the following analysis. Solids recovery in continuous centrifuges should be calculated using Equation 6-1.

poured into 15 test tubes to determine suspended solids by spinning and weighing the settled solids (Vesilind, 1969). The recoveries thus calculated will plot as straight lines versus residence time and centrifugal acceleration on log-probability paper.

The sludge consistency, or how much strength, body or firmness the sludge has, is an indication of its ability to be discharged from the machine. A light fluffy sludge of poor consistency cannot be scrolled, and will therefore escape in the centrate. It is necessary to somehow measure the consistency of the compacted sludge as well as its concentration.

The penetrometer, shown in Figure 6-24, consists of a polished brass rod, 3.0 mm in diameter, and weighing 13.4 g. With a plastic support it provides a means for quantitative measurement of sludge firmness.

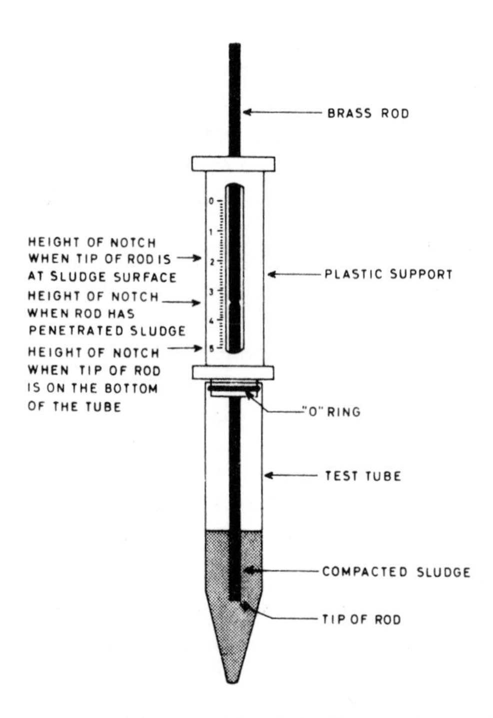

Figure 6-24. Penetrometer used for measuring sludge firmness in test-tube experiments.

After a sample of sludge has been spun a set time at a set centrifugal force, the centrate is poured off and the plastic support slipped on to the top of the tube. The brass rod is then inserted in the plastic support and lowered to the sludge cake level; this level is measured on the scale. The brass rod is then dropped into the sludge from that position and the final height recorded. The penetration is calculated as percent of sludge not penetrated, designated by the symbol P.

Activated sludge, for example, is very difficult to scroll as it has a light fluffy character. Correspondingly, it has very little resistance to penetration. The P value for such a sludge will be almost zero. A raw sludge of good consistency, however, scrolls easily and resists penetration and, therefore, gives high P values.

It can thus be shown that $(C_o - C'_c)/C_o$ is a reasonable measure of the sludge-settling characteristics and that P is a reasonable measure of the ability of the sludge to be discharged from the machine.

By trial and error, it was found that laboratory results could be related to the prototype data on a linear basis if the model is written as

$$\text{Estimated Percent Recovery} = \left(\frac{C_o - C'_c}{C_o}\right)\left(\frac{P}{100}\right)^{0.1} \times 100$$

The centrate clarify, C'_c, is measured at the same centrifugal force and residence time as the prototype. The penetration is measured on a sample spun at the prototype centrifugal force but at an arbitrarily set constant residence time of 60 sec.

The accuracy of the model for predicting centrifuge performance is shown to be within 10%.

Example 6-6. Given the data plotted in Figure 6-25A, estimate the solids recovery for a prototype centrifuge operating at 1,100 gravities with a residence time of 35 sec.

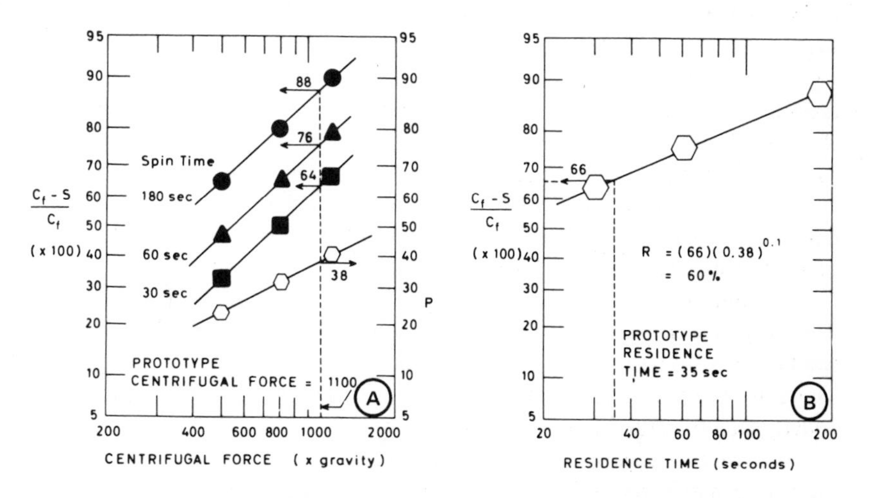

Figure 6-25. Typical data for laboratory test tube centrifugation.

The corresponding values of $(C_o - C'_c)/C_o$ for the three residence times of 30, 60 and 180 sec are read at 64%, 76% and 88%. These values are then plotted against the residence time in Figure 6-16B. If the expected residence time of the prototype centrifuge is 35 sec, the $(C_o - C'_c)/C_o$ at that residence time is 66%. This is the measure of the sludge-settling characteristics.

The P value at the prototype centrifugal acceleration is read as 38%.

$$\text{Estimated Percent Recovery} = \left(\frac{C_o - C'_c}{C_o}\right)\left(\frac{P}{100}\right)^{0.1} \times 100$$

$$= (0.66)\,(0.38)^{0.1} \times 100$$

$$= 60\%$$

Another bench-scale test has been developed by the Wastewater Technology Centre in Burlington, Ontario. In this test, a 50-ml test tube is spun for two minutes at 3,500 rpm, which results in about 1,000 x g. A portion of the sludge cake is scrolled out of the tube with an auger drill bit while the tube is held at a 45° angle. The scrolled-out sludge represents the cake, while the water remaining in the tube is the centrate, and solids analyses can be performed on both (Campbell *et al.*, 1975).

Figure 6-26 represents some comparative data between the bench-scale test and continuous (6- x 12-in.) centrifuge performance. Excellent correlation is noted.

An improved version of the penetrometer has been developed in Denmark (Hansen, 1974) and in France (Colin *et al.*, 1976). The Danish model fits 50-ml centrifuge tubes and has a plastic penetration rod. The French model, is dropped into a 90-mm diameter centrifuge tube with a height of 165 mm which has been spun for 30 minutes at 3,500 gravities. The penetrometer is constructed of aluminum and weighs 35 grams. Typical penetration values are shown in Table 6-9. This device is used to gain qualitative information about sludge centrifugability.

A third laboratory technique, not yet field-tested, is to use a strobe light and actually measure the rate of sludge solids compaction. Such thickening data can be used to size centrifuges in a manner similar to its use in the design of gravitational thickeners (Vesilind, 1974A).

The continuous solid-bowl centrifuge is in fact a highly efficient thickener. The only real difference is that instead of relying on gravity to provide the force, centrifugal acceleration increases the force by over 1,000 times gravity. There is, therefore, reason to believe the thickening theory would apply equally well to centrifuges.

The primary drawback in applying the theory has been the difficulty of measuring batch-thickening velocities at high rotational speeds. This can be overcome by using transparent tubes in laboratory centrifuges and "stopping"

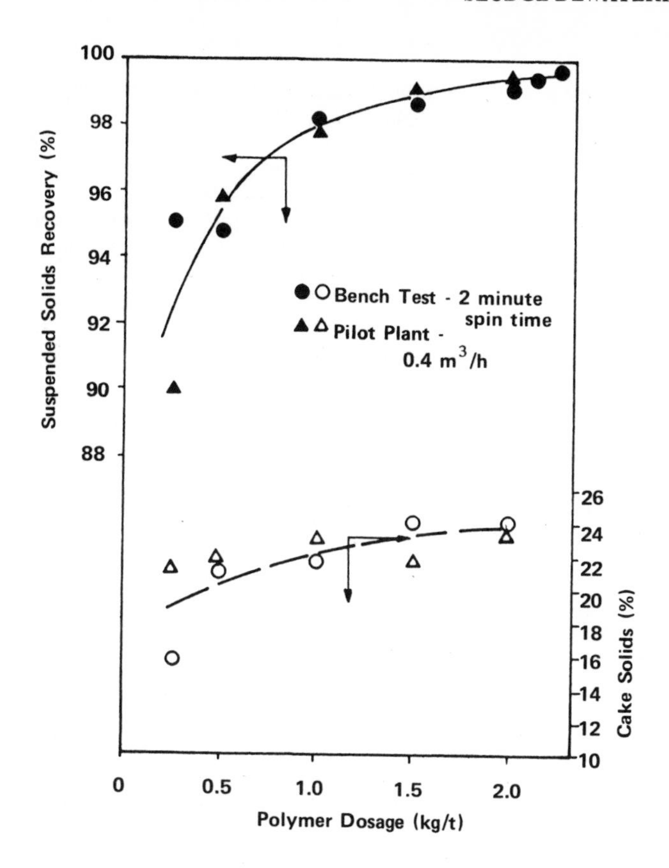

Figure 6-26. Prediction of solid-bowl centrifuge performance from bench tests (Campbell *et al.*, 1976).

their motion by strobe lights. Figure 6-28 is a schematic of such a laboratory set-up. The centrifuge tube is a plastic graduated cylinder with the bottom flange cut off. This settling cylinder is thus already calibrated and has a flat bottom. A hole is cut in the metal tube holder so the sludge in the tube is visible. The strobe is synchronized with the centrifuge to make the tube "stand still" and thus allow for continuous observation of the slurry-liquid interface.

Typical settling curves for a lime slurry, using 10-ml tubes, are shown in Figure 6-29. These curves resemble batch-thickening curves obtained in a conventional settling test.

The data from a series of these tests can be plotted as interface velocity versus solids concentration, and the batch flux is calculated by multiplying velocity by solids concentration. The limiting flux for a desired underflow

Table 6-9. Penetration of Centrifuged Sludge by Standard French Penetrometer (Colin *et al.*, 1976)

Type of Sludge	Penetration in mm	
	Dropped from Surface	Dropped from a Height of 50 mm above Surface
Ferric Hydroxide	15.9	22.7
Anaerobic Digested	7.2	9.9
Aerobic Digested	11.2	16.5
Waste Activated	10.7	17.2

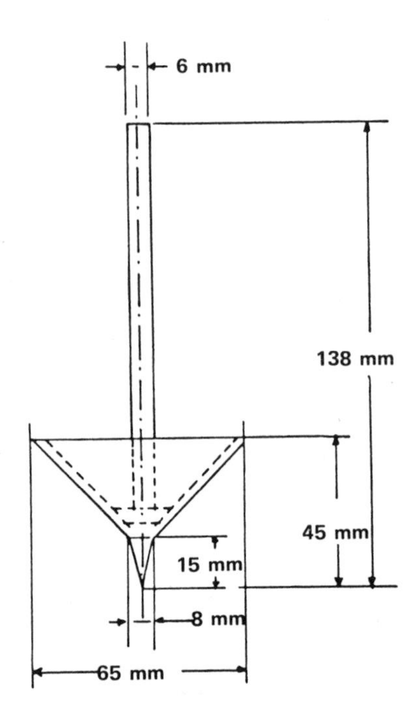

Figure 6-27. Penetrometer used for estimating sludge centrifuge performance (Colin *et al.*, 1976).

solids concentration (or, more appropriately, the sludge cake) can be calculated as discussed in Chapter 5.

The effect of the small test tube diameter on the settling velocity was found by White and Lockyear (1978) to be minimal. They used a test-tube centrifuge to assess the thickenability of sludges and demonstrated that the centrifuge can be a useful tool in estimating practically attainable underflow solids concentrations.

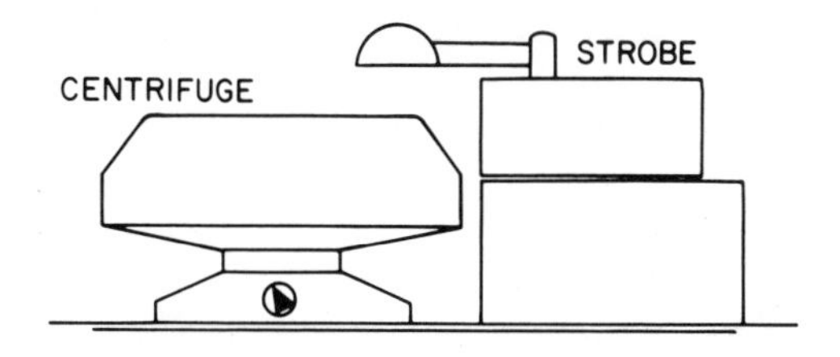

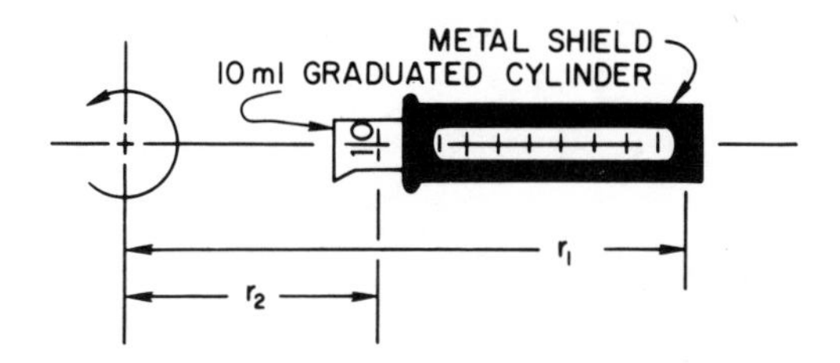

Figure 6-28. Arrangement of strobe light to allow for continuous observation of sludge compaction in a test-tube centrifuge.

Such tests might also be used to develop a parameter for centrifugation which corresponds to specific resistance to filtration for sludge filtration (Vesilind, 1977).

Consider a particle suspended in a liquid in a spinning test tube. This particle has two forces acting on it (if we ignore for a moment interparticle forces). If the rotor turns with an angular velocity ω (radians per second), the particle experiences a centrifugal force. This force can be described as

$$F_c = \omega^2 r(\rho_s - \rho)V$$

where r = radians from the center of rotation to the particle, cm
ρ_s = density of the particle, g/cm^3
ρ = density of the fluid, g/cm^3
V = volume of the particle, cm^3

This force is balanced by the drag force, F_d, which can be written as

$$F_d = C_D \frac{A v^2 \rho}{2}$$

where A = projected area of the particle, cm^2
C_D = drag coefficient
v = particle velocity, cm/sec

Assuming laminar flow (C_D = 24/Reynolds Number) and spherical particle shape,

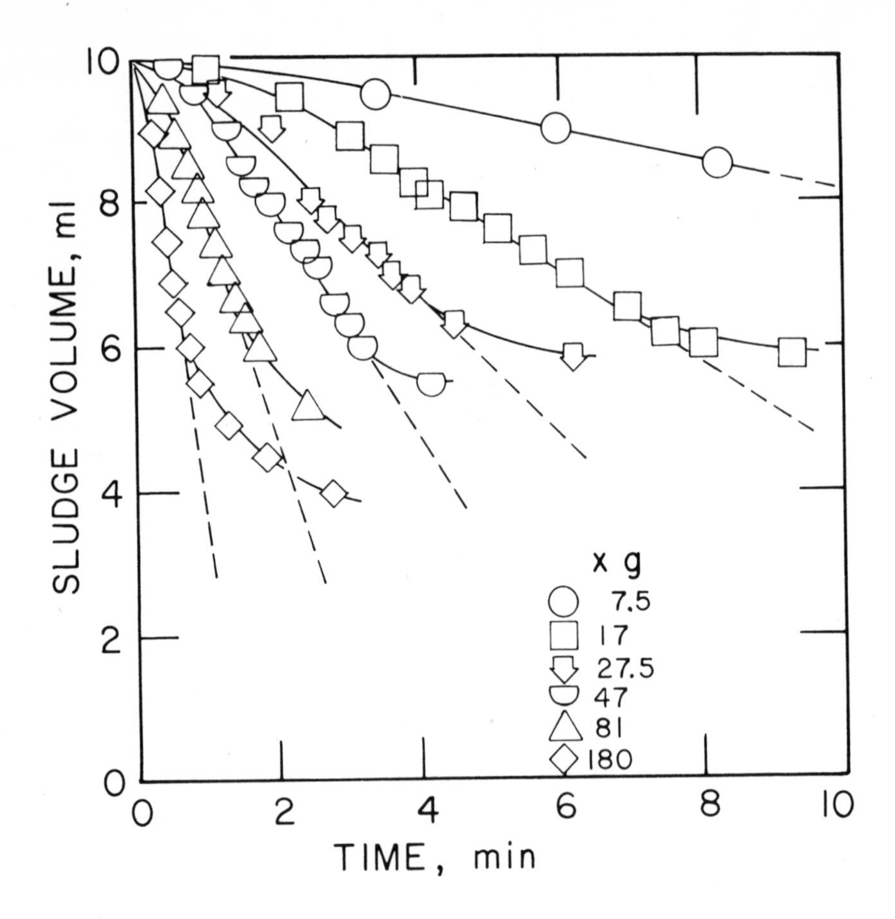

Figure 6-29. Typical settling curves for a mixed digested sludge in a test-tube centrifuge.

$$F_c = F_d$$

can be rearranged to read

$$\frac{v}{\omega^2 r} = \frac{(\rho_s - \rho)d^2}{18\mu}$$

Note that the term on the right-hand side is the well-known Stokes law with the gravity term missing.

We can see that the right-hand side contains variables which are solely properties of the slurry (particles and fluid) while the term on the left-hand side of the equation represents operational variable of the centrifuge.

The term $v/\omega^2 r$ has been used in biochemistry to characterize the settleability of molecules in ultra-centrifuges and is termed the "settling coefficient," S.

In gravitational thickening, we assume that the slurry interface velocity is a function of the solids concentration only, and we can make the same assumption for settling in a test-tube centrifuge, but we encounter an additional problem–the fact that the radius changes as the solids move rapidly outward and thus the centrifuge force imposed on the sludge increases. When the radius is large compared to the total travel distance of the slurry interface, this is not a serious problem. When the radius is small, however, the interface velocity could well be influenced by the increasing radius during settling.

This problem can be eliminated by recognizing that

$$v = \frac{dr}{dt}$$

and by our previous definition,

$$\frac{dr}{dt} = \omega^2 rS$$

Integrating,

$$\ln \frac{r_2}{r_1} = \omega^2 S(t_2 - t_1)$$

where r_1 and r_2 are the position of the interface at times t_1 and t_2, respectively. A graph of $\ln(r_2/r_1)$ versus $(t_2 - t_1)$ yields the slope $\omega^2 S$ and hence S, since the rotational speed is known.

If the settling coeffficient is truly a variable which describes the settleability of a slurry in a centrifugal field (independent of the radius or the speed of rotation), $v/\omega^2 r$ should be independent of $\omega^2 r$. In other words, as either ω or r increases, the settling velocity should increase so as to hold the $v/\omega^2 r$ value unchanged.

Figure 6-30 shows the results on an experiment with activated sludge in which the rotational speed was varied from 240 rpm to 1,150 rpm. The settling coefficient seems to increase slightly with speed, but is essentially constant over a range of centrifugal forces comparable to those in prototypes. The slight variation of S with speed may be explained in part by the assumptions used in its derivation. First, we assume that a single spherical particle was settling in an infinite fluid, in a laminar flow regime. Obviously, hindered settling such as experienced by sludge particles is not independent of concentration, or the external experimental conditions. The small diameter and low initial depth may be especially troublesome artifacts and remain to be investigated.

The "settling coefficient," S, is suggested as a measure of the settleability of a sludge in a centrifuge. The S value is independent of the centrifuge speed and radius, and is thus representative of the sludge characteristics only. This parameter has not yet been applied to centrifuge design.

Although the solids recovery is important, an estimate of the expected cake solids concentration may be even more valuable. This estimation is difficult because recovery varies with cake solids. The best presently available method is to estimate the cake solids at an operational point approaching 100% recovery.

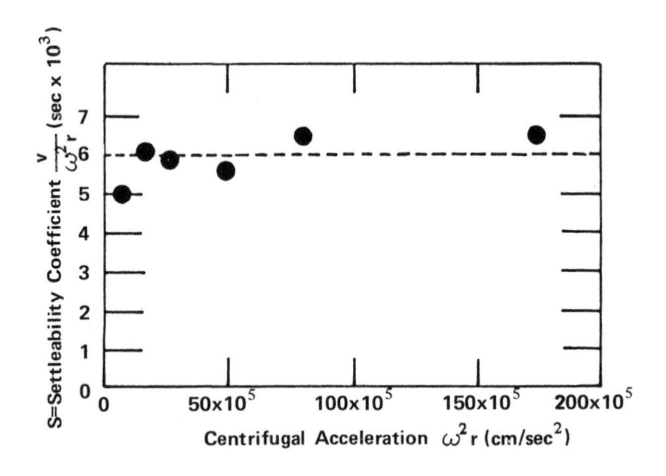

Figure 6-30. Settling coefficient of an activated sludge at different centrifugal accelerations (Vesilind, 1977).

Again, using a test-tube machine, cake solids can be measured for any centrifugal acceleration by allowing the sludge to compact until further compaction is negligible. This can be done readily by placing 10-ml plastic graduated cylinders (with bottom flange cut off) into the test-tube holders and measuring the sludge volume directly by noting the interface height. The cake solids are calculated as

$$C_k = \frac{H_o}{H_F} C_o$$

where H_o and H_F are the initial and final height of the sludge, and C_o is the feed (initial) solids concentration.

Alternatively, larger tubes can be used by first pouring a known volume of sludge into the tube, centrifuging, then measuring the volume of the centrate by pouring it into a graduated cylinder. The difference in volumes is the cake volume.

Typical curves for an activated sludge and a raw sludge are shown in Figure 6-31. Note that the activated sludge would not compact to more than about

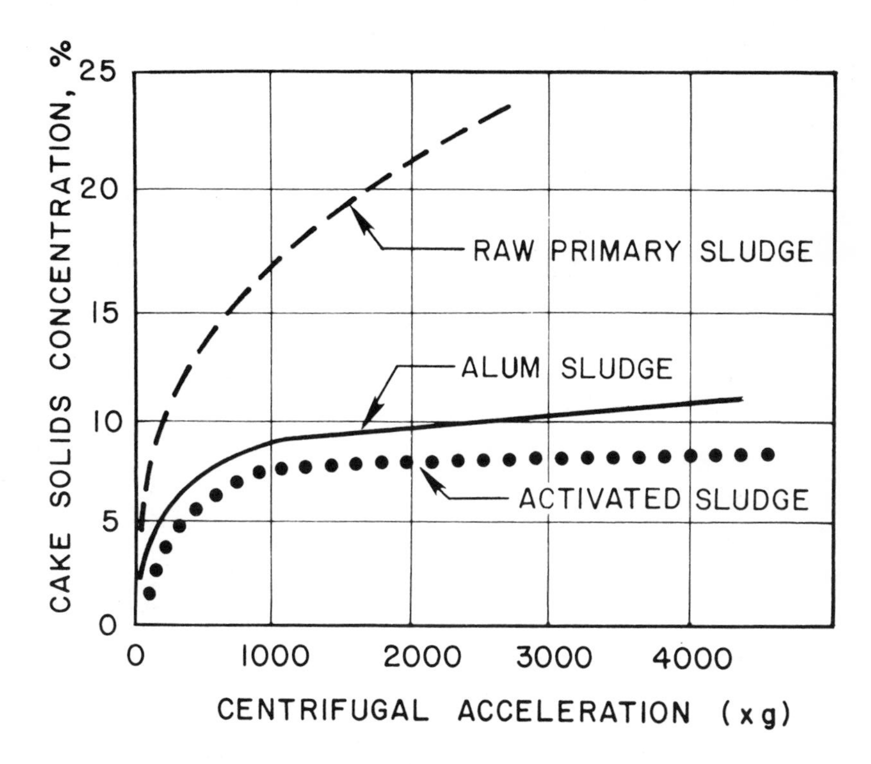

Figure 6-31. Sludge compaction in a test-tube centrifuge.

10% solids, even at very high centrifugal speeds. The raw sludge, on the other hand, seems to compact better and better with higher forces.

In fact, activated sludge seems to have a certain amount of elasticity and actually bounces back after the centrifugal force has been removed. Keith (1978) showed that if the sludge cake height is measured with a strobe light, curves such as Figure 6-32 are obtained for activated sludge. For this sludge, the maximum concentration seems to be about 8.8% solids. However, as the force is removed, the cake solids bounce back to yield only 4.5% solids.

The properties of the sludge which facilitate its removal from the centrifuge by the screw conveyor must also be evaluated. If insufficient solids are removed, the pool will fill up with the solids and no thickening is possible. This is analogous to closing the underflow valve on a gravitational thickener.

The solids movement capacity can be estimated by first assuming that the fraction of the pool depth occupied by the solids is equal to H_F/H_o from the test-tube experiments. If this depth is denoted by y and the radius of the bowl is r_2, the projected area occupied by the solids can be approximated as $2\pi y(r_2)$. The velocity at which the solids are moved in the cylinder is

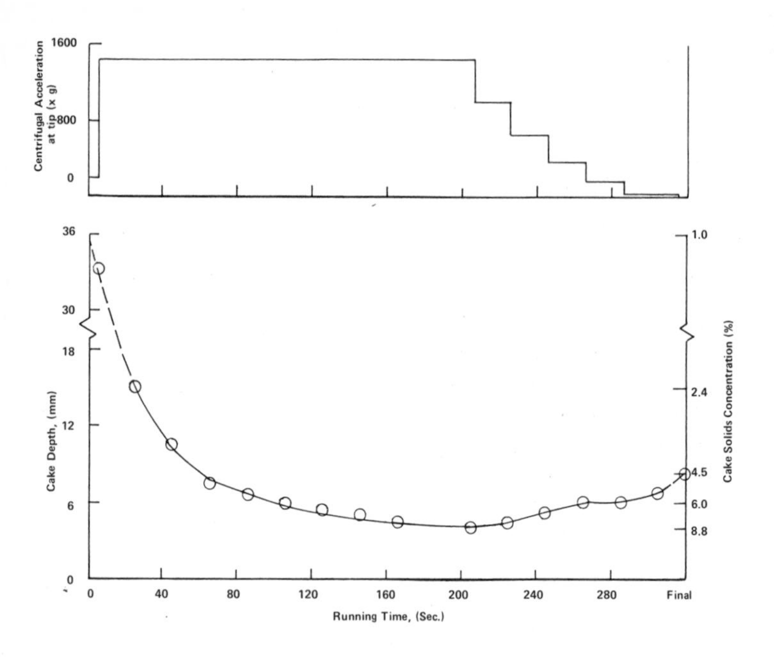

Figure 6-32. Compaction of sludge in a centrifuge tube (Keith, 1978).

$\Delta\omega \times S \times N$, where $\Delta\omega$ is the differential speed between bowl and conveyor in radians per second, S is the pitch of the conveyor blades and N is the number of leads. The total solids throughput is this area times velocity or

$$Q_s = 2\pi y(r_2)(\Delta\omega)SN$$

where Q_s is in units of ft^3/hr. Multiplication by expected cake concentration will yield the more familiar pounds per hour of solids throughput.

This equation assumes no slippage, and this may not be an acceptable assumption. The slippage would depend on the nature of the slurry, rotational speed, conveyor pitch, condition of the blades and beach angle. Slippage factors need to be developed from experience.

Design by Scale-Up from Prototypes. No laboratory simulation can match field testing, and it is recommended that whenever possible prototypes be tested at the plant under actual plant conditions. Sometimes smaller geometrically similar machines can be tested on location, and these results can be scaled up to larger centrifuges.

The first rational attempt to relate the performances of any two continuous geometrically similar centrifuges was by Ambler (1952). His analysis, known as the "Sigma Concept," has been widely used in scaling up (or down) between any two similar machines.

The Sigma Concept. Recall that the Stokes relationship for a particle settling in a continuous fluid is

$$v = \frac{g(\rho_s - \rho)d^2}{18\mu}$$

where ρ_s = particle density
ρ = fluid density
d = particle diameter
μ = viscosity

Assume a particle in a centrifuge (a) attains this terminal velocity immediately, and (b) its settling is slow enough to be in laminar flow regime. (In other words, assume that the Stokes relationship applies.) In a centrifuge, however, gravitational acceleration is replaced by centrifugal acceleration, $\omega^2 r$, where r = radius, ω = radial velocity, radians/sec. Hence, in a centrifuge

$$v = \frac{r\omega^2 (\rho_s - \rho)d^2}{18\mu}$$

The radial distance dr traveled by the particle during time $d\theta$ is

$$dr = v d\theta = \frac{r\omega^2 (\rho_s - \rho)\, d^2}{18\mu}\, d\theta$$

Integrating from r_1 (centerline to pool surface) to r_2 (centerline to bowl)

$$\int_{r_1}^{r_2} \frac{dr}{r} = \frac{\omega^2 (\rho_s - \rho)\, d^2}{18\mu} \int_0^{\theta} d\theta$$

$$(\ln r_2 - \ln r_1) = \frac{\omega^2 (\rho_s - \rho)\, d^2}{18\mu}\, \theta$$

$$\theta = \frac{18\mu}{\omega^2 (\rho_s - \rho)\, d^2} \ln \frac{r_2}{r_1}$$

For the particle to be removed,

$$\theta = \frac{V}{Q} = \frac{\text{volume}}{\text{flow rate}}$$

Inserting gravitational constant,

$$\frac{V}{Q} = \left(\frac{g}{\omega^2} \ln \frac{r_2}{r_1}\right)\left(\frac{18\mu}{g\,(\rho_s - \rho)\,d^2}\right)$$

$$Q = \left(\frac{V\omega^2}{g \ln \frac{r_2}{r_1}}\right)\left(\frac{g\,(\rho_s - \rho)\,d^2}{18\mu}\right)$$

Note that the first term of the above equation has only machine variables, the second only sludge variables. We can define the first term Σ so that $Q = \Sigma v$.

For the same sludge, two similar machines will perform equally if

$$\frac{Q_1}{Q_2} = \frac{\Sigma_1}{\Sigma_2}$$

The above analysis differs a little from Ambler's original, but the end result is equally applicable.

Example 6-7. Using the above expression for Sigma in a solid-bowl machine, estimate the feed rate for centrifuge No. 2 based on the performance of centrifuge No. 1. Centrifuge 1 was found to operate satisfactorily at a flow rate of 0.5 m^3/hr.

	Machine 1	*Machine 2*
Bowl diameter (cm)	20	40
Pool depth (cm)	2	4
Bowl speed (rpm)	4,000	3,200
Bowl length (cm)	30	72

The volume of sludge in the pool can be conveniently approximated as

$$V = 2\pi\left(\frac{r_1 + r_2}{2}\right)(r_2 - r_1)\,L$$

$$V_1 = 2(3.14)\left(\frac{8 + 10}{2}\right)(10 - 8)\,(30)$$

$$V_1 = 3{,}400 \text{ cm}^3$$

$$V_2 = 2(3.14)\left(\frac{16 + 20}{2}\right)(20 - 16)\,(72)$$

$$V_2 = 32{,}600 \text{ cm}^3$$

$$\omega = (\text{rpm})\left(\frac{1}{60}\,\frac{\text{min}}{\text{sec}}\right)\left(2\pi\,\frac{\text{radians}}{\text{revolution}}\right)$$

$$\omega_1 = 4{,}000 \times \frac{1}{60} \times 2 \times 3.14 = 420 \text{ radians/sec}$$

$$\omega_2 = 3{,}200 \times \frac{1}{60} \times 2 \times 3.14 = 335 \text{ radians/sec}$$

$$\Sigma_1 = \frac{(420)^2}{980} \frac{3,400}{\ln\left(\frac{10}{8}\right)} = 2.74 \times 10^6$$

$$\Sigma_2 = \frac{(335)^2}{980} \frac{32,600}{\ln\left(\frac{20}{16}\right)} = 16.8 \times 10^6$$

$$Q_2 = \frac{\Sigma_2}{\Sigma_1} Q_1$$

$$= \frac{16.8}{2.8}(0.5) = 3.2 \text{ m}^3/\text{hr}$$

The problem in applying the Sigma concept to wastewater sludges is that it often does not work. A number of assumptions have been made in this derivation, many of which seriously affect the analysis. It is assumed that Stokes velocity prevails. This is obviously not true because the sludge settles in the hindered settling regime. The compression of sludges is totally ignored. The space occupied by the solids within the machine is not considered. Terminal velocity is assumed for the particles but it is extremely doubtful that any particles attain this terminal velocity because they are continually being accelerated. All particles are assumed to have equal rotational speed, but it is more likely that the particles never do attain this speed and are continually being accelerated. Finally, turbulence is assumed to be negligible and this is obviously not so.

A much greater error in the use of the Sigma concept is that clarification is considered the only major process that occurs within the machine. The movement of solids out of the machine is not included in the analysis. It is entirely possible that centrifuges can be limited not only by clarification but also by their solids throughput.

The Beta Concept. The machine is said to be solids-limiting when its design precludes its ability to discharge a sufficient quantity of solids. It is reasonable to hypothesize that for any two similar solid-bowl centrifuges, solids throughput is proportional to the ratio of the volumes occupied by the solids (Vesilind, 1974B). Ignoring the cone, pool volume can be estimated as $V = \pi DLz$, where z = pool depth, D = bowl diameter and L = bowl (cylinder) length (see Figure 6-33). The volume occupied by the solids is $V_s = \pi DyL$, where y = the depth of solids. The fraction of the pool volume occupied by solids is

$$\frac{V_s}{V} = \frac{\pi DLy}{\pi DzL} = \frac{y}{z}$$

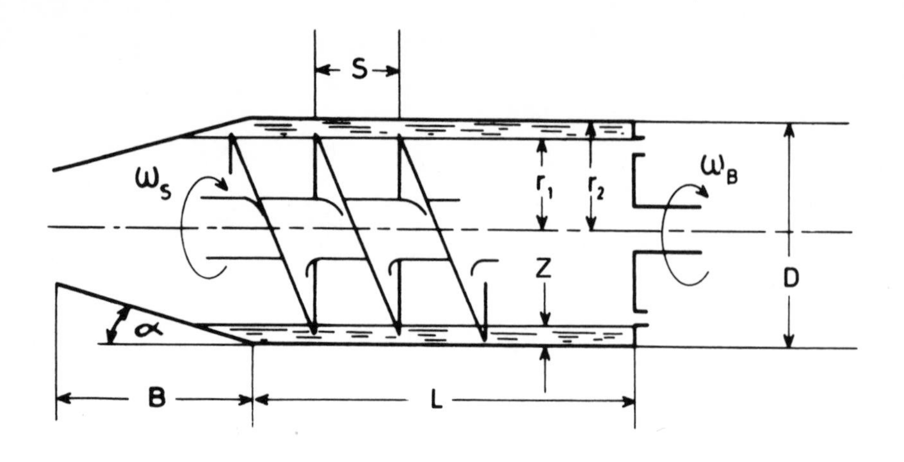

Figure 6-33. Definition of symbols for derivation of Beta.

Assuming no slippage, the travel time for a particle in the bowl is

$$T = \frac{L}{\Delta\omega SN}$$

where $\Delta\omega$ = bowl/conveyor differential speed
S = pitch of blades
N = number of leads

The solids flow into a machine can be approximated as Q_s/γ_c, where Q_s is the solids throughput in PPH, and γ_c is the specific weight of the wet cake. Hence, Q_s/γ_c is in ft^3/hr. The volume occupied by the solids is $V_s = (Q_s/\gamma_c)T$. Substituting,

$$V_S = \frac{Q_S}{\gamma_c}\frac{L}{\Delta\omega SN}$$

The inside bowl wall area is $A = \pi DL$ or $L = A/\pi D$. Substituting,

$$\frac{V_S}{A} = \left(\frac{Q_S}{\gamma_c}\right)\left(\frac{1}{\Delta\omega SN\pi D}\right) = \text{depth of cake} = y$$

From above,

$$\frac{V_s}{V} = \frac{y}{z} = \left(\frac{(Q_s/\gamma_c)}{\Delta\omega SN\pi Dz}\right)$$

The denominator is comprised solely of machine variables, while the numerator contains process variables, Q_s and γ_c. This grouping of machine variables is defined as β, and the scale-up of solids throughput between any two solid-bowl centrifuges is therefore

$$\frac{(Q_s/\gamma_c)_1}{\beta_1} = \frac{(Q_s/\gamma_c)_2}{\beta_2}$$

If the same slurry is used in both machines, and assuming that γ_c is not influenced greatly by differences in gravitational force, then

$$\left(\frac{Q_s}{\beta}\right)_1 = \left(\frac{Q_s}{\beta}\right)_2$$

where Q_s is solids throughput in pounds (or kilograms) per hour and $\beta = \Delta\omega SN\pi Dz$ in ft^3 (or m^3) per hour.

Example 6-8. Using the data in Example 6-7 and the following additional data, estimate the capacity of centrifuge No. 2 based on the performance of No. 1. Centrifuge if No. 1 operated satisfactorily at 60 kg of solids per hr.

	Machine 1	*Machine 2*
Conveyor speed (rpm)	3,950	3,150
Pitch (cm)	4	8
Number of leads	1	1

$$Q_2 = \frac{\beta_2}{\beta_1} Q_1 = \frac{(\Delta\omega SN\pi Dz)_2}{(\Delta\omega SN\pi Dz)_1} Q_1$$

$$= \frac{(50 \times 8 \times 1 \times 3.14 \times 40 \times 4)_2}{(50 \times 4 \times 1 \times 3.14 \times 20 \times 2)_1} (60)$$

$$Q_2 = 480 \text{ kg/hr}$$

This analysis has some obvious limitations. It has been assumed that the density of the submerged packed solids does not differ between the two machines. Laboratory experiments with test tubes support this assumption for most materials. For a spin time of 60 sec, for example, tests with calcium carbonate slurry, biological sludge and suspensions of plastic beads all indicated no appreciable difference in cake depth for centrifugal accelerations ranging from 1,000 x G to 1,500 G. For large differentials and for some materials a variation in compaction with centrifugal force might be expected.

Slippage on the bowl is assumed to be negligible. This might not be true in all cases, especially when a large helix angle (large space between conveyor blades) and/or multiple leads are used. If the formula is considered at face value, simply doubling the lead (N) would double the solids capacity. Obviously, this cannot be expected, since cake dryness will surely suffer. The maximum helix angle and number of leads before performance begins to deteriorate depends on the material handled.

Increasing the conveyor capacity, although an attractive proposition, may also result in the realization of high torque levels in the gear boxes. Present machines can handle up to 50,000 ft lb of torque, but high solids loading can result in gearbox failures.

Another source of error is the interaction between Σ and β. Scale-up of clarification is calculated without consideration for the space occupied by the solids. Similarly, solids scale-up should account for variations in liquid flow and its effects on the depth of the solids mat. It is probably not altogether true, therefore, that there is a sharp division where liquid limits end and solids limits begin. It is more realistic to consider solids and liquid limits as interacting, and to expect a deterioration in performance when either one or both are approached.

The application of this concept is illustrated by the data in Figure 6-34, which shows the effect of polymer dosage on solids recovery for a waste

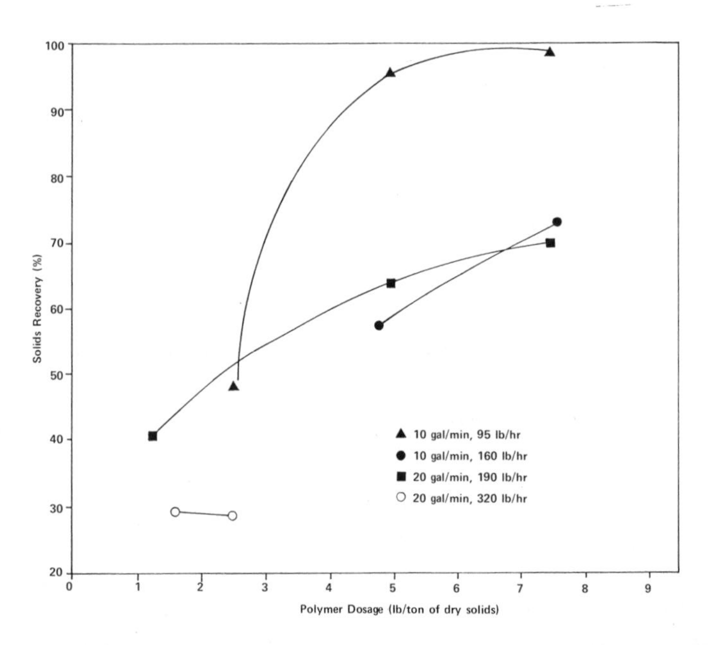

Figure 6-34. Results of sludge centrifugation illustrating solids-limiting performance (Campbell and LeClair, 1974).

activated sludge. For the two series at approximately equal solids feed rates—160 and 190 lb/hr—the results are similar, even though the liquid feed rate for one series was double that of the other—10 gal/min vs 20 gal/min. In this case the machine was clearly *solids limiting.*

Lowering the solids throughput to 95 lb/hr, and maintaining the same liquid feed rate, 10 gal/min, resulted in marked improvement in solids recovery. Conversely, a further increase in solids throughput resulted in further deterioration of performance.

Solids and Liquid Limits (the use of both Σ and β). Data for two materials, activated sludge and calcium carbonate slurry, shown in Table 6-10, are illustrative of how either solids or liquid may limit the scale-up. In this table, the results from the smallest machine are used to predict the performance of the larger machines. Although all these results are single data points and therefore subject to some error, significant conclusions can be drawn.

Table 6-10. The Use of Sigma and Beta in Centrifuge Scale-Up.

	Mach. No.	Solids Recovery (%)	Cake Solids (%)	Sigma Ratio	Beta Ratio	Actual gpm	Est. gpm	Actual lb/hr	Est. lb/hr
Calcium Carbonate	1	89	74	1.0	1.0	7.7		1,450	
	2	92	73	3.11	0.88	9.1	23.8	1,390	1,272
Activated Sludge	1	83	5.8	1.0	1.0	26		78	
	2	83	5.2	2.74	3.30	70	71	251	257
	3	78	5.8	2.52	8.11	70	66	307	663

With reference to the table, the first comparison for calcium carbonate shows that if the Sigma method were used alone, the machine application would have been underdesigned by over 100% (*i.e.*, 23.8 gpm would have been expected of Machine 2, based on Machine 1 tests, but only 9.1 gpm was attained).

The comparison between Machines 1 and 3 of the activated sludge data shows that if the Beta method was used alone, a similar design error would have resulted. The larger machine is clearly limited by hydraulic loading.

The comparison between Machines 1 and 2 for activated sludge shows a situation where both the solids and liquid loadings were limiting simultaneously.

To summarize, the hydraulic capacity of any solid-bowl centrifuge (Machine 2) can be estimated on the basis of data for one geometrically similar (Machine 1), operating on the same slurry, by the relationship

$$(Q_L)_2 = \left(\frac{Q_L}{\Sigma}\right)_1 (\Sigma)_2$$

where

$$\Sigma = \frac{V\omega^2}{g \ln\left(\frac{r_2}{r_1}\right)}$$

The solids handling capacity can be estimated as

$$(Q_S)_2 = \frac{(Q_S)_1}{\beta_1} (\beta)_2$$

where

$$\beta = \pi \Delta\omega D S N z$$

The lowest rate thus determined will limit performance and must be used for scale-up.

One centrifuge manufacturer provides the customer with a graph showing the feed solids concentration versus "capacity" of the machine in m^3/hr as shown in Figure 6-35. Such a curve in fact acknowledges the existence of both a liquid and a solids capacity for a centrifuge.

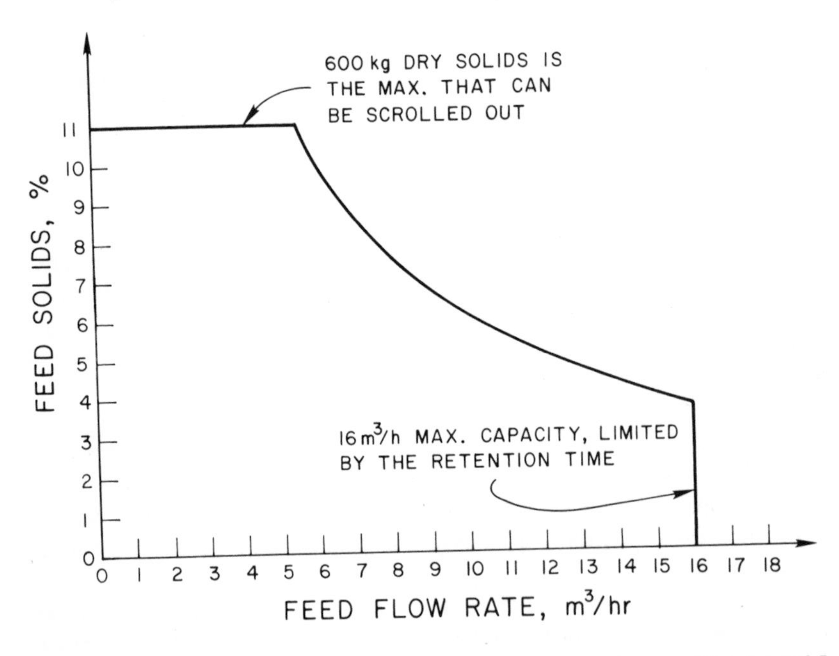

Figure 6-35. A capacity curve used by one centrifuge manufacturer (Courtesy of I. Kruger A/S).

Summary

The centrifuge has a great many advantages over other dewatering methods. In addition to lower capital cost, a reasonable relative operating cost can be achieved especially if labor costs are taken into account. It is not necessary for a machinist to keep watch over an operating centrifuge, as is required with vacuum filters.

A centrifuge is an enclosed unit and thus odor-free; it can be used for dewatering such materials as heat-treated sludges, which are notoriously obnoxious. Centrifuges can be accommodated in small buildings or for that matter, outside, as is the case in California and other parts of the U.S. Chemicals are not necessary for operation but are generally recommended for continuous good operation and to avoid the buildup of fine solids in a treatment plant. A centrifuge is a flexible unit as noted previously; it is easy to start (by pressing a button) and to clean (by running clean water through the machine). The latter seldom takes more than a few minutes, compared to an hour for cleaning a vacuum filter.

Centrifuges do have some problems, however. Chemicals are usually needed for operation and this can be costly. There have been instances where trash has become embedded in the sludge and clogged the machine very badly, causing many hours of downtime. Maintenance, especially with the wear and tear on the conveyor, can also lead to substantial costs. The life of a conveyor varies from about 1,000 to over 4,000 hours, depending on the type of sludge handled.

Other types of machines have been proposed for wastewater treatment but none has shown the versatility and operational characteristics of the solid-bowl machine. It seems reasonable to assume that this machine will be used in wastewater treatment for many years to come.

OTHER MECHANICAL DEWATERING METHODS

Other centrifugal devices have been used for sludge thickening and/or dewatering. The cyclone is a simple one, consisting of a cone into which the sludge is pumped tangentially (Figure 6-36). The heavy sludge solids tend to move to the inside wall, down, and drop out the bottom while the clean liquid moves up and out the center. Cyclones are quite useful to recover heavy materials and have found a use in degritting sludge prior to centrifugation.

In recent years, the disc centrifuge, originally devised for the separation of cream from milk, has been applied to thicken activated sludge. The disc machine, shown in Figure 6-37A, has the advantage of being fairly compact and efficient in its solids recovery of slow solids feeds due to high hydraulic capacity.

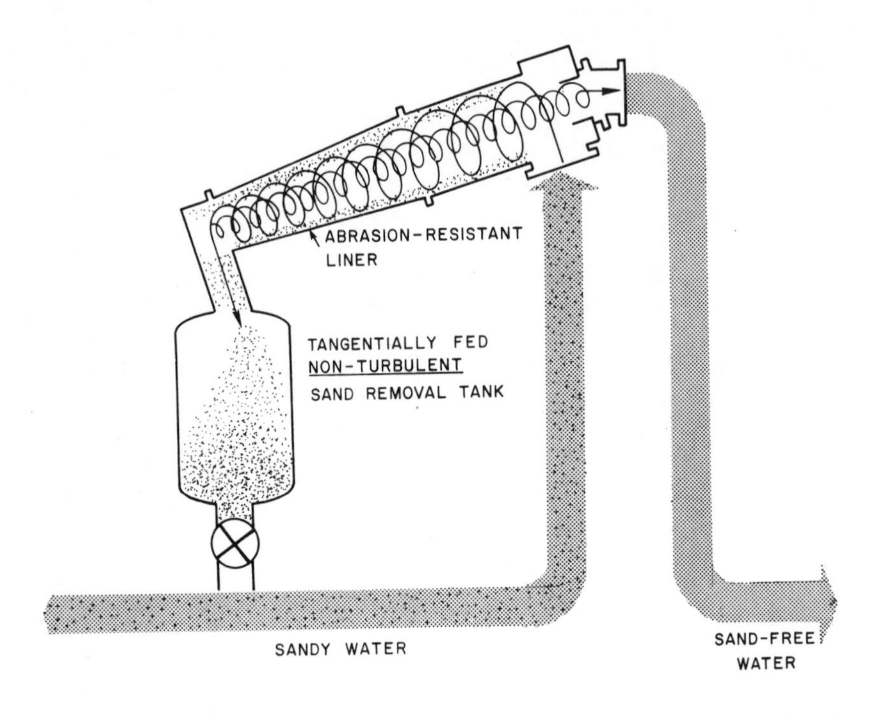

Figure 6-36. Cyclones used for degritting (Courtesy of Krebs, Equipment Engineers, Inc.).

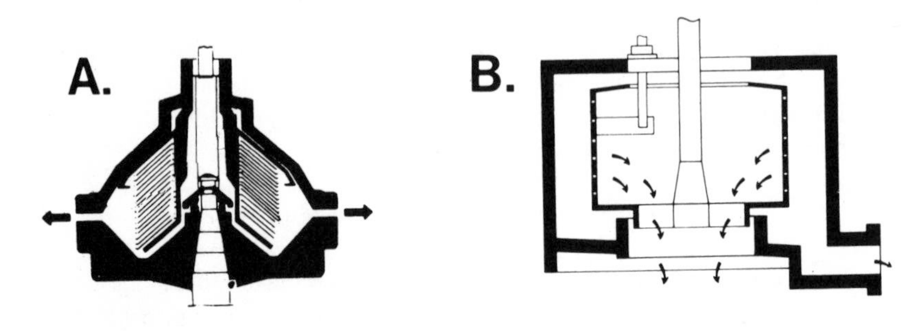

Figure 6-37. Disc and basket centrifuge (Courtesy of DeLaval Separator Co.).

The solids are settled outwardly and the lighter solids separated by settling against the underside of the discs, gaining in density and also sliding outward. The clarified liquid flows out through the disc stack.

The solids are removed through nozzles in the bowl. These nozzles can be so constructed to open intermittently, expel the solids and close before the liquid can escape, or a portion of the solids can be recycled through the machine, thus maintaining a high solids inventory. The latter arrangement results in better solids control as well as permitting the use of larger nozzles, thus reducing the chance of clogging. The efficiency of solids separation for the disc machine can be related to the SVI, just as for the solid bowl centrifuge (Albertson and Vaughn, 1971).

The downfall of the disc machine in dewatering wastewater sludges has been the clogging problem. Trash in the sludge and unexpected power failures can cause clogged innards, and cleaning the machine is a tedious process. The stack must be disassembled one disc at a time—a process that can take an entire working day.

Another centrifuge of some interest in wastewater treatment is the basket centrifuge machine shown in Figure 6-37B. The basket is simply a vertical rotating bowl without a conveyor. Periodically, as the solids collect on the sides of the bowl, a scoop comes by and scrapes out the solids as the cake. Because of this cyclic operation it is advisable to have a mechanic available at all times, although the manufacturers insist such supervision is unnecessary because the machine is totally automated.

For a basket centrifuge, the Sigma value, defined and developed using the same arguments as for the solid-bowl machine, can be written as (Ambler, 1952)

$$\Sigma = \frac{2\pi H\omega^2}{g}\left(\frac{3}{4}r_2^{\,2} + \frac{1}{4}r_1^{\,2}\right)$$

where H = height of bowl
ω = radial velocity of the bowl
r_2 = radius of bowl
r_1 = radius to liquid layer

In addition to the nonideality conditions involved in the application of sigma to solid-bowl machines, a further complication also arises because the effective volume of the bowl decreases with time.

The basket centrifuge, by virtue of its basic simplicity, can be fairly inexpensive and simple to operate. For example, the Lavin centrifuges (Hatboro, MA) range from 12- to 20-inch diameter machines, have nominal feed capacities of 500 to 3,000 gallons per hour and achieve 2,200 gravities, but yet are light enough to be portable.

New and imaginative devices for sludge dewatering are being introduced. One device combines the operation of a vacuum filter with a belt press; others use the screw press principle to provide pressure. There is not yet sufficient operating data on which to judge the future applicability of these devices to sludge dewatering.

CONCLUSIONS

Wasetwater sludges have been difficult to dewater mainly because only two basic differences between solid and liquid have been used–density and size. The solids in wastewater have specific gravities in the range of only 1.05 (dangerously close to 1.00). Similarly, there are amorphous compressible particles formed in large disturbable flocs, making filtration difficult. Neither process viewed in this light looks very promising, yet we have been using them for years.

Dewatering of sludges is often desirable and even necessary. However, it must be emphasized that dewatering is simply a preparation for ultimate disposal, and may not be necessary at all, as illustrated by the following horror story.

In one treatment plant, a vacuum filter was designed so that the sludge cake was placed in a hopper through which it was to drop into a waiting truck. Unfortunately, the engineer designed the hopper sides at an angle such that the sludge cake would not drop into the truck. As a result, the operator used a firehose to flush the solids down into the truck. The sludge now contained too much water, so he switched to a tank truck to transport the now liquid sludge. Then, to make the situation even more ridiculous, the tank truck dumped the sludge into a quarry already half full of water (Lamb, 1973)!

This example of when *not* to dewater a sludge illustrates the state of mind of many design engineers, those who consider dewatering a necessary step in sludge treatment. Obviously, this is not so, and it is good practice to always analyze the advisability of dewatering sludge rather than do it as a "knee-jerk" reaction (Dick, 1973).

REFERENCES

Adrian, D. D., P. A. Lutin and J. H. Nebiker (1968). "Source Control of Water Treatment Waste Solids," Dept. of Civil Enginering, Univ. of Massachusetts, Amherst.

Albertson, O. E., and E. J. Guidi (1967). "Centrifugation of Waste Sludges," *J. Water Poll. Control Fed.* 41:4.

Albertson, O. E., and D. R. Vaughn (1971). "Handling of Solids Wastes," *Chem. Eng. Prog.* 67:9.

Ambler, C. M. (1952). "The Evaluation of Centrifuge Performance," *Chem. Eng. Prog.* 48:3.

Austin, E. P. (1978). "The Filter Belt Press–Application and Design," *Filtration & Separation (Br.)* July/Aug.

Bakerville, R. C., A. M. Bruce and M. C. Day (1978). "Laboratory Techniques for Predicting and Evaluating the Performance of a Filterbelt Press," *Filtration & Separation (Br.)* Sept./Oct.

Baskerville, R. C., and R. S. Gale (1968). "A Simple Automatic Instrument for Determining the Filterability of Sewage Sludges," *Water Poll. Control (Br.)* 67:233.

Bennett, E. R., D. A. Rein and K. D. Linstedt (1973). "Economic Aspects of Sludge Dewatering and Disposal," *J. Environ. Eng. Div., ASCE* 99:55.

Burd, R. S. (1968). "A Study of Sludge Handling and Disposal," FWPCA (EPA) Pub. WP-20-4, Washington, DC.

Campbell, H., and B. P. LeClair (1974). "Effects of Control Variables and Sludge Characteristics on the Performance of Dewatering and Thickening Devices," *Proceedings, Sludge Handling and Disposal Seminar,* Toronto (Ottawa: Environmental Protection Service).

Campbell, H. W., R. W. Kuzyk and G. R. Robertson (1978). "The Use of Pulped Newsprint as a Conditioning Aid in the Vacuum Filtration of a Municipal Sludge," *Prog. Water Tech. (Br.)* 10(5/6):79.

Carman, P. C. (1933). "A Study of the Mechanism of Filtration," *J. Soc. Chem. Ind. (Br.)* 52:280T.

Carnes, B. A., and J. M. Eller (1972). "Characterization of Wastewater Solids," *J. Water Poll. Control Fed.* 44:8.

Christensen, G. L., W. R. Elliott and W. K. Johnson (1976). "Interactions Between Sludge Conditioning, Vacuum Filtration, and Incineration," *J. Water Poll. Control Fed.* 48(8):1955.

Coackley, P., and B. R. S. Jones (1956). "Vacuum Sludge Filtration," *Sew. Ind. Wastes* 28:8.

Colin, F. *et al.* (1976). "Characterisation des Boues Residuaires," Summary of French contributions to the European COST 68 Project, Ministry of the Environment, Paris.

Dick, R. I. (1973). "Sludge Handling & Disposal—State of the Art," *Ultimate Disposal of Wastewaters and Their Residuals.* (Raleigh, NC: Water Research Institute, North Carolina State University).

DiGregorio, D., and G. L. Shell (1976). "Dewatering Chemical-Primary Sludges," *J. Environ. Eng. Div., ASCE* EE5:1087.

Gale, R. S. (1967). "Filtration Theory with Special Reference to Sewage Sludges," *Water Poll. Control (Br.)*, p. 6.

Gale, R. S. (1970). "Studies on the Vacuum Filtration of Sewage Sludges," *Water Poll. Control (Br.),* p. 514.

Gale, R. S. (1971). "The Calculation of Theoretical Yields of Rotary Vacuum Filters," *Water Poll. Control (Br.)* p. 114.

Gale, R. S., and R. C. Baskerville (1970). "Polyelectrolytes in the Filtration of Sewage Sludges," *Filtration & Separation (Br.)* Jan/Feb., p. 1.

Gerlich, J. W., and M. D. Rockwell (1973). "Pressure Filtration of Wastewater Sludge with Ash Filter Aid," EPA-R2-73-231, Washington, DC.

Hansen, J. (1974). Danish Technical University, Lyngby. Personal communication.

Jones, B. R. S. (1956). "Vacuum Sludge Filtration," *Sew. Ind. Wastes* 28:9.

Keith, F. W., Jr. (1978). "Thickening Waste Activated Sudge by Centrifugation," Proceedings, Engineering Foundation Conference, Asilomar, CA.

Lamb, J. C. (1973). Discussion at the Conference on Ultimate Disposal of Wastewaters and Their Residuals, Durham, NC.

Levin, P. (1971). "Dewatering of Activated Sludge by Cellulose Sponge Adsorption and Expression," PhD Thesis, Illinois Institute of Technology.

Lewin, V. H., and M. H. E. Sharkey (1966). "Sewage Sludge Dewatering," *Water Waste Treatment J.* 11:19.

Martel, C. J., and F. A. DiGiano (1978). "Production and Dewaterability of Sludge from Lime Clarified Raw Wastewater," Dept. of Civil Engineering, Univ. of Massachusetts, Amherst.

Metcalf and Eddy. *Wastewater Engineering.* (New York: McGraw-Hill Book Company, 1973).

Nebiker, J. H., T. G. Sanders and D. D. Adrian (1968). "An Investigation of Sludge Dewatering Rates," *Proc. 23rd Ind. Waste. Conf.,* Purdue Univ., West Lafayette, IN.

Nissen, J. A., and P. A. Vesilind (1973). "Biological Sludge Preservation," *Environ. Syst. Eng. Series, No. 13,* (Durham, NC: Duke University).

Nyberg, B. E. (1970). "Experiences with Dewatering with the Belt Filter Press," *Vatten* (in Swedish) 3:255.

Parker, D. G., C. W. Randall and P. H. King (1972). "Biological Conditioning for Improved Sludge Filterability," *J. Water Poll. Control Fed.* 44:11.

Penman, A., and D. W. Van Es (1974). "Winnipeg Freezes Sludge, Slashes Disposal Costs 10 Fold," *Civil Eng.* 43:11.

Petersen, B., and P. Hansen (1974). "Floc Strength and Turbidity Variations During Anaerobic Storage of Aerobic Sludge," unpublished paper, Dept. of San. Eng., Technical University of Denmark, Lyngby.

Puolanne, J. (1977). "Tutkimuksia Jätevesilietteen Kunnostuksesta ja Koneellisesta Kuivauksesta," Vesihallitus (National Board of Waters) Report No. 124, Helsinki, Finland.

Randall, C. W. (1969). "Are Paved Drying Beds Effective for Dewatering Digested Sludge?" *Water Sew. Works* 3:10.

Randall, C. W., J. K. Turpin and P. H. King (1971). "Activated Sludge Dewatering: Factors Affecting Drainability," *J. Water Poll. Control Fed.* 43:1.

Ronen, M. (1978). "Characterization of Lime Sludges from Water Reclamation Plants," D. Sc. Dissertation, Univ. of Pretoria, South Africa.

Schnittger, J. R. (1970). "Integrated Theory of Separation for Bulk Centrifuges," *Ind. Eng. Chem.* 9:3.

Schultz, S. E. (1967). "Dewatering of Domestic Waste Sludges by Centrifugation," PhD Thesis, Univ. of Florida, Gainesville.

Schwoyer, W. L., and L. B. Luttinger (1973). "Dewatering of Water-Plant Sludges." *JAWWA* 65:6.

Shepman, B. A., and C. F. Cornell (1956). "Fundamental Operating Variables in Sewage Sludge Filtration," *Sew. Ind. Wastes* 28:12.

Smith, J. E. (1973). Ultimate Disposal Research Program, EPA. Personal communication.

Swanwick, J. D. (1972). "Theoretical and Practical Aspects of Sludge Dewatering," 2nd European Sewage and Refuse Symposium (EAS), Munich.

Swanwick, J. D. (1974). "Summary for the Use of Standard Stirring Test," Minutes from the COST-68/2 meeting, L. Ulmgren (Ed.), Scandiaconsult Int., Stockholm.

Swanwick, J. D., W. J. Fisher and M. Foulkes (1972). "Some Aspects of Sludge Technology Including New Data on Centrifugation," *Water Poll. Manual (Br.)*.

Swanwick, J. D., F. W. Lussignea and R. C. Baskerville (1962). "Recent Work on the Treatment and Dewatering of Sewage Sludge," *Adv. Water Poll. Res. (Br.)*.

Tabasaran, O. (1971). "The Dewatering of Digested Sludge Using Synthetic Filtering Agents," *Water Res.* 5:61.

Vesilind, P. A. (1969). "Estimation of Sludge Centrifuge Peformance from Laboratory Tests," Report T3/69, Norwegian Institute for Water Research, Oslo.

Vesilind, P. A. (1970). "Estimation of Sludge Centrifuge Performance," *J. San. Eng. Div., ASCE* 96:SA3.

Vesilind, P. A. (1973A). Discussion of "Economic Aspects of Sludge Dewatering," by E. R. Bennett, D. A. Rein and K. D. Linstedt. *J. Environ. Eng. Div., ASCE* 99:EE5.

Vesilind, P. A. (1973B). "Solid Waste Laboratory Manual," *Environ. Sys. Eng.* Ser. No. 7, Dept. of Civil Eng., Duke Univ., Durham, NC.

Vesilind, P. A. (1974A). "Estimating Centrifuge Capacities," *Chem. Eng.* 81:7.

Vesilind, P. A. (1974B). "Scale-up of Solid Bowl Centrifuge Performance," *J. Environ. Eng. Div., ASCE* 100:EE2.

Vesilind, P. A. (1977). "Characterizing Sludge for Centrifugal Dewatering," *Filtration & Separation (Br.)* Mar./Apr., p. 115.

Vesilind, P. A., and J. E. Loer (1970). "Centrifugal Activated Sludge Thickening," unpublished.

Villiers, R. V., and J. B. Farrell (1977). "A Look at Newer Method for Dewatering Sludge," *Civil Eng.* Dec., p. 66.

Water Pollution Control Federation (1959). "Sewage Treatment Plant Design," Manual of Practice No. 8.

Water Pollution Control Federation (1969). "Sludge Dewatering," Manual of Practice No. 20.

Water Pollution Research (1971). H. M. Stationery Office, London.

Water Pollution Research (1972). H. M. Stationery Office, London.

Weston, Roy F., Inc. (1971). "Process Design Manual for Upgrading Existing Wastewater Treatment Plants," *EPA Technology Transfer,* Contract 14-12-933.

White, M. J. D., and C. F. Lockyear (1978). "A Novel Method for Assessing the Thickenability of Sludges," *Proc., Int. Assoc. of Water Poll. Res.*, Stockholm.

Wilhelm, J. H. (1978). "The Use of Specific Resistance Data in Sizing Batch-Type Pressure Filters," *J. Water Poll. Control Fed.* 50:1.

Wuhrmann, K. A. (1977). "A More Precise Method of Determination of Specific Resistance to Filtration," *Water Poll. Control (Br.)*, p. 377.

Zingler, E. (1970). "Significance and Limits of the Buchner Funnel Filtration Test," Proc. 5th Int. Conf. on Water Poll. Research, II-31, San Francisco.

PROBLEMS

1. Calculate the number of times a sand bed could be filled each year given the conditions in Example 6-1, except that the sand bed is covered.
2. Estimate the required sand bed area for a city of 200,000 people which had a secondary plant with anaerobic digestion.
3. A filter leaf test's data are as follows:

   Feed solids = 6%
   Cake solids = 25%
   Area of filter leaf = 300 cm^2
   Cycle time = 30-60-60 sec as submerged, drying and off
   Filtrate solids = 800 mg/l
   Depth of a cake on filter leaf = 1 cm

   Calculate the expected filter yield and the solids recovery.
4. A Buchner funnel filtration test is conducted with a pressure of 6×10^4 N/m^2, an area of 400 cm^2 and a feed solids concentration of 6%. The filtrate volumes are as follows:

Time (min)	Filtrate Volume (ml)
0	0
1	10
2	20
3	29
4	37
5	43
6	48

   Calculate the specific resistance and the filter yield if the same cycle time as in Problem 3 is used. Assume viscosity as that of water.
5. A sludge centrifuge is obtaining high solids recovery but a wet cake. The operator wants to increase cake dryness. What changes in the centrifuge operation should be made to achieve this?

6. A test-tube centrifuge, operating at 500 rpm and with a radius from center to top of tube and bottom of a 15-ml tube of 6 and 12 cm respectively, yields the following results:

Time (sec)	Volume of Sludge (ml)
10	13
20	11
30	8
40	7
50	7

Calculate the "settling coefficient," S.

7. How fast should a sand grain (specific gravity of 2.6, diameter of 1 mm) settle in a centrifuge tube spinning at 200 rpm, with a 15-cm radius? As this sand grain moves to the bottom of the tube, does its velocity increase or decrease? Why would its actual velocity at any given point be lower than the theoretical?

8. Two geometrically similar machines have the following characteristics:

	Machine 1	Machine 2
Bowl diameter (cm)	25	35
Pool depth (cm)	3	4
Bowl length (cm)	30	80

If the flow rates are 0.6 and 4.0 m^3/hr, and the first machine operates at 1,000 rpm, what speed should machine 2 be run at in order to achieve similar operation?

9. Given the data shown in Figure 6-35, how might it be possible to enhance the performance of this machine?

10. A centrifuge experiment produced the following results:

Food Solids Concentration (mg/l)	Flow Rate (m^3/min)	Solids Recovery (%)	Cake Solids (%)
2,000	4	70	20
4,000	4	68	22
6,000	4	42	25
2,000	8	38	33
4,000	8	36	30
6,000	8	21	29

Discuss the capacity of this machine relative to solids and liquid loading.

CHAPTER 7

SLUDGE CONDITIONING

Disposal of wastewater sludges is greatly facilitated if the sludge is previously treated or *conditioned.* The purpose is usually to increase the efficiency of dewatering or thickening, although sterilization and odor control can also be important.

CHEMICAL CONDITIONING

Theory

It is not possible in this limited space to review fully the theory of how chemicals influence sludge dewatering. This information can be found in two articles, by O'Melia (1969) and by Tenney *et al.* (1970).

Briefly, it is thought that two mechanisms are necessary to condition sludge sufficiently: one is the neutralization of the charge while the other is the bridging of individual particles into a floc structure. The classical double-layer model is an attempt to explain the charge neutralization. As shown in Figure 7-1, a negatively charged particle tends to collect a layer of positively charged ions which travel with the particle and are fixed to it. Around this layer is a loose layer of negative and positive ions, which can be substituted by other ions. Two electrical charges affect the particle, one of which is the force due to repulsion of like charges, and the other is the attractive force, called the London or van der Waals force. The van der Waals force is extremely strong, but acts only at very close range.

Summing these two forces, as shown in Figure 7-1, results in an energy hill of repulsive forces that must be overcome if any two similar particles are to unite into a larger one. This can be achieved by substituting trivalent aluminum or trivalent iron ions in the outer layer. If these ions are substituted, the net repulsive force due to the negative charge is decreased; if

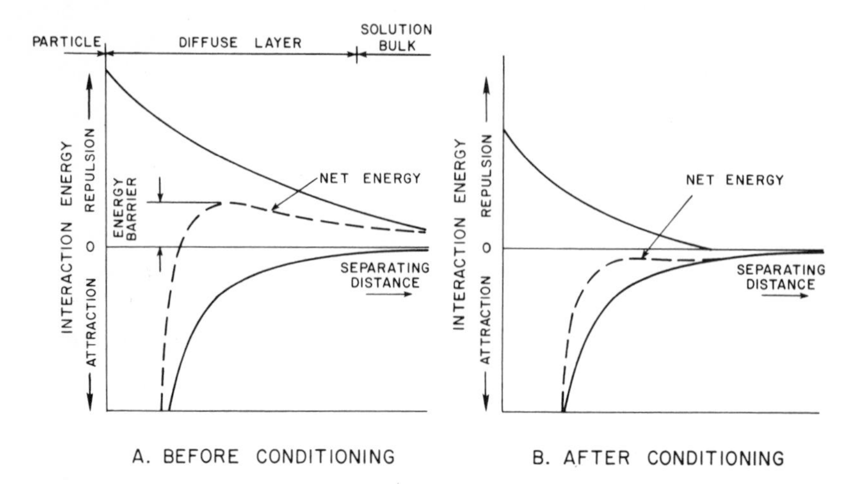

Figure 7-1. Representation of the classical double-layer theory of particle coagulation (after O'Melia, 1970).

this decrease is sufficient, the particle will no longer repulse similar particles but instead be attracted by them. Such a charge neutralization will then enable the particles to agglomerate into fewer large particles and to decrease the surface area of solids. The formation of such flocs makes the sludge easier to dewater mechanically.

This model works rather well for some inorganic sludges and for clay in water treatment. However, for organic sludges in wastewater treatment operations, the model does not fully explain observed results. Because of this deficiency, the bridging model has been used to explain how chemicals will flocculate biological sludges.

The basic premise of the chemical bridging model is that flocculants such as metal hydroxides and organic polyelectrolytes form long molecules that attach themselves to the sludge particles and thus bridge the gap between particles and draw them together, creating the strong floc lattice structure necessary for efficient dewatering.

Inorganic Chemical Conditioners

Some of the earliest chemicals used in sludge conditioning were ferric and aluminum compounds, especially ferric chloride and aluminum sulfate. The double-layer theory assumes that the ionized iron (III) and aluminum (III) will enter the outer layer of the particle, neutralize it, and make it possible for the particles to flocculate. It is true, however, that the iron (III) and aluminum (III) ions do not really exist in an aqueous solution. They form complexes with water molecules such as $Fe(H_2O)_6{}^{+++}$ and $Al(H_2O)_6{}^{+++}$ and, in addition, both compounds form hydroxides with very low solubilities that act as the bridges between the particles. These hydroxides are long, sticky molecules with complex structure, not simple Al $(OH)_3$ or Fe $(OH)_3$. These hydroxides are soluble at both low and high pH. Within their optimum pH range, defined as the range of least solubility, they undergo rapid hydrolytic reaction and form the hydropolymers. At lower pH values, Fe (III) and Al (III) form free ions; above these optimum values, they are negatively charged hydroxyl species. The optimum pH range for Fe (III) is between 6 and 7 and for Al (III), between 4.5 and 5.5 (Tenney *et al.*, 1970).

When iron compounds are used, it is usually necessary to raise the pH with lime to make the reaction efficient and produce the necessary hydroxides.

Organic Chemical Conditioners.

The other group of sludge conditioning chemicals are the organic polyelectrolytes or polymers. Most polymers contain one kind of monomer which itself contains some ionized groups such as a carboxyl, amino or other. If these ionizable groups have a positive charge, the polymer is called a cationic polymer. A negative charge makes them anionic polymers and nonionic polymers have no charge. Examples of these three polymers are shown in Figure 7-2. These polymers must perform the same two functions as the inorganic chemicals, that is, neutralize the charge and bridge the particles. Cationic polymers are the most common in wastewater treatment because their positive charge easily matches the negative charge usually found on the particle; the bridging is then rather simple. Anionic and nonionic polymers are not as easily attached to the particles and thus their action is less efficient.

It is possible to use anionic polymers effectively, however, by first adding cations such as hydrogen ions to the sludge, thus reducing the electrostatic interference to the anionic polymer. Since anionic polymers are often less expensive than other types, it may be economical to lower the pH of the sludge, then use an anionic polymer. It has also been found that anionic polymers work very effectively in conjunction with cationic forms, and they also have been used with inorganic chemicals such as iron and aluminum.

Polyelectrolytes for sludge conditioning are available as liquid, powder or pellets. The polymers are purchased in concentrated form and mixed with water before use. It has been found that polymer concentrations as low as 0.01% perform efficiently in sludge conditioning. Presumably, at such dilute concentrations the polymer molecules have a chance to stretch out and perform at maximum efficiency.

Although the liquid form is the easiest to mix with water, it is expensive because of high transportation costs. The most expensive form is pellets, which are also easy to mix. Powders can be the most difficult types of

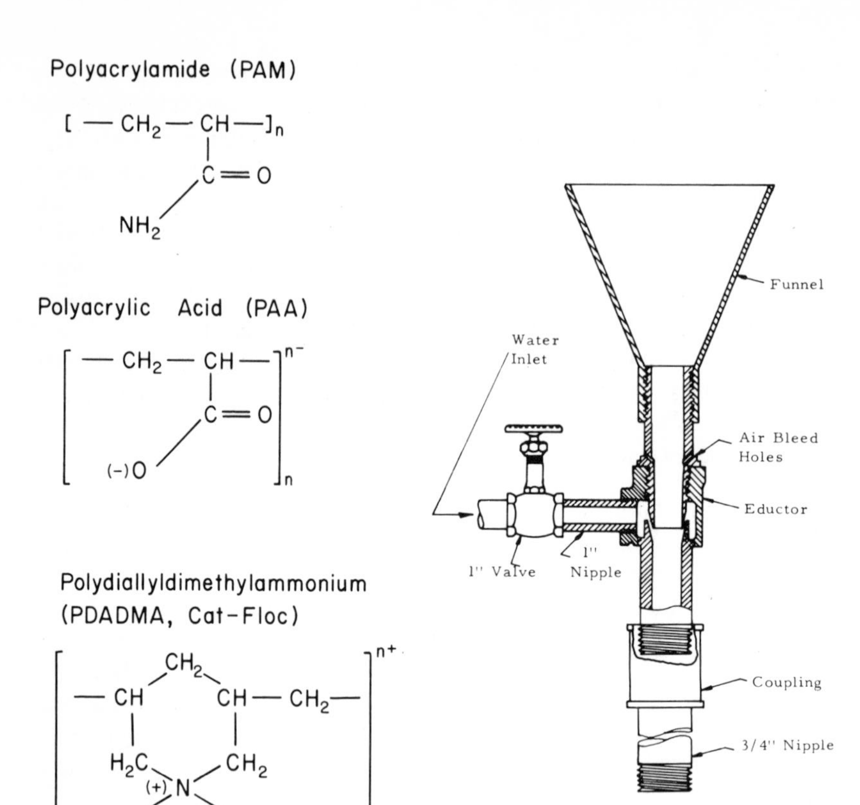

Figure 7-2. Three types of organic polyelectrolytes–cationic, anionic and nonionic (after O'Melia, 1969).

Figure 7-3. Aspirator used for mixing polymers (courtesy of Hercules, Inc.).

polymer to mix, but they are also the least expensive. However, powder polymers can cause many operational problems by making floors extremely slippery and by necessitating the wearing of dust filters to prevent inhalation during handling.

Polymers are often prepared in stock solutions of 1-5% that are further mixed with water before introduction to a dewatering unit. In mixing the powder form, a bag of polymer cannot simply be poured into a mixing tank and stirred, because the powder will form a gelatinous coating around itself and remain in large globs for a long time resisting all types of mixing. If these globs attach themselves to the shafts of the mixing propellers, considerable damage may be done to the mixers.

One solution to the problem of dissolving powders is the jet aspirator, shown in Figure 7-3. The polymer is placed in the funnel and water is pumped at a high velocity through an aspirator at the tip of the funnel. This creates a suction and draws out a small amount of polymer into a violent water vortex, effectively preventing the formation of gelatinous globs.

Polymers have a great many advantages over inorganic sludge conditioners including safety in handling and ease of operation. Another advantage of the polyelectrolytes is that if the sludge is to be incinerated, lower cake solids concentrations will be acceptable since sludge dewatered with organic polymers has better thermal characteristics than with inorganic conditioners. This is because organic polymers are themselves combustible and result in less ash. Swanwick *et al.* (1972) suggests the following necessary solids requirements for autothermal combustion (Table 7-1):

Table 7-1. Solids Concentrations Necessary for Self-Sustaining Combustion (50% Excess Air)

Exit Temperature (°C)	Solids Concentration (%) Required for Conditioning Method Shown	
	Polyelectrolyte	Lime and Copperas
430	17.0	24.3
850	24.8	34.8

Incidentally, it has been found that the cost of chemical conditioning may be decreased by simply switching chemicals periodically. It seems that if only one chemical is used for a long time, it allows certain kinds of fine solids to escape dewatering and these are recycled back into the plant. If the type of chemical is changed, say from a cationic to an anionic polymer, the original fine solids may be efficiently removed and, or course, a different kind of fine solid may be recirculated back to the plant. Some operators have found it advisable to set a cycle of about four months with three different types of chemicals.

Screening Chemical Conditioners

The best method to evaluate the worth of a chemical as a conditioner is to use it on full-scale equipment in the treatment plant. This can be a tedious and expensive process due to the large number of commercially available chemicals. Several quick and easy tests have been developed to screen chemicals in the lab and thereby narrow the number of products to be tested in full-scale experiments.

One of these is the capillary suction time (CST) discussed in Chapter 6. Each test takes only a few minutes and the data are quantitative (seconds).

A simple jar test is perhaps the most widely used method to evaluate chemical conditioning. The standard jar test mechanism is employed, with one-liter samples of sludge into which various chemicals at different concentrations are introduced. Following a short period of rapid mix, the sludge is flocculated for a few minutes then allowed to settle. The clarity of the liquid and the apperance of the floc structure is an indication of the chemical's ability to condition the sludge. This test does not yield quantitative data, but it does give enough qualitative information to allow quick screening.

In one study, (Hansen, 1974) the minimum chemical required in a jar test to produce a floc which could not be improved upon by higher polymer dosage (called the "point of flocculation") was correlated with the minimum polymer dosage required in a prototype centrifuge. This suggests that an experienced engineer could well define quantitatively the required polymer dosage in a centrifuge by a simple jar test.

Rotating viscometers can also be used to judge the effect of polyelectrolytes on sludge, specifically the rate of floc formation and strength of the sludge. Colin (1974), reporting on work performed at the Institut de Recherches Hydrologiques in Nancy, stated that if a paddle-type rotational viscometer is inserted in a sludge sample and spun at a speed sufficient to mix the suspension, and then polymer is added to the sludge, a distinctive curve of torque, as registered on the viscometer versus time, will be obtained. Figure 7-4 is a typical curve, showing that there is an initial increase in torque after polymer addition, corresponding to floc formation. As shearing continues, some of this floc is eventually broken, resulting in decreased torque. The curve finally flattens out as an equilibrium is reached. Higher speeds will decrease the time of floc formation.

Petersen (1973) proposed the simplest way to evaluate polyelectrolytes: take an empty beer bottle, fill it with sludge, and add a certain quantity of a polyelectrolyte. After gently shaking the bottle pour a small amount of sludge onto a white porcelain surface. It is then possible to visually examine the sludge and see the type of floc formed. A large floc is not necessarily the best for dewatering since agitation would tend to break these flocs apart. A floc that is small and very tight (in other words, a strong floc) is the desired result. Petersen has found this test gives a fairly accurate reading of required chemical dosage for centrifuges.

HEAT TREATMENT

As early as 1900, W. K. Porteous found that heat applied to sludges seemed to aid the subsequent separation of solid and liquid. He developed what is now known as the Porteous process, which involves heating a sludge

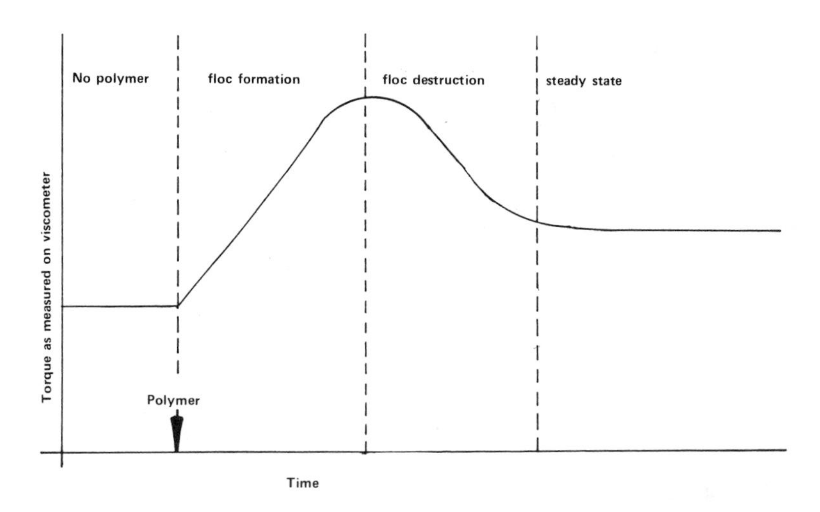

Figure 7-4. As polymer is added to sludge in a rotating viscometer, floc formation occurs, followed by floc destruction and a steady-state condition (Colin, 1974).

by steam injection in a batch process. This operation was not successful, however, from the commercial standpoint and was discontinued. In 1950, W. E. Farrer Ltd. bought out the Porteous Process and developed a continuous system now marketed as the Farrer System.

As noted in Chapter 2, biological sludge is composed mostly of organisms that contain water in several different forms. These cells contain water sufficiently well-bound to the cell that mechanical dewatering methods will not separate the water from the cell. Activated sludge, for example, could therefore be concentrated only to about 10% solids, even at extremely high forces. With chemical conditioning, the concentration can be increased only to about 15% solids as the ideal. The high moisture content makes the incineration of activated sludge impractical.

Heat, applied at sufficient quantities and for sufficient length of time, will destroy the cell and allow all the water bound in the cell to escape. The cells are, in fact, lysed or broken open.

The basic process of heat treatment is very simple. Two variables are heat and time. The temperature in most processes ranges from 180°C-200°C (about 380°F) with a residence time of 20-30 min. The simplest construction is a pipe containing the sludge pumped under pressure, with a pressure-reducing valve on the end as shown in Figure 7-5. The pressure for the heat treatment is usually around 19 kg/cm^2.

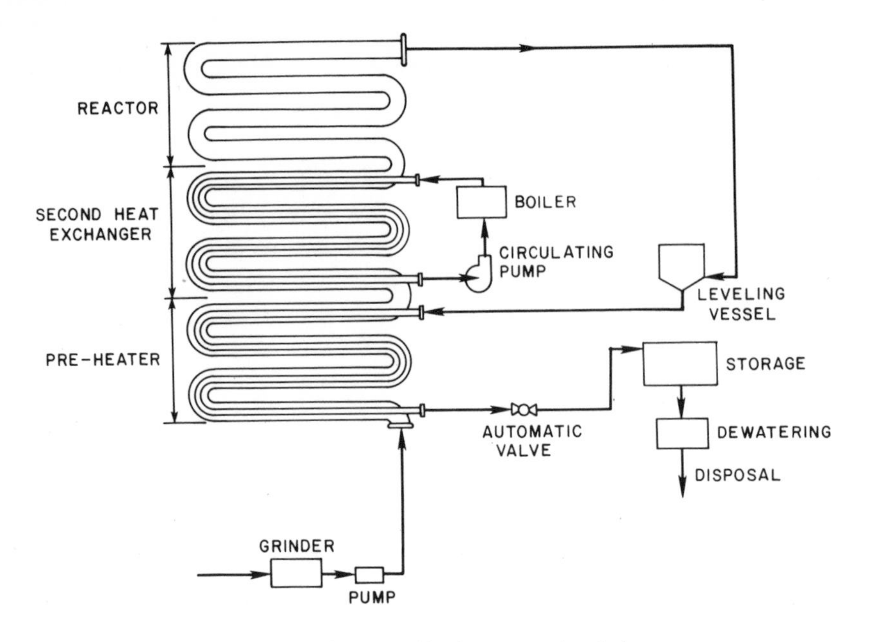

Figure 7-5. Heat conditioning system for sludge.

It is possible with laboratory studies to evaluate the time and temperature required for decreasing resistance to dewatering. Typical curves are shown in Figure 7-6.

One serious problem with heat treatment is the concentrated effluent produced from the operation. It has been observed that about 6,000 mg/l COD is produced for every 1% feed solids. Recycling this effluent to the treatment plant can result in substantial additional biological load on the plant. If the oxygen demand is exerted slowly, the recipient might be harmed. As a result, if heat treatment is installed in a plant, its effect on the overall plant operation must be considered. It has been suggested that for domestic wastewater treatment plants the design capacity must be increased by about 15% to allow for this additional load.

Experiences with a continuous low-pressure heat treatment system in Pudsey, UK, has been graphically described (Hirst *et al.*, 1971). After 18 months' operation, the problems and costs were significant and total costs exceeded $40 per ton of dry solids processed. In 1970, Swanwick of the British Water Pollution Control Laboratory accurately expressed the view of many sanitary engineers when he noted: "A few years ago, much interest and promise were shown with heat treatment and sludge pressing, but lately there is less enthusiasm for this type of plant."

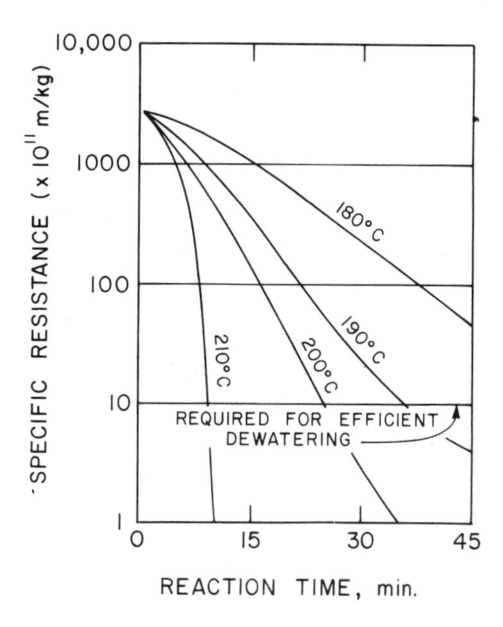

Figure 7-6. Typical curves from laboratory experiments for evaluating the effect of heat conditioning on sludge filtration.

ELUTRIATION

Elutriation is the washing of sludge to remove constituents that would interfere with subsequent dewatering operations. It is commonly used following anaerobic digestion. The potential problem with elutriation is that too much solid material may be returned with the washwater and that these fine solids could build up in the treatment plant.

A typical elutriation operation is a multistage countercurrent process. The ratio of washwater to sludge is approximately 2:1. The tanks act as thickeners and should, in fact, be so designed. From experience, for digested sludge elutriation tanks are designed at approximately 2 $kg/hr/m^2$ (10 $lb/ft^2/day$) of solids load.

Elutriation also tends to remove the nitrogen from the sludge and thus make it less valuable as a fertilizer. If the sludge is to be used further as a soil conditioner, elutriation is not recommended.

Elutriation, by lowering the alkalinity and removing the fine solids, lowers the chemical demand of the sludge. For example, a well-operated elutriation process will lower the ferric chloride demand by as much as 50%, but the fine solids buildup in the plant can still be a potential problem.

Such a problem occurred in the Washington, DC, Blue Plains Plant where conversion of the plant to a secondary activated sludge process resulted in the production of a very dirty elutriate—and, of course, the predictable build-up of fine solids, upsets in the unit operations, and increased liquid-solids separation costs. The problem was finally stabilized (if not solved) by adding chemical flocculants to the elutriation tanks. Similar experiences were reported in Richmond, California, and Toronto, Ontario.

OTHER METHODS OF CONDITIONING

Observation has shown that if sludge in drying beds is allowed to freeze during the winter, its dewatering characteristics will improve once it has thawed. Laboratory studies have shown that freezing can, in fact, lower the resistance to filtration considerably. This realization caused the predictable rush to produce freeze-conditioning systems for sludge dewatering. Unfortunately, the freezing process must be very slow so the water within the cells is allowed to crystallize and squeeze the solids into compact granules thereby making the sludge more easily dewatered once thawed. All practical commercial freezers were too fast and did not allow for the necessary crystallization. Although mechanical freezing does not hold much promise for the future, the use of natural cold for freeze conditioning has been successful (Mahoney and Duensing, 1972).

It has also been suggested that sludge conditioners such as newspaper pulp (Carden and Malina, 1968) and fly ash (Moehle, 1967) can be used to condition sludge prior to dewatering.

Smith *et al.* (1972) recognized that ash from sludge solids incineration could be beneficial in dewatering. The idea was attractive because transportation was not involved—the ash is produced at the plant.

A full-scale plant in Cedar Rapids, Iowa, uses fly ash from both power generation and sludge combustion to precoat pressure filters and as a sludge conditioner. The digested secondary sludge dewaters to 48% solids, with impressive filter yields (Gerlich and Rockwell, 1973).

At Indianapolis, raw primary and secondary sludges are successfully dewatered on vacuum filters using incinerator fly ash for conditioning. The ratio of fly ash to dry sludge solids (weight:weight) is about 1:2. The filtration characteristics before and after conditioning with fly ash are shown in Table 7-2 (Smith, 1974).

Table 7-2. Filtration Performance Before and After Fly Ash Conditioning

	Before	After
Filter Yield, lb solids/ft^2/hr	1 to 2	5.5
Cake Solids, %	15	33
Polymer Requirement, lb/ton of solids	15	6
Cake Release	poor	excellent

These data are somewhat misleading, however. With an ash:sludge solids ratio of 1:2, the sludge cake produced contains one-third fly ash, and the true filter yield (lb sludge solids removed as cake/ft^2/hr) is only 2/3 x 5.5 = 3.6 lb/ft^2/hr. Care must be taken not to confuse the dewatering of the conditioner with sludge solids removal.

CONCLUSIONS

The wastewater treatment plant operator often has a wide latitude in determining how efficient the solids removal system will be. Unfortunately, it is possible to operate many dewatering systems at low solids recoveries simply by inadequately conditioning the sludge. This saves money (since sludge conditioning can be expensive) relative to other plant operating costs.

It cannot be stressed enough, however, that saving money by lowering dewatering efficiency (lower solids recovery) is false economy. The fine solids in the dewatering effluent will recirculate within the plant and accumulate, making all other unit operations more difficult and, eventually, more expensive.

Operators are often besieged by local officials to reduce the expenses of wastewater treatment. They have a difficult time convincing politicians that 95% recovery is necessary and that they must purchase the required chemicals. There is overwhelming evidence, both from domestic and industrial wastewater treatment plants, that it is absolutely necessary to remove the fine solids and to avoid their buildup. This requires sludge conditioning, and the corresponding operating expense.

REFERENCES

Carden, C. A., and J. F. Malina (1968). "Effect of Waste Paper Additions on Sludge Filtration Characteristics," *Cent. Res. Wat. Res., No. 24.* Univ. of Texas, Austin.

Colin, F. (1974). "Study of Sludge Centrifugability," Minutes of COST-68/2 meeting, Dubendorf, Switzerland.

Gerlich, J. W., and M. D. Rockwell (1973). "Pressure Filtration of Wastewater Sludge with Ash Filter Aid," *EPA-R2-73-231* Washington, DC.

Hansen, P. H. (1974). "Laboratory Scale Characterization and Full Scale Centrifugation of Sewage Sludge," unpublished paper, I. Kruger A/S, Copenhagen.

Hirst, G., K. G. Mulhall and M. L. Hemming (1971). "The Sludge Heat Treatment Plant of Pudsey," *Water Poll. Control (Can.)*, 71.

Mahoney, P. F., and W. Duensing (1972). "Precoat Vacuum Filtration and Natural Freeze Dewatering of Alum Sludge." *J. Am. Water Works Assoc.* 64:6.

Moehle, F. W. (1967). "Fly Ash Aids in Sludge Disposal," *Environ. Sci. Technol.* 1:374.

O'Melia, C. R. (1969). "A Review of the Coagulation Process," *Public Works* 100:5.

Petersen, Aksel (1973). I. Kurger, A/S, Copenhagen. Personal communication.

Smith, J. E., Jr. (1974). Ultimate Disposal Research Program, EPA. Personal communication.

Smith, J. E., Jr. *et al.* (1972). "Sludge Conditioning with Incinerator Ash," *Proceedings of the 27th Industrial Waste Conference, Purdue University* (West Layfayette, IN: Purdue Research Foundation).

Swanwick, J. D., W. J. Fisher and M. Foulkes (1972). "Some Aspects of Sludge Technology Including New Data on Centrifugation," *Water Pollution Manual (Br.),* HMSO.

Tenney, M. W., W. F. Echelberger, J. T. Coffey and T. J. McAloon (1970). "Chemical Conditioning of Biological Sludges for Vacuum Filtration," *J. Water Poll. Control Fed.* 42:R1.

CHAPTER 8

SLUDGE DRYING AND COMBUSTION

Thickening and dewatering are only partially effective in separating the water from sludge particles. Thermal energy is required to obtain dry sludge. The sludge drying process yields a dry product high in volatile solids, while combustion involves complete oxidation of the volatile matter and the production of an inert residue.

SLUDGE COMBUSTION

In Europe, as well as North America, sludge incineration seemed, for a while, to be the ideal solution to the problems of increased sludge production. Through the early 1970s, the trend to incineration in the United States was rapidly increasing and it was projected that by 1985 as much as 35% of the sludge would be incinerated. This trend has now been reversed, however, with probably less than 10% of the sludge in the U.S. being incinerated in 1985 (Farrell, 1974).

The reason for this abrupt reversal is the energy crunch. Almost all incinerators require auxiliary fuel and, as the price of this fuel increased dramatically, the economics of incineration were reversed. Although most wastewater sludges are in large measure organic in nature, they will sustain combustion only if sufficient quantities of water have been removed. The removal of this water, however, is expensive and difficult.

In addition to the moisture content, the heat value of sludges also depends on the elemental composition. As seen by some typical data in Table 8-1, the composition can vary considerably. In most applications, the heat value of the sludge is insufficient and auxiliary fuel is required. The amount of this auxiliary fuel will define the efficiency of the combustion process.

Various methods for calculating the heat value of a sludge have been proposed.

Table 8-1. Elemental Composition of Sewage Sludge (EPA, 1973)

Elemental Composition	Sludge No. 1	No. 2	No. 3	No. 4
Carbon (%)	63.4	65.6	55.0	51.8
Hydrogen (%)	8.2	9.0	7.4	7.2
Oxygen (%)	21.0	20.9	33.4	38.0
Nitrogen (%)	4.3	3.4	3.1	3.0
Sulfur (%)	2.2	1.1	1.1	Trace
Volatile (%)	47.9	72.5	51.4	82.0
VSS (Btu/lb)	12,840	12,510	10,940	8,990
SS (Btu/lb)	6,160	9,080	5,620	7,380

One of the earliest methods of calculating heat value was the Dulong formula (Corey, 1969), based on the composition of carbon, hydrogen and oxygen. The heat value is calculated as

$$\text{Btu/lb} = 145.4\ C + 260 \left(H - \frac{O}{8} \right) + 41\ S$$

where C, H, O and S are the weight percentages in the sludge of carbon, hydrogen, oxygen and sulfur.

Owen (1957) pointed out that this formula can give unsatisfactory answers and suggested that calorimeter experiments be used.

Fair *et al.* (1968) suggested an empirical formula for the fuel (heat) volume of sludge:

$$Q_H = A\left[\frac{100\ C_V}{100 - D}\right] - B\left[\frac{100 - D}{100}\right]$$

where Q_H = heat value in Btu/lb solids
C_V = volatile solids, percent
D = dosage of inorganic chemicals used in dewatering the sludge as % of weight of sludge (D = 0 for organic polymers)
A = empirical constant, 107 for activated sludge, 131 for raw primary
B = empirical constant, 5 for activated sludge, 10 for raw primary.

Other equations have been suggested by Niemitz (1965) and Kempa (1970). They are, respectively,

$$Y = 83.3X - 1{,}089$$

and

$$Y = 53.5Z + 365$$

where Y = heat value in kcal/kg total solids
X = loss at ignition of a sludge sample (500°C as % of TS)
Z = loss at ignition (% of TS, temperature unspecified)

Studies in Canada (Shannon *et al.*, 1974) have shown that the caloric value can be closely approximated using only the volatile solids content, as shown in Figure 8-1. The equation for this empirical relationship is

$$\text{Btu/lb} = 122\,(C_V) - 660$$

where C_V = percent volatile solids.

If at all possible, a bomb calorimeter should be used to determine the heat value. A bomb calorimeter is a massive gun-metal cylinder fitted with a screwed cover, capable of withstanding high pressures. The cylinder is immersed in an insulated water bath and the material in question is burned in this cylinder in an atmosphere of pure oxygen, to assure complete combustion. The heat of combustion is measured by the temperature rise of the calorimeter and the surrounding water.

The rule of thumb for estimating the fuel value of sludges is to expect 10,000 Btu/lb of dry volatile solids (5,600 cal/g). By comparison, coal has the heat value of approximately 14,000 Btu/lb and oil approximately 19,000

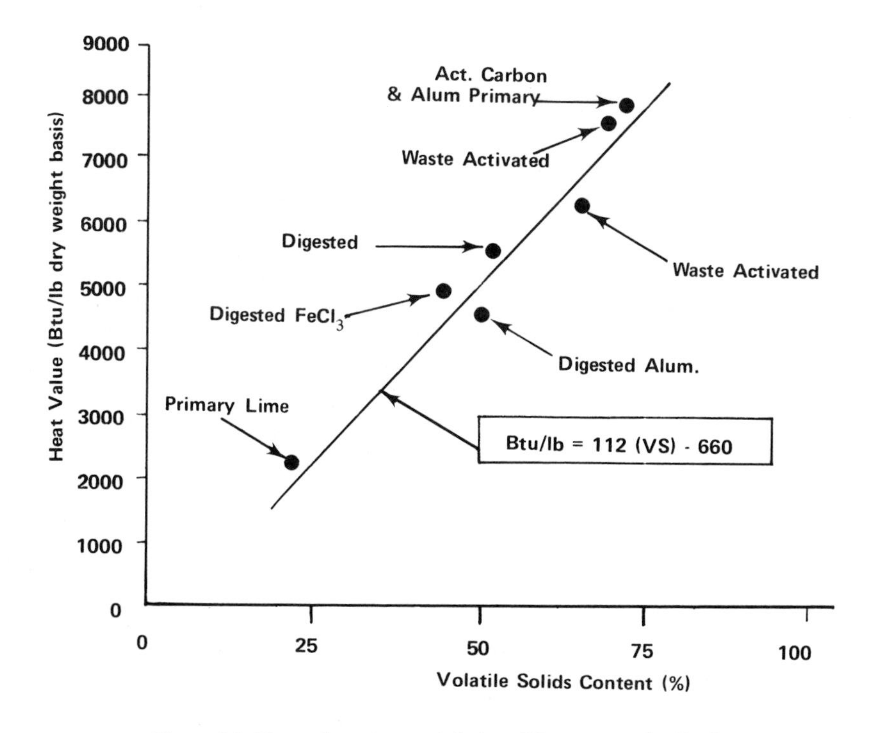

Figure 8-1. Heat value of several sludges (Shannon *et al.*, 1975).

Btu/lb. It is necessary, however, to realize that the fuel value for sludge thus calculated is for dry volatile solids, and sludge very seldom has a dry solids content of more than 20 or 25%, or a volatile solids content of greater than 70%. Thus the actual fuel value of sludge might be 10,000 x 0.2 x 0.7 = 1,400 Btu/lb wet sludge (800 cal/g). Wet sludge thus has a relatively low fuel value, and additional fuel is frequently required for its combustion.

The solids concentration (or more appropriately, the volatile solids concentration) required to sustain combustion is graphically shown in Figure 8-2. This shows that a primary sludge with 60% volatile solids must have at least a 28% solids concentration before combustion can be sustained without the use of auxiliary fuels.

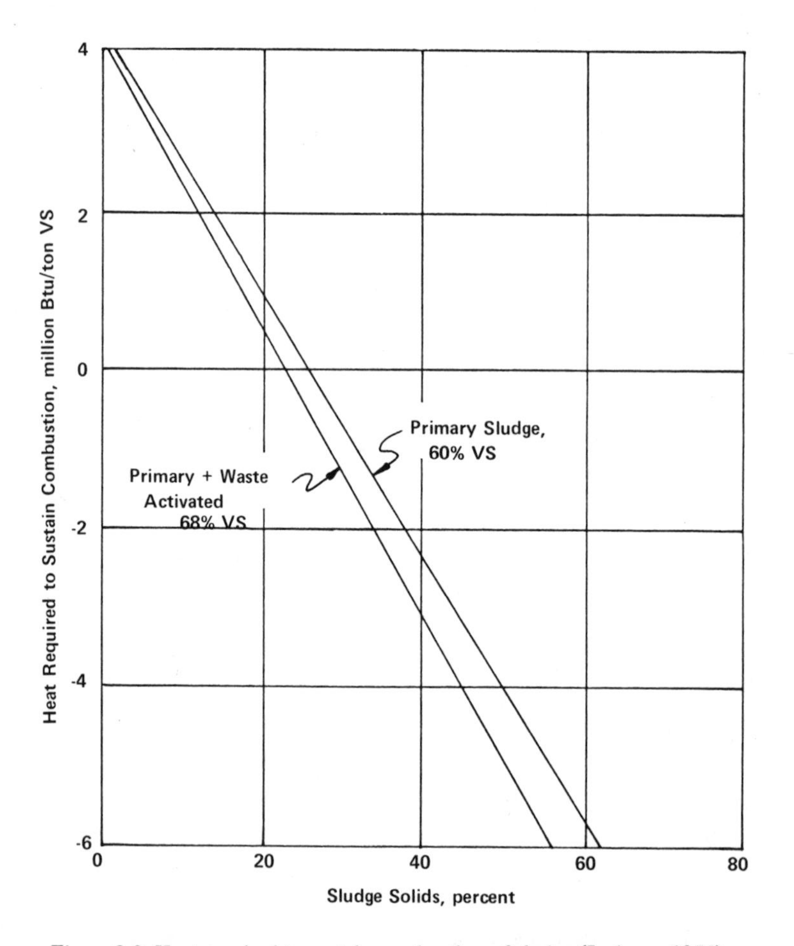

Figure 8-2. Heat required to sustain combustion of sludge (Jackson, 1975).

Example 8-1. If a sludge with a 60% VS and a solids concentration of 20% is to be incinerated, calculate the required auxiliary fuel oil per dry ton of sludge.

From Figure 8-2, at 20% solids, an additional 0.9 million Btu is required per ton of VS.

If oil has 19,000 Btu/lb, the incinerator will require

$$\frac{900{,}000 \text{ Btu}}{19{,}000 \text{ Btu/lb}} = 47.4 \text{ lb fuel oil per ton of VS}$$

The sludge is 60% VS, so the fuel requirement is

$$\frac{47.4}{0.60} = 79 \text{ lb fuel oil per ton of dry solids}$$

One way to increase the fuel value of sludge is to add a combustible material as a dewatering aid. Shredded newspaper, discussed in Chapter 6, shows considerable promise. Another approach is to use powdered coal, although coal does not seem to have any appreciable beneficial effect on the filter yield (Hathaway and Olexsey, 1975).

The two types of incinerators in wide use are the multiple-hearth incinerator and the fluid-bed incinerator. The multiple hearth (Figure 8-3) consists of several tiers through which the sludge is dropped. The upper layers of the incinerator are used for the vaporization of moisture and cooling of the exhaust gases. Volatile gas and solids are burned in the intermediate layers while the lower layers are used for the slow-burning compounds and the cooling of the ash. The temperatures within the sludge incinerator are approximately 550°C on the top, 900°C-1,000°C in the middle, and 350°C in the bottom hearths.

The fluid-bed incinerator shown in Figure 8-4 consists of a bed of sand into which sludge is introduced. The upward flow of gas suspends the sand so it is not necessary to provide grates for the burning. The sand is preheated to approximately 1,500°F and the violent boiling eliminates any need for a mixing device. The grinding and turbulent motion within the incinerator requires only 20% excess air, a savings over the multiple-hearth incinerator. The problems of odors can be eliminated by maintaining the exit temperature at greater than 800°C.

A third kind of incinerator used in some treatment plants is the rotary drum. This consists of a revolving drum through which the sludge is moved, again without the use of grates or other mixing devices. The temperature within the drum is approximately 850°C.

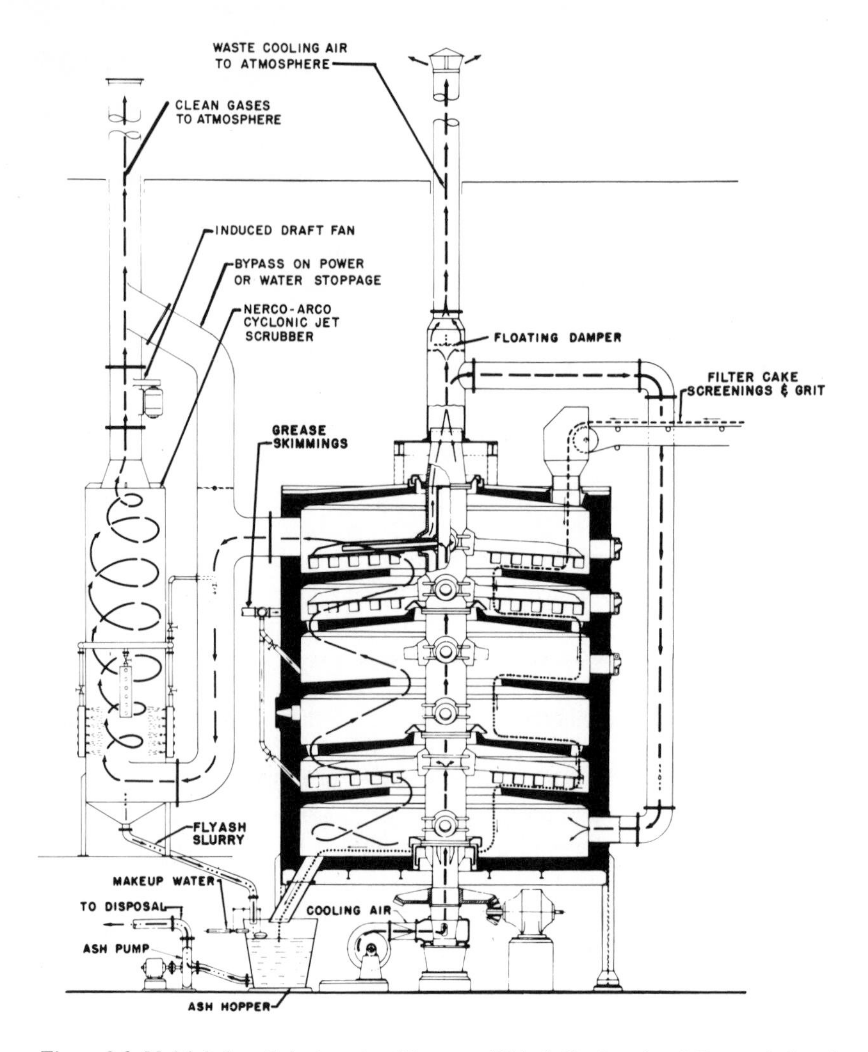

Figure 8-3. Multiple-hearth incinerator (Courtesy Nichols Engineering & Research Corp.).

An improved version of the rotary drum uses a tangential air supply that creates vortex air patterns, thus insuring complete and rapid combustion.

Successful incineration is possible if the following are provided in adequate supply:

Time
Temperature
Turbulence
Oxygen
Constant and adequate feed

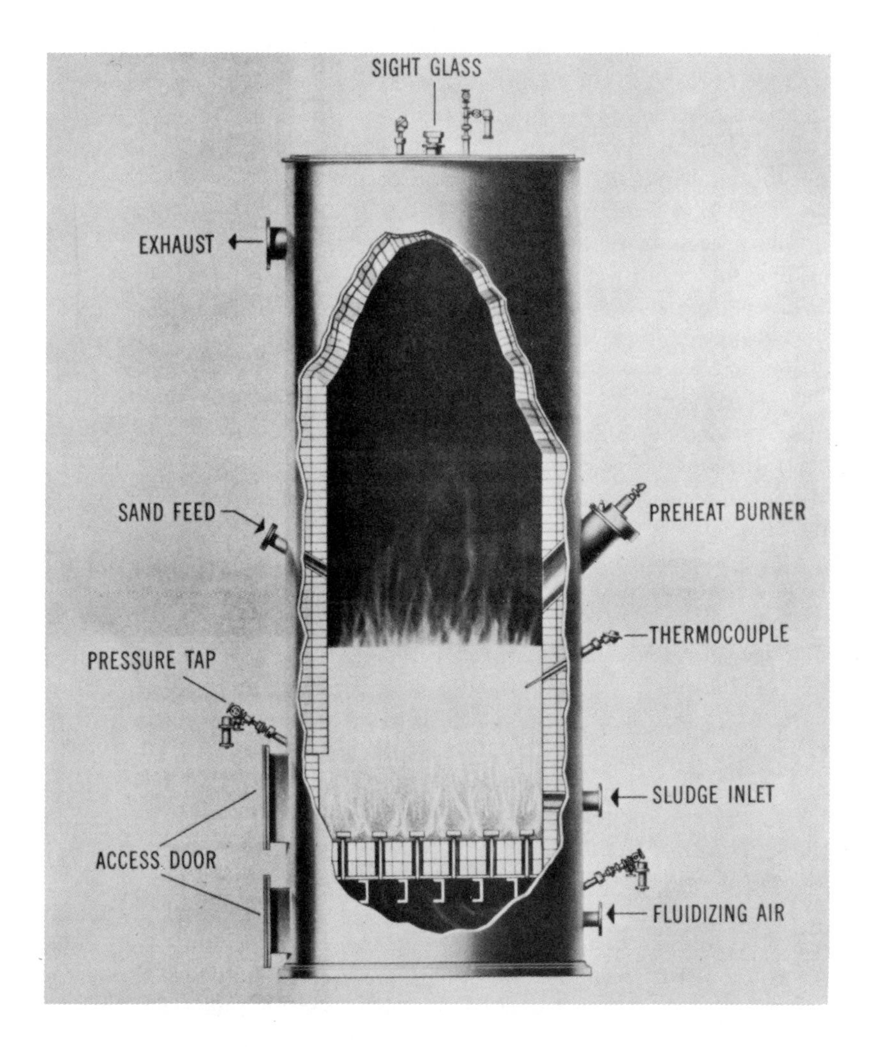

Figure 8-4. Fluidized-bed incinerator (courtesy of Dorr-Oliver Inc.).

Oxygen is, of course, supplied as air. All incinerators require more air than might be theoretically required. This *excess air* is needed because the process is not totally efficient. Most incinerators operate at 50-100% excess air. Although a large amount of excess air enhances combustion, it also quenches the reaction and a significant portion of the heat produced is used to heat this air. In addition, a larger volume of air means a larger volume of gases to be treated in the air pollution control system.

Combustion time is controlled by the feed rate and by mechanical devices such as the speed of the rabble arms in the multiple-hearth incinerator. Turbulence is necessary to bring the solids in contact with the oxygen.

The temperature can be controlled by varying the feed rate, the air flow, and the auxiliary fuel. Most incinerators operate at around 1,000°C to insure the elimination of odors.

The process of a sludge incinerator involves determining the allowable loading, the quantity of excess air and auxiliary fuel. Most designs are based on the performance of individual incinerators, and generalization is difficult. For example, the amount of excess air necessary can vary between 20 and 200%.

Incinerators must be able to meet local and national air pollution regulations. In addition to the emission of odor, incinerators without air pollution controls emit large quantities of fly ash but only small amounts of sulfur and nitrogen oxides. Most manufacturers now advertise that with proper controls, their incinerators will not emit more than 0.2 lb of particulates per 1,000 lb of stack gas and 50% excess air. This is within the Los Angeles air pollution code—the strictest in the U.S.

The reduction in pollutants is not free, however, and the cost of incineration has increased considerably. Such devices as secondary combustion, electrostatic precipitation, adsorption and scrubbing all have been used on incinerators, and have been known to as much as double the cost of the process.

Liao and Pilat (1972) found that for a fluid-bed incinerator, the concentrations of gaseous pollutants such as NO_X, CO and SO_2 were well within the standards. However, uncontrolled particulate emissions were high and had to be reduced with a scrubber.

Recently, concern has also been expressed about the heavy metals and refractory organic toxins which might be emitted to the atmosphere during incineration. Farrell and Salotto (1973), however, found that pesticides and polychlorinated biphenyls (PCB) are destroyed in multiple-hearth and fluidized-bed incinerators, and that most heavy metals are deposited in the ash. Mercury was the exception, being totally vaporized into the stack gases.

The form in which the metal is found in the sludge could influence its fate. For example, if cadmium is present in the sludge in solution as cadmium chloride, it could volatilize upon incineration. On the other hand, if it is present as a precipitated hydroxide, it would probably decompose to the oxide and exit with the bottom ash.

In a series of tests on sludge incinerators, an EPA study (1974) showed that if an incinerator is equipped with a wet scrubber, only about 1.6 grams of mercury would be emitted to the atmosphere per dry metric ton of sludge solids incinerated. This represented about 10% of the mercury in the sludge, with about 60% being deposited in the ash and the remainder removed with the scrubber water. The 1.6 grams per dry metric ton is well within EPA's Clean Air Act Amendments of 1970 for mercury emissions (EPA, 1975).

There does not seem to be a great difference in the metal emissions with the type of incinerator used. Fluidized-bed incinerators seem to yield emission data similar to multiple-hearth models (Copeland, 1975).

INCINERATION OF SLUDGE WITH MUNICIPAL REFUSE

An old concept, the co-disposal of waste sludge and municipal refuse, has recently received attention, due mainly to increased environmental restrictions on the disposal of both of these urban effluent streams.

A rough idea of the efficiency of incinerating sludge and refuse together can be seen in Figure 8-5 which illustrates the required sludge moisture content for sustaining combustion (Culp *et al.*, 1976). In this figure, it is assumed that sludge has a heat value of 10,000 Btu/lb VS, and refuse has a heat value of 4,750 Btu/lb, with a moisture content of 25%.

Example 8-2. A sludge with 50% VS and 20% TS is to be incinerated with refuse. Find the dry tons of sludge allowed per ton of refuse incinerated.

From Figure 8-5 at a moisture content of 80%, the sludge to solid waste ratio is about 4.3. Hence 4.3 tons of sludge can be incinerated per ton of refuse.

Unfortunately, the operational problems such as poor burnout and ignition with co-disposal of sludge and refuse have forced all municipalities using this process to abandon the program (Sussman, 1977; Niessen *et al.*, 1976). As one city engineer graphically stated: "We tried it, and it put the fire out." Most problems arise from lack of controlled feed and lack of proper design for sludge combustion.

Two newer techniques for achieving co-disposal have been introduced recently, and both seem to hold promise.

The first technique uses the heat from the incinerator to dry the sludge; the dried sludge is then burned in the incinerator. Rotary driers are commonly used for the sludge-drying operation and the dry powdered sludge is burned in suspension.

The second method is to treat the refuse to increase its heat value and change its physical character so it may be burned in incinerators such as the multiple-hearth or fluidized-bed (Vesilind *et al.*, 1979). The refuse processing would commonly include shredding and air classification to remove the light organic fraction. This "fluff" has a low moisture content (about 15%) and a

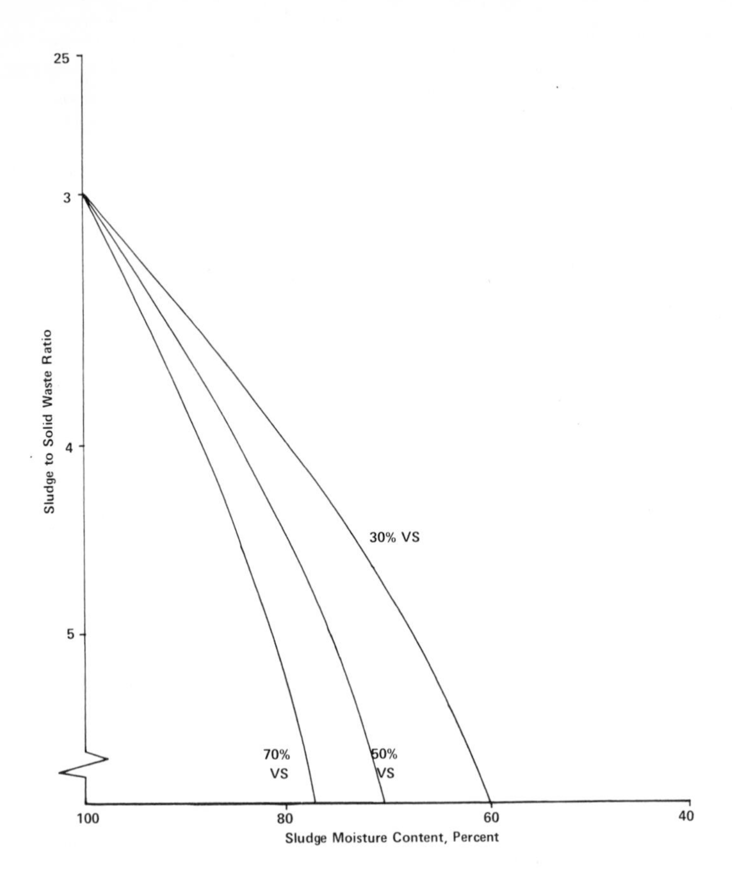

Figure 8-5. Combustion of sludge and solid waste (Culp, Wesner and Culp, 1976). (Assumptions: Heat value of sludge = 10,000 Btu/lb VS; heat value of solid waste = 4,750 Btu/lb; moisture of solid waste = 25%; heat required to evaporate water = 2,100 Btu/lb water.)

high fuel value (5,000 to 7,000 Btu/lb) and does not contain large inorganics that could cause operating problems or produce clinker in the furnace. Such operations could well produce exportable energy as heat or steam (Bracken *et al.*, 1977).

In practice, the two approaches to achieving more efficient combustion with refuse and sludge (*i.e.*, dry the sludge or treat the refuse) boil down to five techniques, as listed in a thorough study of worldwide co-disposal facilities (Niessen *et al.*, 1976).

1. Combustion of dewatered sludge and shredded- and air-classified refuse in multiple-hearth incinerators. These systems have been successful in Europe.
2. Combustion of dewatered sludge and shredded- and air-classified refuse in a fluidized-bed incinerator. Although no plants using this technique presently exist, the concept seems feasible.
3. Combustion of refuse and use of the heat to dry the sludge, which would then be fired in the same combustion chamber. The sludge is predried in a process step prior to combustion using the heat produced in its combustion. This technique seems to be the most widely used method of co-disposal.
4. Spraying wet sludge into the combustion chamber. All of these systems have been abandoned.
5. Use of pyrolysis (or more accurately starved-air combustion) in co-disposal. This system is described below in more detail.

PYROLYSIS

There is no clear line between incineration and true pyrolysis. Technically, pyrolysis is thermal decomposition in the absence of oxygen. Many so-called pyrolysis systems are, however, starved-air incinerators.

The major difference between incineration and pyrolysis is that the latter produces three potentially useful energy-rich products: a gas (mostly methane), a liquid and a char. Depending on how the process is carried out (temperature, time, etc.) varying amounts of these three end products can be produced. Incineration systems where the air flow is restricted and an elevated temperature maintained so as to achieve combustion of the gases within the combustion chamber—although an equally effective process—cannot thus be classified as true pyrolysis. Nevertheless, the lack of adequate alternative nomenclature requires that in this discussion we consider pyrolysis any thermal decomposition system which does not have excess air (over the stoichiometric requirement).

Although pyrolysis has received the most attention as a method for processing refuse, it is now being seriously considered a method of sludge disposal as well. A typical pyrolysis system for sludge uses a multiple-hearth incinerator as the basic reactor and an afterburner where the offgases are combusted in order to increase the temperature of the emissions for odor control. Due to stricter air pollution requirements, the low air volumes used in pyrolysis make this process an attractive alternative for large systems.

Starved-air combustion, according to Sieger (1978), has many advantages over ordinary incineration, such as:

- greater feed capacity since the amount of air needed is about 30% of stoichiometric, as opposed to 200% for incinerators, thus reducing air velocities and allowing larger feed volumes for comparable airflows;
- savings in auxiliary fuels, since the offgases use combustible and are burned in a separate afterburner;

- easier operational control since the main hearth reaction is controlled by oxygen availability, and thus fluctuations in feed will not cause large temperature changes;
- reduced emissions, since less air is used, resulting in lower velocities and less entrained particulates (50% less than incineration).

Starved-air combustion also has disadvantages, such as additional equipment (afterburner and scrubbing), corrosion problems due to the highly reactive nature of the gas entering the afterburner, and additional controls and instrumentation.

On balance, the potential of sludge pyrolysis using the starved-air principle seems promising. An additional incentive for some communities would be the possibility of co-firing sludge- and refuse-derived fuel (the organic fraction of municipal refuse) in the same unit. In one large facility, the use of refuse-derived fuel as the auxiliary source of energy, at ratio of refuse: sludge of 1:2 or a wet basis, achieved autogenous combustion and produced enough power to run the refuse processing facility and the wastewater treatment plant (Sieger, 1978).

WET OXIDATION

Wet oxidation, known also as wet combustion, is the burning of wet organics under high temperature and pressure. Since this is a wet process, prior sludge dewatering is not required. The efficiency of the process can be measured in terms of COD reduction. Since process can be run at various temperatures and pressures, the reduction in COD achieved is dependent on the temperature and pressure of operation, as shown in Table 8-2. The production of ash in the high oxidation mode (high temperature and pressure) equals that obtained with incineration.

The process is controlled by four parameters: temperature, air supply, pressure and feed solids concentration. The effect of temperature and pressure has been noted above. The air supply is necessary in order to provide complete oxidation, but too much air reduces the thermal efficiency. These systems are capital-intensive and have, in the past, experienced problems with corrosion and safety. They seem to be most applicable for industrial waste treatment.

Table 8-2. Reduction in COD in Wet Oxidation (Ewing and Culp, 1976)

Type of Oxidation	Reduction in COD (%)	Temperature (°F)	Pressure (psi)
Low	5	350-400	300-500
Intermediate	40	450	750
High	92-98	675	1,650

SLUDGE DRYING

Dried sludge can be a very good fertilizer and soil conditioner. Unfortunately, the cost of drying sludge is high, and the economics of the entire operation–treating, drying and marketing–make such enterprises unattractive.

The types of driers available include the flash dryer, screw-conveyor, multiple-hearth, rotary and atomized-spray towers. The flash driers are the most common in wastewater treatment. This process, pictured in Figure 8-6, is a recirculating device by which sludge is pumped into a warm column of air and the water vapor driven off. Some of the sludge is recycled to help the incoming wet sludge.

Another sludge dryer which may have potential is the rotating-shelf dryer. These resemble the multiple-hearth furnace in that the material is transferred downward from tray to tray. The trays rotate at the same speed, about 1 rpm,

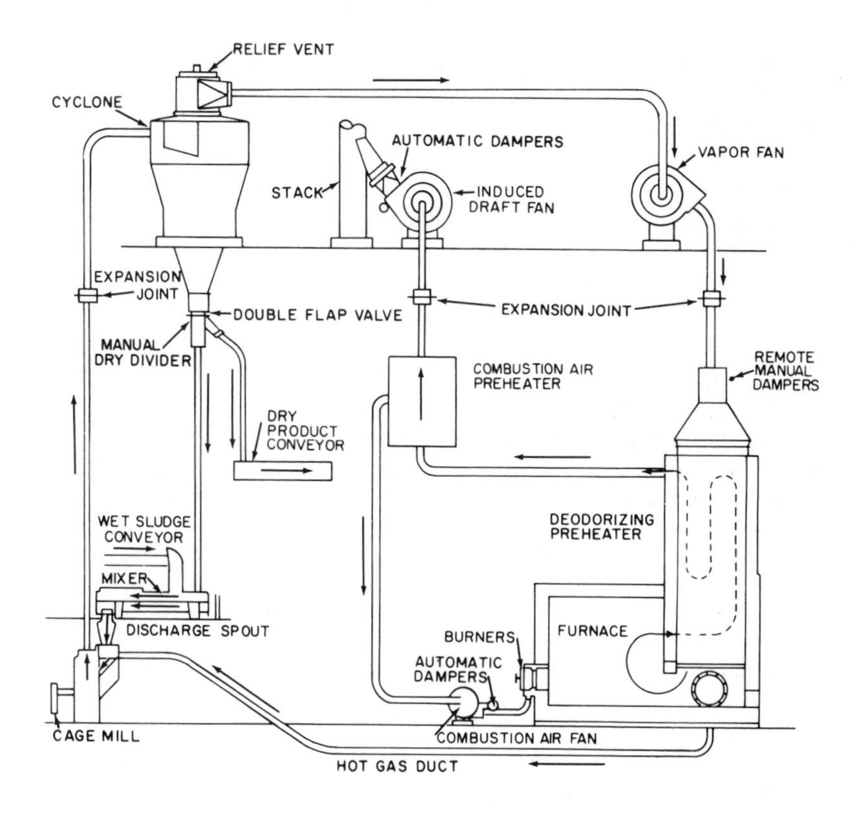

Figure 8-6. Flash drying system (Courtesy of Combustion Engineering).

and stationary wiper arms push the material through slots onto the tray below. Solids retention time in the dryer ranges from 0.5 to 2 hr, and can be controlled by adjusting the speed of the trays (Lee, 1976).

A third drying system is a proprietary method employing oil mixing, evaporation and dewatering. Marketed as the Carver-Greenfield process, it is widely used in the food and agricultural industry.

Other hardware is available also (Fischer, 1975), although operating data on wastewater sludges are scarce.

Sludge drying operations have had some safety problems, due mostly to the formation of a highly flammable dust. The product from such a drying process, an effective fertilizer and soil conditioner, must be sold to make the operation economically feasible. Few plants have been able to sell enough to make it worthwhile.

CONCLUSIONS

Sludge combustion is reasonable only in situations where no other disposal method is practicable. The costs are high, and the operation can be troublesome. Most operations require auxiliary fuel and, if this is in short supply, the operation can be forced to shut down.

Sludge drying is not widely used because the end product must be sold to make the operation economical. Additional problems of sludge drying are safety and the need for oil or gas as the fuel.

REFERENCES

Bracken, B. D., G. A. Horstkotte and T. D. Allen (1977). "Full Scale Testing of Energy Production from Solid Waste," *Proceedings*, Third Nat. Conf. on Sludge Management Disposal and Utilization, Miami Beach, FL.

Copeland, B. (1975), "A Study of Heavy Metal Emissions from Fluidized Bed Incinerators," in *Proceedings of the 30th Industrial Waste Conference, Purdue University* (Ann Arbor, MI: Ann Arbor Science Publishers, Inc.).

Corey, R. C., Ed. (1969). *Principles and Practice of Incineration* (New York: Wiley-Interscience).

Culp, Wesner and Culp (1976), "Energy Conservation in Municipal Wastewater Treatment," EPA Contract 68-03-2185.

Environmental Protection Agency (1973). Technology Transfer Seminar, Atlanta.

Environmental Protection Agency (1974). "Background Information on National Emission Standards for Hazardous Air Pollutants–Proposed Amendments to Standards for Asbestos and Mercury," EPA 450/2-74-009a, October, Washington, DC.

Environmental Protection Agency (1975). "Air Pollution Aspects of Sludge Incineration," EPA 625/4-75-009, June, Washington, DC.

Ewing, L. J., and R. L. Culp (1976). "Total Cost of Heat Treatment of Wastewater Sludges," *Proceedings*, National Conference on Sludge Management Disposal and Utilization, Miami Beach, FL.

Fair, G. M., J. C. Geyer and D. A. Okun (1968). *Water and Wastewater Engineering* (New York: John Wiley & Sons, Inc.).

Farrell, J. B. (1974). "Overview of Sludge Handling and Disposal," *Proceedings*, National Conference on Sludge Management, Pittsburgh, PA.

Farrell, J. B., and B. V. Salotto (1973). "The Effect of Incineration on Metals, Pesticides, and Polychlorinated Biphenyls in Sewage Sludge," in *Ultimate Disposal of Wastewaters and their Residuals,* Research Inst., N.C. State Univ. (Raleigh, NC: Water Research Institute, North Carolina State Univ.).

Fischer, R. (1975). "Drying of Residential Sludge," *Proceedings*, Int. Conf. and Tech. Exhib., Conversion of Refuse to Energy, Montreaux, Switzerland.

Hathaway, S. W., and R. A. Olexey (1975). "Improving the Fuel Value of Sewage Sludge," *News of Environmental Research in Cincinnati,* EPA Municipal Environmental Research Lab., Cincinnati, OH (October 31).

Jackson, J. (1975). "Sludge Incineration: Present State-of-the-Art" WWEMA Ad Hoc Committee on Sludge Incineration, Washington, DC.

Kempa, E. S. "Combustion Heat from Sewage Sludge," *Gas Wasser-Fach* (In German) 111:10.

Lee, D. A. (1976). "Operation and Control Aspects of Convection-Type Rotating Shelf Dryers," *Bulletin, Am. Ceramic Soc.* 55(5):498.

Liao, P. B., and M. J. Pilat (1972). "Air Pollutant Emissions from Fluidized Bed Sewage Sludge Incinerators," *Water Sew. Works,* 9:2.

Niemitz, W. (1965). "The Caloric Value of Sewage Sludge and its Relation to Other Sludges," *Gas Wasser-Fach* (In German) 106:1392.

Niessen, W., A. Daly, E. Smith and E. Gilardi (19760. "A Review of Techniques for Incineration of Sewage Sludge with Solid Wastes," EPA-600/2-76-288, December, Washington, DC.

Owen, M. B. (1957). "Sludge Incineration," *J. San Eng. Div., ASCE,* 83: SA1.

Shannon, E. E., D. Plummer and P. J. A. Fowlie. "Aspects of Incinerating Chemical Sludges," *Proceedings*, Sludge Handling and Disposal Seminar, Toronto, Ontario; Environmental Protection Service, Ottawa.

Sieger, R. B. (1978). "Sludge Pyrolysis: How Big a Future?" *Civil Eng.,* May, p. 88.

Sussman, D. (1977). "Co-Disposal for Solid Wastes and Sewage Sludge," EPA Technology Update, Office of Solid Waste Management Programs, Washington, DC.

Vesilind, P. A., A. E. Rimer and G. W. Pearsall (1979). *Unit Operations in Resource Recovery Engineering* (Englewood Cliffs: Prentice Hall).

PROBLEMS

1. Use the Dulong formula to calculate the heat value of the four sludges described in Table 8-1, and compare your answers to the heat values tabulated.
2. Plot the Niemitz and Kempa equations for heat value on a copy of the curve in Figure 8-1. Comment on any differences.
3. Suppose Sludge No. 1 in Table 8-1, at 25% solids, is to be combusted, how much fuel oil will be needed in order to sustain combustion?
4. Suppose a sludge at 20% solids and 60% volatile fraction is to be combusted with a refuse-derived fuel. How much refuse is needed per ton of wet sludge?
5. If the production of organic refuse is approximately 2.5 lb/capita/day, how dry must a sludge be (assuming 55% volatiles) in order for a municipality to burn all of its refuse and sludge together? Assume that the wastewater treatment plant produces raw primary and waste activated sludge, and does not have a digester.
6. If the only reaction in combustion was $C + O_2 \rightarrow CO_2$ (it obviously isn't), calculate the theoretical quantity of air necessary to combust Sludge No. 1 in Table 8-1.

CHAPTER 9

CHEMICAL SLUDGES

Because of the increase in industrial and other exotic waterborne wastes, and an increasing need to remove pollutants other than solids and BOD, conventional treatment is becoming less capable of protecting the water quality of the recipients. Accordingly, physical-chemical methods are often the only practical means of treatment and have many advantages over biological processes. Physical-chemical treatment operations, alone or in conjunction with biological treatment, can remove organic materials and solids as well as or better than biological units alone and, in addition, nutrient removal is greatly enhanced. But the sludges produced from such operations are not typical wastewater treatment sludges, and therefore deserve special attention. This chapter is devoted to a discussion of these increasingly important products of wastewater treatment.

In addition to their concern about chemical sludges from wastewater treatment, sanitary engineers are increasingly involved with the problem of waste sludge disposal from water treatment plants. Since 1884, when aluminum sulfate was first used for treating surface water supplies, the method of sludge disposal has been simply to flush it into the nearest watercourse. In light of the increasingly stricter controls on industrial wastes, this disposal method has become increasingly untenable (Lamb, 1969). The federal government now views water plant sludges as a type of industrial waste and requires discharge permits for its disposal.

Most often, water treatment sludges are waste alum or lime (from lime softening). They are universally difficult to handle and dewater. They have the advantage, however, of being relatively inert, and, therefore, seldom require biological treatment. But as with wastewater sludges, the composition of these water treatment sludges varies from plant to plant, necessitating individual attention (Schwoyer and Luttinger, 1973).

ALUM SLUDGES

The usual method of disposing sludges from water treatment is to wait for the settling tanks to fill up with sludge (usually from one to six months), then empty the sludge into the nearest stream or other waterway. This disposal method is not harmful to the aquatic environment when sufficiently diluted. New regulations, however, are forcing many plants to seek alternative means of disposal.

These sludges are usually fairly high in moisture content—98% or more. The BOD is low, about 50 mg/l, but the COD is fairly high, from 500-1,500 mg/l. The pH of waste alum sludge is about 6, and approximately 40% of the solids are volatile. The major effect of these waste alum discharges is the formation of mudbanks along the stream.

One of the main problems of waste alum sludge is its very high moisture content. Thickeners have been used in some places to reduce this and make the sludge more efficient to handle. Tube settlers have been introduced for waste alum sludge and appear to be quite effective in increasing the solids concentration.

Dewatering waste alum sludge is difficult. The specific resistance of alum sludges is about 10 to 40 x 10^{12} m/kg, which is approximately the same as activated sludge. Interestingly, the specific resistance of alum sludge decreases with increasing solids concentration (Glenn *et al.*, 1973). Alum sludges at even high solids concentrations behave as a liquid, with Newtonian flow characteristics (Gates and McDermott, 1968).

Centrifugal dewatering is possible with high polymer dosages. Nielsen *et al.* (1973) found that 2 lb polymer/ton of solids had almost no effect on solids recovery, but the addition of 1 lb/ton more polymer resulted in a jump to better than 90% solids recovery. In other words, the recovery-polymer dose curves were very steep. Cake solids of about 15% were obtained which was considered acceptable since a drying system followed centrifugation.

Pressure filters are used to dewater alum sludges in a number of cities, with lime conditioning to aid the dewatering (Inhoffer and Doe, 1973).

The dual-cell gravity solids concentrator has also been used successfully in dewatering both alum and lime-softening sludges, using polymers for conditioning (Schwoyer and Luttinger, 1973).

One possible method of disposal is to discharge waste alum sludge into sanitary sewers. Although the total effects of such discharges have not been fully investigated, especially in small treatment plants, experiences in some larger cities have shown little detrimental effect. In fact, such discharges may have some beneficial effect as a coagulation aid in primary clarification.

Freezing has been used successfully in Europe for this purpose (Doe *et al.*, 1965); however, this method becomes economical only when no other means of alum sludge disposal is available.

Alum has also been used in wastewater treatment plants for phosphorous removal. The desired reaction produces an insoluble phosphate. Simplified,

$$Al^{+++} + PO_4^{\equiv} \longrightarrow AlPO_4 \downarrow$$

Experiences of dosing existing biological plants with alum are limited, but all indications show that excellent phosphorous removal is attained and the sludge volume is not greatly increased. The total weight of sludge is doubled, however, and this should be considered in plant design and operation.

Anaerobic digestion of alum sludge from wastewater treatment operations likewise needs more work. One report is that the sludge has an adverse effect on the digestion process and that the digested sludge will not dewater well on drying beds (Brown, 1973).

Some typical sludge production and loading figures for alum sludges (as well as lime and ferric chloride) in physical-chemical wastewater treatment plants are shown in Table 9-1 (EPA, 1973). Additional performance data on sludge production may be found in the EPA manual on upgrading wastewater treatment plants (Weston, 1971).

Theoretically, it should be possible to recover alum by raising the pH according to the following reaction:

Table 9-1. Some Chemical Wastewater Sludge Handling Results (EPA, 1973)

	Ferric Chloride	Alum	Hydrated Lime
Approximate Chemical Dose, mg/l	120	150	460
Sludge Production			
Wastewater, sludge produced, lb/mil gal	700	700	700
Chemical, sludge produced, lb/mil gal	700	500	6,300
Allowable solids loading, on a gravity thickener, lb/day/ft^2	15	5	43
Underflow solids, from gravity thickening, % solids	30	15	200
Vacuum Filtration			
Conditional chemical for vacuum filtration, type	$Ca(OH)_2$	$Ca(OH)_2$	None
Conditioning chemical dose, % by wt	20	20	0
Filter yield, lb/hr/ft^2	1.2	0.8	10
Filter cake moisture content, % by wt	80	80 80	60

Note: lb/mil gal x 120 = kg/m^3
lb/day/ft^2 x 0.203 = kg/hr/m^3
lb/hr/ft^2 x 4.9 = kg/hr/ft^2

$$Al(OH)_3 + 3H^+ \longrightarrow Al^{+++} + 3H_2O$$

The problem here is with impurities and the buildup of color and iron compounds.

When alum is added to wastewater, the two main precipitates formed are aluminum phosphate and aluminum hydroxides/oxides. The calculation of sludge production thus requires the estimation of these two precipitates.

Example 9-1. Assume an alum dosage of 13 mg/l as Al^{+++} to achieve a 0.9 mg/l phosphorus residual for an incoming waste with 6.3 mg/l P concentration. Calculate the production of alum sludge.

Phosphorus removal = 6.3 - 0.9 = 5.4 mg/l.
The removal equation with gram molecular weights is

$$Al^{+++} + PO_4^{\equiv} \longrightarrow AlPO_4 \downarrow$$

$$(27) \quad (95, \text{ or } 31 \text{ as P}) \rightarrow (122)$$

or 31 mg of P results in the formation of 122 mg of $AlPO_4$.

$$5.4 \text{ mg/l P removed} \times \frac{122 \text{ mg } AlPO_4}{31 \text{ mg P}} = 21.3 \text{ mg/l}$$

Hence solids production due to P removal is 21.3 mg/l.
The amount of Al^{+++} used for the production of $AlPO_4$ is

$$5.4 \text{ mg/l P} \times \frac{27 \text{ mg } Al^{+++}}{31 \text{ mg P}} = 4.7 \text{ mg/l } Al^{+++}$$

Excess Al^{+++} or 13 - 4.7 = 8.3 mg/l must have gone into the formation of aluminum oxides/hydroxides. The simplest thing to assume is that

$$Al^{+++} + 3OH^- \longrightarrow Al(OH)_3 \downarrow$$

$$(27) \ (3 \times 17 = 51) \rightarrow (78)$$

and so the production of $Al(OH)_3$ is

$$8.3 \text{ mg/l } Al^{+++} \times \frac{78 \text{ mg } Al(OH)_3}{27 \text{ mg } Al^{+++}} = 24 \text{ mg/l}$$

The total solids production is therefore $[AlPO_4 + Al(OH)_3] = 21.3 + 24 \cong 45$ mg/l.

IRON SLUDGES

Both ferric and ferrous compounds can be used to precipitate impurities in wastewater. The sludges thus formed are surprisingly soft and fluffy and difficult to dewater to more than 10 or 12% solids. Such sludges still behave as liquids.

The recovery of iron from such operations is theoretically possible but not economical.

The production of iron sludge can be calculated as shown in the following example.

Example 9-2. Assume a dosage of 15 mg/l Fe^{+++} to achieve a 0.9 mg/l P level for a water having 6.3 mg/l P. Calculate the solids production.

The equations and gram molecular weights are

$$Fe^{+++} + PO_4^{\equiv} \longrightarrow FePO_4\downarrow$$

$$(56)\ (95, \text{ or } 31 \text{ as P}) \longrightarrow (151)$$

$$Fe^{+++} + 3(OH)^- \longrightarrow Fe(OH)_3\downarrow$$

$$(56)\ (51) \longrightarrow (107)$$

as in the alum example above, the solids due to the $FePO_4$ precipitate are

$$5.4 \text{ mg/l P removed} \times \frac{151 \text{ mg } FePO_4}{31 \text{ mg P}} = 26.3 \text{ mg/l}$$

The Fe^{+++} used in $FePO_4$ formation is 5.4 x 56/31 = 9.8 mg/l.

The excess Fe^{+++} is 15 - 9.8 = 5.2 mg/l,
and the Fe $(OH)_3$ formed is

$$5.2 \text{ mg/l } Fe^{+++} \times \frac{107 \text{ mg } Fe(OH)_3}{56 \text{ mg } Fe^{+++}} = 9.9 \text{ mg/l}$$

The total solids is therefore 26.3 + 9.9 ≅ 36 mg/l.

LIME SLUDGES

Lime, as CaO (quicklime) or $Ca(OH)_2$, can be used for removing many of the impurities in wastewater. By adding sufficient quantities of lime, the pH can be raised to about 11.5, and calcium carbonate, metal hydroxides and phosphates are precipitated. The phosphorous is precipitated as mostly calcium hydroxyapatite $Ca_5(OH)(PO_4)_3$. The small quantities of aluminum, magnesium and manganese oxides aid in the removal of silt and other impurities. Example 9-3 shows how sludge quantities with lime addition can be estimated. These calculations are a good first guess. Due to the complex nature of wastewaters the only acceptable means of estimating sludge production is by laboratory and pilot studies.

Example 9-3. Assume raw wastewater suspended solids = 250 mg/l, raw wastewater $PO_4^{\equiv}$ = 34.5 mg/l, the hardness of the raw wastewater is 150 mg/l and of the treated effluent is 185 mg/l, both as $CaCO_3$, and that all suspended solids are removed. Assume also that the lime reacts with phosphates to form only $Ca_5(OH)(PO_4)_3$ and the remainder precipitates as $CaCO_3$. For a lime [$Ca(OH)_2$] dose of 200 mg/l, assume that all phosphate is removed as $Ca_5(OH)(PO_4)_3$. The concentration of this precipitate is

$$\frac{\text{mol wt of } Ca_5\,(OH)\,(PO_4)_3}{\text{mol wt of } (PO_4)_3} \times PO_4 \text{ conc.} = \frac{502}{285} \times 34.5 = 61 \text{ mg/l}$$

Some of the lime contributed to the increased hardness in the effluent,

$$185 - 150 = 35 \text{ mg/l as } CaCO_3$$

The Ca lost in the effluent is

$$35 \times \frac{\text{mol wt of Ca (40)}}{\text{mol wt of } CaCO_3 \text{ (100)}} = 14 \text{ mg/l}$$

The Ca in $Ca_5(OH)(PO_4)_3$ is

$$61 \times \frac{200}{502} = 24 \text{ mg/l}$$

The total Ca added was

$$200 \times \frac{40}{74} = 108 \text{ mg/l}$$

The Ca in the sludge must therefore be

$$108 - (24 + 14) = 70 \text{ mg/l}$$

The $CaCO_3$ formed is

$$70 \times \frac{100}{40} = 175 \text{ mg/l}$$

Sludge composition:

(a) raw wastewater solids	=	250 mg/l
(b) $Ca_5\ (OH)\ (PO_4)_3$	=	61 mg/l
(c) $CaCO_3$	=	175 mg/l
		486 mg/l

which translates to 4,050 lb solids/mil gal of wastewater treated.

It is possible to recover lime from the sludge produced in lime precipitation. One method is to centrifuge the slurry at low solids recoveries so as to remove only the heavy $CaCO_3$. The magnesium hydroxide and other light solids are then dewatered in a second centrifuge (Figure 9-1). The $CaCO_3$ cake is then recalcined with the addition of heat according to the following reaction:

$$CaCO_3 \xrightarrow{\Delta} CaO + CO_2$$

The recalcination process is fairly simple because the cakes produced in centrifuges and vacuum filters are quite dry, with solids ranging from 40-50%. Lime sludges with a high pH, however, have proved difficult to dewater.

The recalcining process can be conducted in a rotary kiln or in a fluidized-bed furnace. In both cases, the CO_2 produced can be used to dissolve some of the hydroxides or for recarbonizing the finished water to bring the pH down.

Parker *et al.* (1973) showed that the addition of lime to the primary clarifier produced a thick sludge that centrifuges could easily classify. Approximately 90% of the calcium carbonate fed to the first centrifuge was recovered in the cake, while 50-75% of the other solids were rejected as the centrate. The calcium carbonate slurry was subsequently dewatered to 50% solids and incinerated. The classification curves, based on 20 equilibrium runs, are shown in Figure 9-2.

Verbaan (1975) confirmed most of the above findings and also showed that adding polymer increased the recovery of the non-$CaCO_3$ constituents, which is, of course, contradictory to classification.

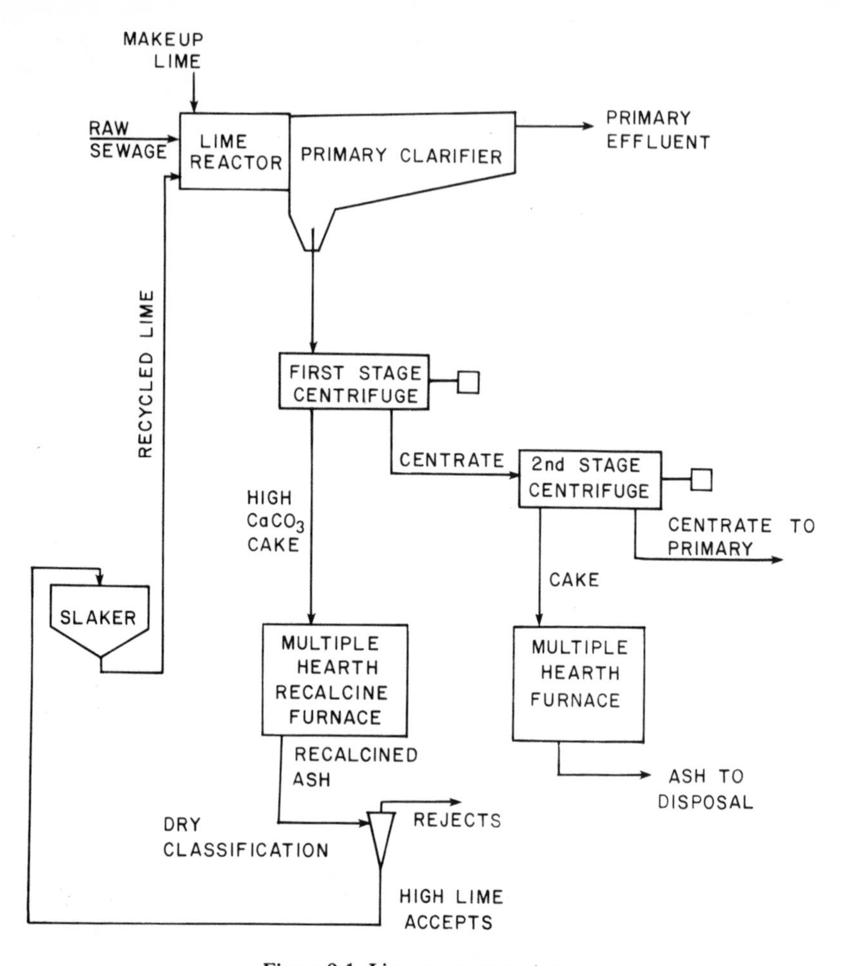

Figure 9-1. Lime recovery system.

The lime produced in recalcining is CaO, or quicklime, a dangerous compound. It is often slaked by adding water,

$$CaO + H_2O \longrightarrow Ca(OH)_2$$

and this hydrated lime is much safer to handle.

Quicklime can also be used in the dewatering or drying of biological sludge by mixing the lime and sludge in a common concrete mixer. The above reaction is exothermic, and thus the sludge is dried and disinfected as a result of the high temperatures produced (approaching 100°C). This process, used in

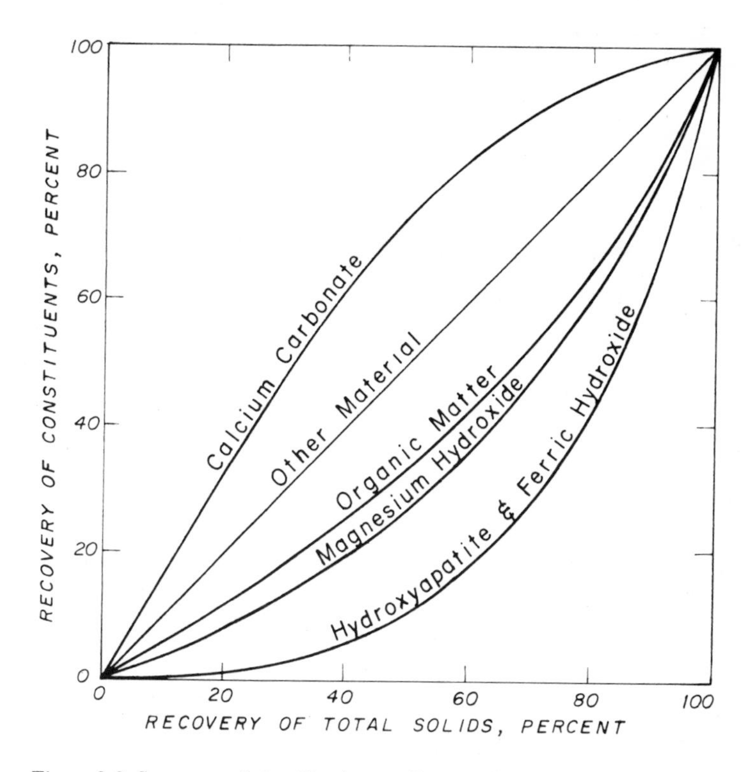

Figure 9-2. Summary of classification studies on a lime sludge (Parker, 1973).

in some European treatment plants, yields a product which is marketable as a soil conditioner, especially where the soil is acidic, lacks permeability and has poor water-holding capacity.

At pH levels above 11.0, lime-sludge settling and dewatering properties are greatly dependent on the concentration of magnesium. Figure 9-3, for example, shows the sludge solids concentration in a clarifier underflow as a function of magnesium (Ronen and Halbertal, 1975). The effect of magnesium on filterability is shown in Table 9-2 (Hawkins *et al,* 1974). As a point of reference, sludge filtration becomes quite difficult if the specific resistance exceeds about 10^{12} m/kg. Thus a magnesium ion concentration of greater than perhaps 25 mg/l will make the filtration of a lime sludge difficult.

FLUE-GAS DESULFURIZATION SLUDGES

A chemical sludge not usually considered a product of wastewater treatment is power plant flue-gas desulfurization sludge. By 1980 it is estimated

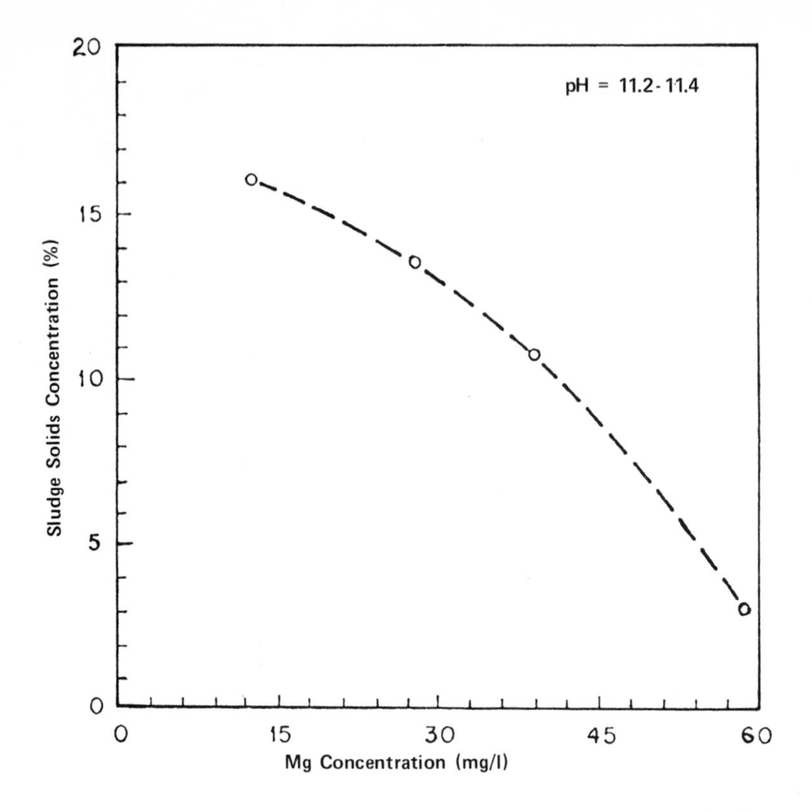

Figure 9-3. Underflow sludge solids on a lime sludge are strongly influenced by the magnesium concentration (Ronen and Halkertal, 1975).

Table 9-2. Effect of Mg^{++} on Lime Sludge Specific Resistance (Hawkins *et al,* 1974)

Mg^{++} (mg/l)	Specific Resistance (m/kg)
0	1.4 to 2.1 x 10^{11}
12	1.6 x 10^{11}
36	2.1 x 10^{12}

that electric utilities will be faced with the disposal of 100 million tons per year of wet scrubber sludge (Haas and Taylor, 1974). Twenty years of production at this rate will result in a disposal area of 144 square miles at a depth of 10 feet. The obvious solution to this problem is the proper utilization of this material. Unfortunately, the solid waste produced with lime scrubbing is mostly calcium sulfate;

$$CaCO_3 + heat \rightarrow CaO + CO_2$$

$$CaO + SO_2 + \frac{1}{2} O_2 \rightarrow CaSO_4$$

$$CaO + SO_3 \rightarrow CaSO_4$$

which has negligible market value. It is therefore discarded into the environment, often with thickening and lagooning as the concentration processes.

Other types of processes for flue-gas desulfurization are the magnesium oxide scrubber, double alkali, sodium carbonate and organic reagents. Some of these systems produce elemental sulfur as the end product, while others produce dilute sulfuric acid or hydrogen sulfide. The technology of flue-gas desulfurization is changing rapidly, mostly prompted by stringent government standards for SO_2 emissions.

METAL HYDROXIDES

The ordinary method of treating industrial wastes containing high concentrations of heavy metals is to increase the pH and precipitate out the resulting hydroxides/oxides. Figure 9-4 is a solubility diagram for chromium, and shows that the optimum pH for precipitating chromium hydroxide is about 8. These hydroxides are then settled and thickened prior to dewatering, reprocessing or ultimate disposal.

The thickening characteristics of metal hydroxide sludges depend substantially on a number of variables (Vesilind and Hayward, 1974) including

- pH of slurry
- time since original pH change
- method of pH change (how fast, and if the desired pH was attained by first increasing to a higher pH and then decreasing it)
- temperature
- stirring
- turbulence during mixing

All these variables can be controlled in the laboratory, but they would be difficult to translate into prototype design.

Some preliminary results of thickening tests with metal oxide slurries, conducted at about pH 9 without stirring and equal concentrations, are shown in Figure 9-5. Lead and chromium hydroxides seem to settle well, while nickel and zinc settle quite poorly. There is obviously a great variability in the settling rates among different metal hydroxides. Much work remains to be done in this area.

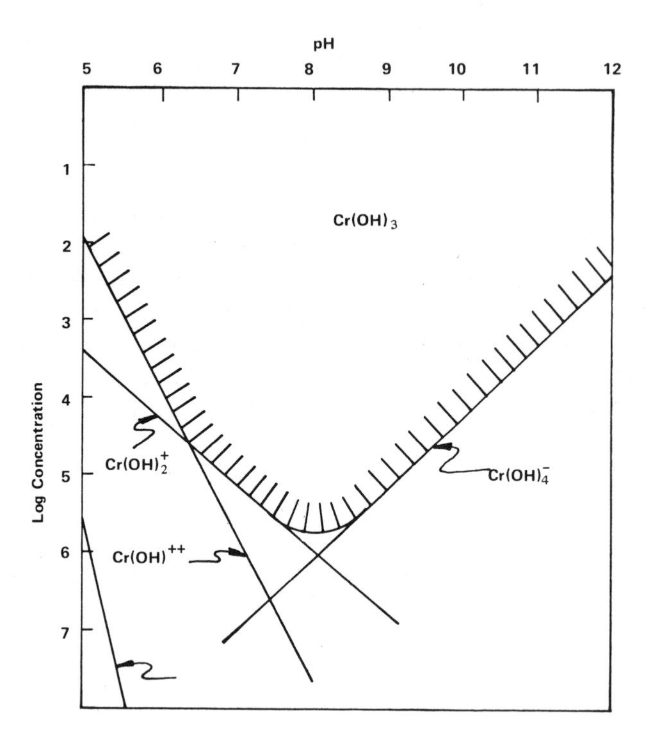

Figure 9-4. Solubility diagram for chromium.

ULTIMATE DISPOSAL OF INDUSTRIAL SLUDGES

In addition to chemical sludges formed in water and wastewater treatment many industries produce waste sludges that must be disposed of. Industrial sludges are often of little value as soil conditioners and, in fact, are often highly toxic. The ultimate disposal of such sludges has been a headache for years, and increasing pressure by government agencies and private groups has reduced disposal alternatives.

The origins of these sludges vary with the industries producing them. For example, metal-finishing plants often produce sludges high in zinc, chromium and other heavy metals. Such sludges cannot be disposed of in the ordinary manner, and are usually either lagooned or removed by a private contractor (final fate unknown).

For industries near the ocean, barging has been a much-used method and, in spite of governmental concern, will probably continue for some time. Economically, however, its usefulness is limited to coastal areas.

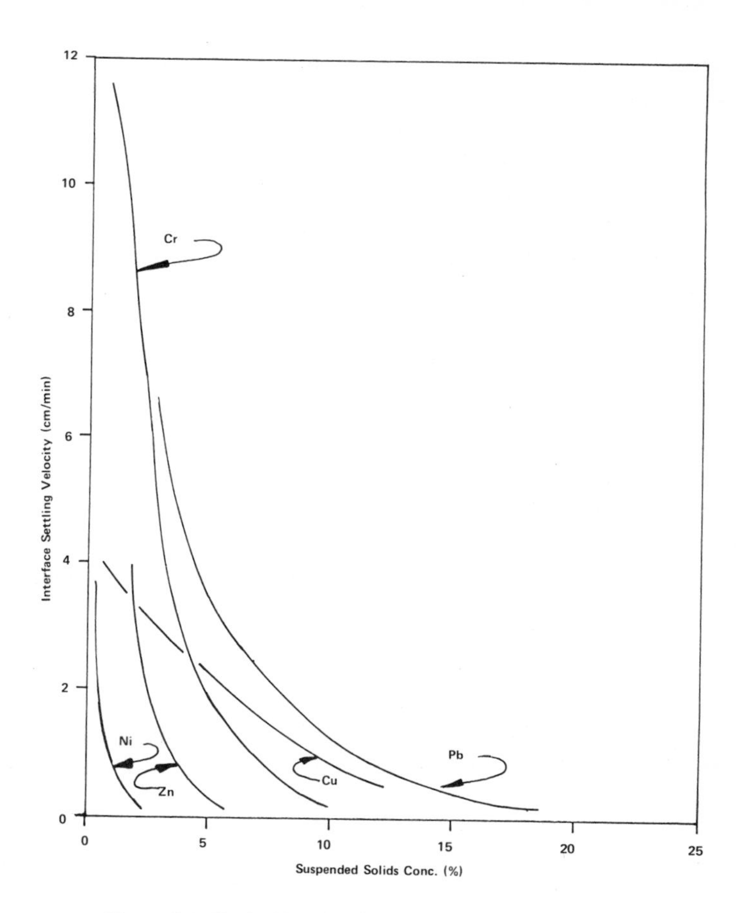

Figure 9-5. Typical batch-thickening data for hydroxides.

Liquid wastes have also been injected into deep wells, a method originally developed by the petroleum companies that pumped wastewaters back into empty wells from which gas or oil had been extracted, thus restoring the geological formations to equilibrium pressure. This has been viewed as truly ultimate disposal and is often the least expensive alternative. The disadvantage is twofold. First, the waste pumped into the ground can and probably will ultimately find its way back to the surface. Second, if contained below, it would be difficult to reach if this ever became necessary. Deep-well injection is thus viewed with considerable alarm by many engineers and environmentalists.

Some sludges can be dewatered by the methods described in Chapter 6 and thus disposed of in landfills if the liquid fraction can be disposed of as readily.

Unfortunately, many industrial sludges such as metal oxide slurries are difficult to dewater, and the dewatered cake seldom is transportable.

One interesting alternative, presently used by several large industrial firms, is chemical fixation (Conner, 1972). The process is nonthermal and operates at atmospheric pressure. A proprietary compound is mixed with the sludge and the mixture is pumped to a lagoon where, in a few days, the liquid gels to a solid. This solid has been shown to be stable under normal environmental conditions, with minimal leaching of metals into water.

CONCLUSIONS

With increased production of exotic industrial wastes, physical-chemical treatment alone or in tandem with biological treatment, will be widely used in the future. The sludges produced in such operations are usually difficult to handle but less decomposable than biological sludges. In addition, chemical sludges are much more predictable than biological sludges, and it should be possible to operate a chemical treatment plant to produce a sludge with specific characteristics. This is difficult to do in a biological plant.

Ultimate disposal of industrial sludges can be accomplished in many ways. One philosophy must be followed after a method is selected: The waste must be considered a potential resource as well as a toxin. Acceptance of this basic premise makes some methods of disposal unacceptable, since the waste cannot be readily recovered. Deep-well injection and ocean disposal into deep ocean bottoms are two methods that should be avoided. Containment and control until better disposal (or recovery) methods are available is far better than risking serious environmental damage or dispersing a potential raw material.

REFERENCES

Brown, J. C. (1973). "Alum Treatment of High Rate Trickling Filter Effluent at Chapel Hill, N.C.," in *Ultimate Disposal of Wastewaters and their Residuals,* (Raleigh, NC: Water Research Institute, North Carolina State Univ.).

Conner, J. R. (1972). "Ultimate Liquid Waste Disposal Methods," *Plant Eng.* (October 19).

Doe, P. W., D. Benn and L. R. Bays (1965). "Sludge Concentration by Freezing," *Water Sew. Works* 11:112.

Environmental Protection Agency (1973). Technology Transfer Seminar, Atlanta.

Gates, C. D., and R. F. McDermott (1968). "Characterization and Conditioning of Water Treatment Plant Sludge," *J. Am. Water Works Assoc.* 60:7.

Glenn, R. W., J. F. Judkins and J. M. Morgan (1973). "Filterability of Water Treatment Plant Sludges," *J. Am. Water Works Assoc.* 65:6.

Haas, J., and R. Taylor (1974). "Potential Uses of the Byproducts from Lime/Limestone Scrubbing of SO_2 From Flue Gases," Presented at annual meeting, AIME.

Hawkins, F. C., J. F. Judkins and J. M. Morgan (1974). "Water Treatment Sludge Filtration Studies," *J. Am. Water Works Assoc.* 60:11.

Inhoffer, W. R., and P. W. Doe (1973). "Design of Wastewater and Alum Sludge Disposal Facilities," *J. Am. Water Works Assoc.* 65:6.

Lamb, J. C. (1969). "Disposal of Wastes from Water Treatment Plants," Am. Water Works Assoc. Research Foundation Report.

Nielsen, H. L., K. E. Carns and J. N. DeBoice (1973). "Alum Sludge Thickening and Disposal," *J. Am. Water Works Assoc.* 65:6.

Parker, D. S., F. J. Zadick and K. E. Train (1973). "Sludge Processing for Combined Physical-Chemical-Biological Studies," Office of Research and Development EPA-R2-73-250, Washington, DC.

Ronen, M., and Y. Halbertal (1975). "Lime Treatment of Oxidation Pond Effluents in a Solid-Contact Type Reactor Clarifier," Mekorot Water Co., Tel Aviv, Israel.

Schwoyer, W. L., and L. B. Luttinger (1973). "Dewatering of Water-Plant Sludges," *J. Am. Water Works Assoc.* 65:6.

Verbaan, B. (1975). "Processing and Utilization of Chemical Sludges Derived from Wastewater Reclamation Water Treatment and Industry," NIWR Internal Report (Restricted), Pretoria, South Africa (as reported in Ronen, 1978).

Vesilind, P. A., and P. Hayward (1974). "Thickening of Metal Hydroxides," Presented at annual meeting, ACS, Chicago.

Weston, R. F. Inc. (1971). "Upgrading Existing Wastewater Treatment Plants," EPA Technology Transfer Contract 14-12-933.

PROBLEMS

1. A thickener is to process metal hydroxides to 10% solids as the underflow concentration. Using the data in Figure 9-5, calculate the relative thickener areas for lead, copper, chromium and zinc hydroxides.

2. If a coal contains 5% sulfur, how much limestone ($CaCO_3$) will theoretically be necessary to remove the sulfur from 1 tonne of coal? How many dry tonnes of $CaSO_4$ will be produced? If a power plant consumes 1,000 tonnes of coal per hour, how much $CaSO_4$ will be produced in a year, both in terms of volume and weight?

3. An industrial waste stream at 10 m^3/min contains 57 mg/l of phosphorus. This is to be treated with alum so as to achieve an effluent phosphorus of less than one mg/l. How much alum is needed and how much alum sludge will be produced?

4. Suppose lime is used as the only treatment for a raw domestic sewage. Estimate, using reasonable values, the lime requirement to achieve a phosphorus concentration of less than one mg/l, and estimate total solids production.
5. A wastewater requires 20 mg/l alum as Al^{+++} to treat a secondary effluent. The plant influent has 10 mg/l phosphorus and achieves 90% of removal. Estimate the quantity of alum sludge produced.

CHAPTER 10

ULTIMATE DISPOSAL IN THE MARINE ENVIRONMENT

The sludges produced from wastewater treatment must be disposed of in one of three places–in the sea, on the land or in the air. Disposal in space is a fourth alternative, but at present is too expensive.

Disposal in the air is in reality not ultimate disposal but rather temporary storage, because most waste will eventually settle to the earth. Even if strict air pollution controls are adhered to, residue still must be disposed of on the land or in the sea. Sludge combustion is thus only a means of volume reduction, not a method of ultimate disposal. Wastewater solids can thus be truly disposed of only on land or in the marine environment. This chapter is a discussion of ultimate disposal in large bodies of water.

EXPERIENCES IN SLUDGE DISPOSAL IN THE MARINE ENVIRONMENT

Coastal cities have been disposing of sludge in the sea for many years. Philadelphia, for example, has been dumping primary digested sludge by barge since 1961. The Hyperion Plant in Los Angeles pumps digested sludge seven miles into the sea. The city of Miami has been dumping sludge in the Miami Channel, which unfortunately passes close to the city of Palm Beach, north of Miami. Some of Miami's sludge ends up on Palm Beach's beaches, much to the disgust of tourists and resort owners. New York has been dumping about five million cubic yards of sludge per year for five decades into an area twelve miles offshore.

METHODS OF SLUDGE DISPOSAL IN THE OCEAN

The two principal methods of moving sludge to the disposal site are by barge or by pipeline. The barges may be self-propelled or, as is the case in some European countries, converted oil tankers. The common method of sludge transport, however, is towed barge. The use of long pipelines became popular a few years ago, especially on the west coast of the United States. They owe their popularity in part to a few consulting engineers and pipe manufacturers who pressed for the construction of these long submerged pipelines over the continental shelf and into the deep parts of the ocean. Since the deep parts are so far removed from land, they believe that filling them up with sludge will have little environmental impact. What is not considered, however, is that deep waters have a very fragile ecology, and any change in their environment such as the introduction of sludge, can have a disastrous effect on marine life. Such long pipelines should therefore be avoided at all costs. If sludge disposal in the sea is necessary, the disposal site should have strong currents to provide rapid dilution and adsorption into the marine environment.

ANALYSIS OF OUTFALLS

Most receiving waters must satisfy quality standards. These are usually in terms of such things as coliforms, turbidity, grease and floating objects. Although diffusers minimize the problem of local high concentrations of pollutants, potential degradation of water must nevertheless be estimated.

The important variables in the analysis of outfall performance are waste decay, initial dilution and waste dispersion (Burchett *et al.*, 1967).

Waste Decay

The pollution of coastal waters is most often judged by the concentration of coliforms. The coliform dieoff has been estimated using the equation

$$\frac{dC}{dt} = -k_b C$$

where C = concentration of coliforms
t = time
k_b = dieoff constant

The constant k_b is difficult to evaluate, and a conservative estimate is to let $k_b = 0$, assuming no dieoff. Integrated,

$$C = C_0 e^{-k_b t} \qquad 10\text{-}1$$

where C is the bacterial concentration at time t and C_0 is the initial concentration. The value of k_b depends on factors such as salinity, temperature, etc.

Kehr and Butterfield (1943) found that k_b = 2 $days^{-1}$ at 25°C, 1.0 $days^{-1}$ at 15°C and 0.5 $days^{-1}$ at 5°C in nonsaline water. Experiences with California outfalls, however, indicate that the dieoff in saline water is probably much greater. Camp (1963) has calculated that for secondary sludge, k_b would be about 4 $days^{-1}$, and for primary sludges, k_b would be between 7 and 16 $days^{-1}$.

Initial Dilution

A projection of the diffuser pipes perpendicular to the current direction is a measure of the initial *sewage field*. The concentration of coliforms at the diffusers can be approximated by

$$C_0 = \frac{V_x bh}{Q}$$

where h is the average thickness of the sewage field, Q is the rate of sewage discharge, b is the length of the diffusers and V_x is the velocity of the current (Figure 10-1). The depth h is, of course, a function of thermal stratification and the depth of the outfall.

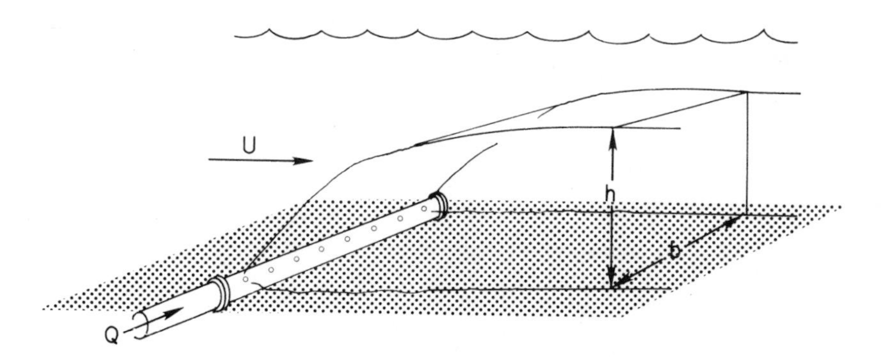

Figure 10-1. Definition of terms for dilution of sewage from a diffuser.

Waste Dispersion

There does not seem to be a clearly superior method of calculating the extent of the diffusion. A number of methods have been proposed but the one suggested by Brooks (1960) appears to be quite practical.

According to Brooks, the schematic of the sewage field can be represented as in Figure 10-2. In this figure, Z is the width of the field as it is expanding

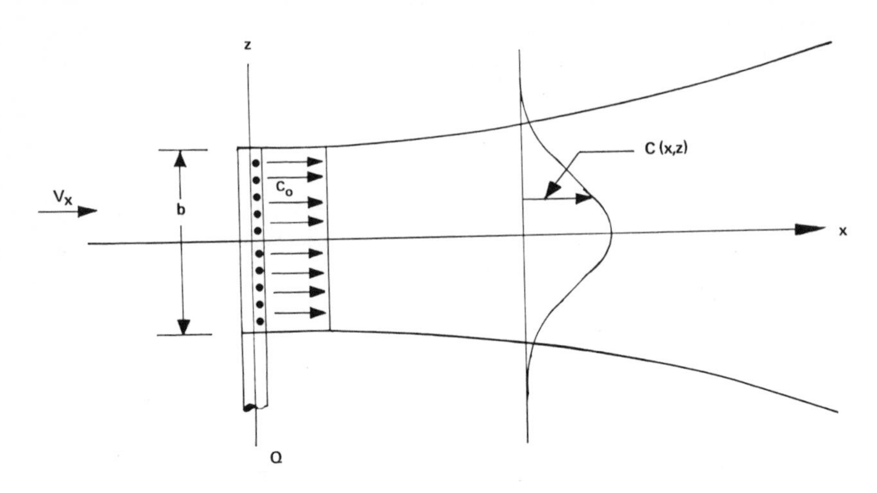

Figure 10-2. Definition sketch of a sewage plume produced by a submerged diffuser.

from the initial width of the diffuser b. Assuming no decay of coliforms (no dieoff), the basic differential equation describing steady-state conditions as assumed by Brooks is

$$\frac{\partial}{\partial z} - \epsilon \frac{\partial C}{\partial z} + V_x \frac{\partial C}{\partial x} = 0$$

where ϵ = eddy diffusivity, m^2/sec
C = concentration of coliforms at any distance or time
V_x = current and plume velocity, m/sec

From experiments by Pearson (1956) and Gunnerson (1958), Brooks has concluded that the eddy diffusivity varies with the 4/3 power of Z, or

$$\epsilon = 0.01\ Z^{4/3}$$

and by extrapolation,

$$\epsilon_0 = 0.01\ b^{4/3}$$

where ϵ_0 is the initial diffusivity at the discharge point.

Brooks has solved the differential equation for the maximum concentration at the plume centerline (z = 0) as

$$C = C_0 \text{ erf} \left[\frac{1.5}{\left(1 + \frac{8 \epsilon_0 x}{V_x b^2}\right)^3 - 1} \right]^{1/2}$$

where C = concentration of coliforms at the centerline, distance x downstream, coliforms/100 ml
C_0 = concentration of coliforms after the initial concentration, coliforms/100 ml
x = distance downstream, m
b = width of diffuser, m
ϵ_0 = initial diffusivity, m^2/sec
V_x = velocity of current, m/sec
erf = error function, defined as

$$\text{erf } X = \frac{2}{\sqrt{\pi}} \int_0^X e^{-v^2} dV$$

The coliform organisms injected into the ocean water die, however, and thus the concentration decreases by dieoff as well as by dilution. Incorporating the dieoff of coliforms into the diffusion equation, Brooks suggested that at $z = 0$,

$$C = C_0 \cdot 10^{-k_b t} \text{ erf} \left[\frac{1.5}{\left(1 + \frac{8 \epsilon_0 x}{V_x b^2}\right)^3 - 1} \right]^{1/2}$$

The solution of this equation is rather cumbersome, and a nomograph (Figure 10-3) can be used to facilitate solution.

Example 10-1: A diffuser produces a plume with an initial coliform concentration of C_0 of 1×10^6 per 100 ml. The diffuser length, b, is 30 m, and the current velocity, V_x, is 0.1 m/sec. What would be the maximum coliform concentration at the shore 915 m away if the current is directly onshore?

Using Figure 10-2, which is in English units,

x = 915 x 3.28 = 3,000 ft
b = 30 x 3.28 = 98.4 ft
V_x = 0.1 x 3.28 x 60 = 19.68 ft/min

Entering Figure 10-3 at V_x = 19.68 ft/min and connecting x = 4,920 ft, transcribes a point on the turning line. Connecting b = 98.4 ft and the point

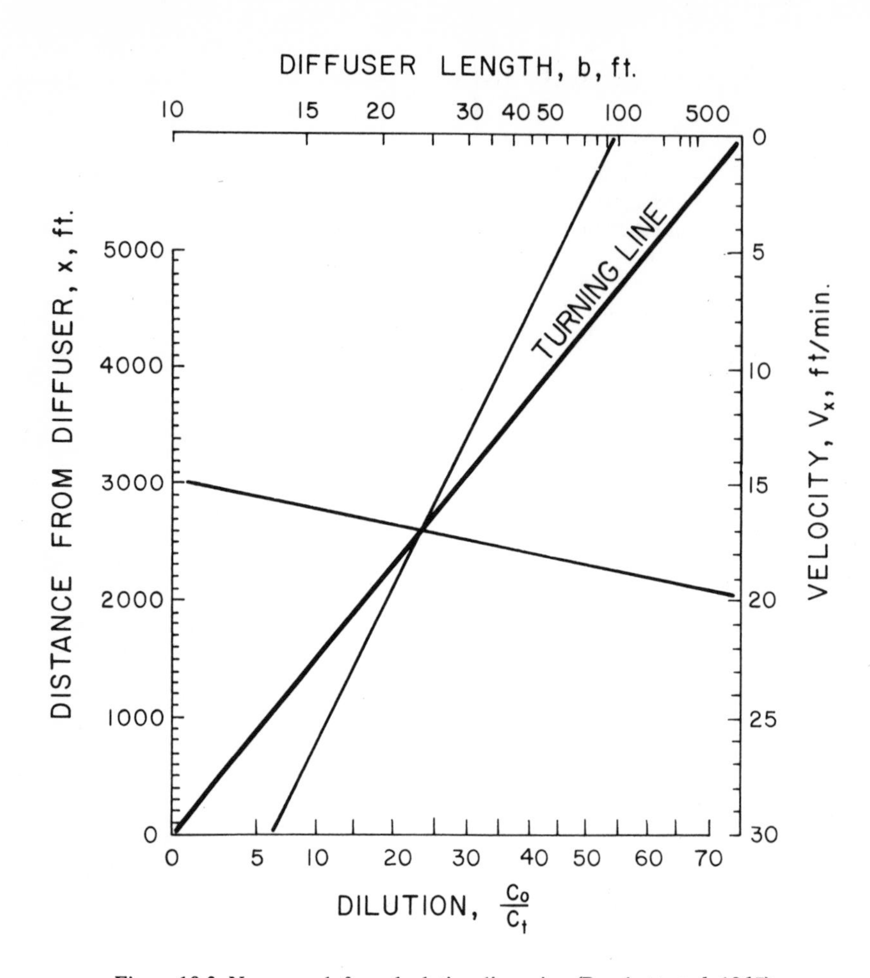

Figure 10-3. Nomograph for calculating dispersion (Burchett *et al.* 1967).

on the turning line, we find $C_0/C = 7$. Thus the maximum concentration of coliforms on the shore can be expected to be

$$C = \frac{C_0}{7} = \frac{1 \text{x} 10^6}{7} = 143{,}000 \text{ coliforms/100 ml}$$

In addition to dilution, the coliforms will die off. The time to reach shore is

$$t = \frac{3{,}000}{19.68} = 152 \text{ min} = 0.105 \text{ days}$$

and from the previous discussion,

$$k_b = 7\ \text{day}^{-1}$$

Hence

$$10^{-k_b t} = 10^{-7 \times 0.105} = 0.184$$

The maximum expected concentration of coliforms on the beach is thus

$$143{,}000 \times 0.184 = 26{,}000\ \text{coliforms/100 ml}$$

Alternatively, an approach suggested by Pomeroy (1960) is to estimate the coliform bacteria concentration in the surf zone by

$$N = \frac{KQ^2}{YL^2}$$

where N = number of bacteria found at the shore station of highest pollution in terms of the arithmetic average or the 80 percentile.
Q = average sewage flow, m^3/day
Y = depth of discharge, m
L = distance from end of the outfall to the shore, m
K = constant. This varies with topography and oceanographic conditions. A value of 0.01 has been found applicable for several outfalls off the California coast.

Hydraulic Design of Outfalls

Practical consideration limits the velocity in the pipeline to between 2 and 10 ft/sec (0.6 and 3 m/sec) (Burchett *et al.* 1967). The diffuser should be placed perpendicular to the prevailing current. When the currents are not predominant in any one direction, a Y-shaped arrangement is used. Usually 10-15 ft (3-4.5 m) of diffuser length is provided per million gallons per day of design flow. Port diameters can range between 9 and 3 in. (23 and 8 cm) with spacing of about 10 ft (3 m).

Rawn *et al.* (1960) described a method for the hydraulic design of diffusers. The rate of discharge, q, from an orifice or port in the side of a pipe is

$$q = C_D a \sqrt{2gE}$$

where C_D = discharge coefficient
a = area of port
E = total head in the pipe

If the port furthest from the shore is numbered 1, the flow through this port is

$$q_1 = C_D \, a_1 \sqrt{2g \, E_1}$$

$$= C_D \frac{\pi}{4} d_1{}^2 \sqrt{2g \, E_1}$$

where d = port diameter

The velocity in the pipe immediately preceding the port is

$$V_1 = \frac{q_1}{\frac{\pi}{4} D^2}$$

where D = diameter of pipe

If the pressure difference between the inside and outside of the pipe immediately upstream of the first port is denoted by h_1, the total head at the first port is

$$E_1 = h_1 + \frac{V_1{}^2}{2g}$$

The discharge coefficient can be approximated from Figure 10-4 (McNown and Hsu, 1951). For the first port with a sharp edge, $C_D = 0.61$.

It is necessary to select E_1 and then calculate q_1 and V_1. Proceeding to port 2, the total head is calculated as

$$E_2 = E_1 + h_f + \frac{\Delta s}{s} \Delta z_1$$

where h_f = head loss due to friction between the two ports
s = specific gravity of the fluid in the pipe
Δs = difference in specific gravities between the seawater and the fluid
Δz = difference in elevation

The term $(\Delta s/s) \Delta z$ can be called the *density head*. Since the pertinent pressure at any point in the diffuser is the pressure differential between the fluid inside the diffuser and seawater outside at the level of the port, a decrease in depth of the ports along the diffuser (Δz) increases the pressure differential between the fluid inside and the seawater outside the diffuser (Rawn *et al.*, 1960). Seawater specific gravities range from 1.01 to 1.03.

Continuing to the next port

$$q_2 = C_D \, a_2 \sqrt{2gE_2}$$

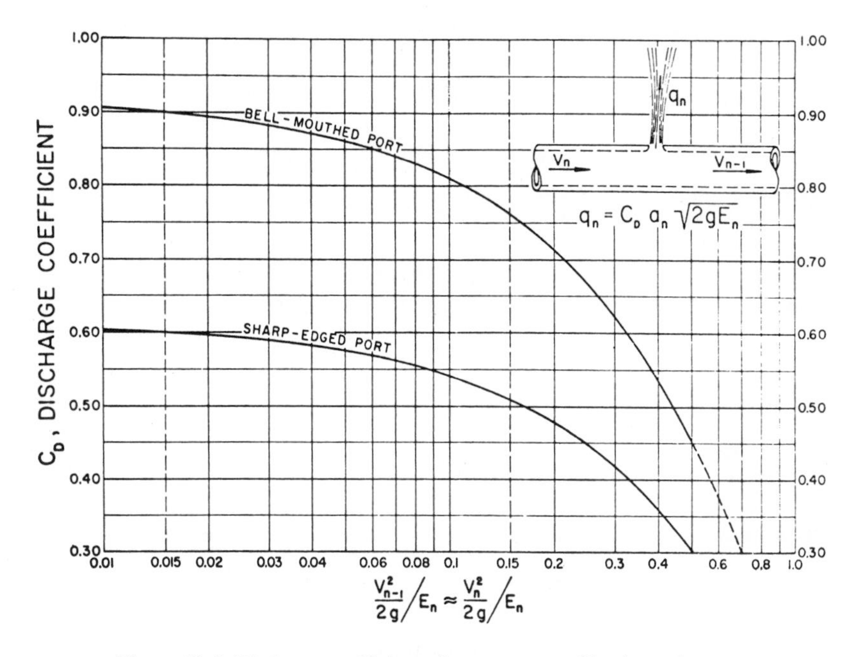

Figure 10-4. Discharge coefficients for two types of horizontal ports (McNown and Hsu, 1951).

and the velocity in the pipe is

$$V_2 = V_1 + \frac{q_2}{\frac{\pi}{4} D^2}$$

This procedure, continued step by step to the last port, is obviously cumbersome and demands a computer-aided solution. The first step is illustrated below.

Example 10-2: A 2-ft diameter diffuser is to discharge 10 mgd. The sharp-edged diffuser ports, one on each side of the pipe, are spaced at 10-ft intervals and have a diameter of 6 in. We need to find the length of the required diffuser and of the head loss within the diffuser. Assume the following: E_1 (total head at first port, farthest from the shore) = 100 ft, Δz_n (the change in elevation between any two points) = 0.5 ft, and $\Delta s/s$ (difference in specific gravities/specific gravity of the sludge) = 0.025. The flow through the first port is

$$q_1 = C_D a_1 \sqrt{2g\,E_1}$$

$$= C_D \frac{2\pi}{4} d_1{}^2 \sqrt{2g\,E_1}$$

$$= \frac{(0.61)(2)(3.14)(0.5)^2}{4} \sqrt{2(32.2)100}$$

$$= 1.92 \text{ cfs}$$

The velocity within the pipe between ports 1 and 2 is

$$V_1 = \frac{q_1}{\frac{\pi}{4} D^2} = \frac{1.92}{\frac{3.14(2)^2}{4}}$$

$$= 0.611 \text{ ft/sec}$$

Proceeding to port 2,

$$E_2 = E_1 + h_{f_1} + \frac{\Delta s}{s}(\Delta z)_1$$

The head loss due to friction is calculated as

$$h_{f1} = f \frac{L_1}{D} \frac{V_1{}^2}{2g}$$

where L_1 = length of pipe = 10 ft
f = 0.022 (from a Moody diagram of Darcy-Weisbach friction factor vs Reynolds Number)

$$h_{f_1} = 0.022 \frac{10}{2} \frac{(0.611)^2}{(2)(32.2)} = 6.4 \times 10^{-5} \text{ ft}$$

$$E_2 = 100 + 0.000064 + 0.025\ (0.5)$$

$$\cong 100$$

The ratio $(V_1{}^2/2g)/E_2 = 0.0005$, hence $C_D = 0.61$ from Figure 10-4. The flow from the second ports is then

$$q_2 = C_D\, a_2 \sqrt{2g\,E_2}$$

$$q_2 = 1.92 \text{ cfs}$$

and

$$V_2 = V_1 + \Delta V = V_1 + \frac{q_2}{\frac{\pi}{4} D^2}$$

$$= 0.611 + \frac{1.92}{\frac{3.14\,(2)^2}{4}}$$

$$= 1.22 \text{ ft/sec}$$

This procedure is then continued until the summation of the flows from individual ports equals 10 mgd. The total head at that port minus E_1 equals the head loss within the diffuser.

ADVERSE EFFECTS OF OCEAN DISPOSAL

In coastal waters, stratification may prevent the mixing of sludge within the main body of water. If barge disposal is used, the wind-mixed surface layer may entrap the low-density sludge plume and carry it intact for long distances, and perhaps back to shore. Floating materials (oil, grease, other sewage artifacts) swept back to the shore discourage recreational fishing and swimming.

Large slugs of sludge can seriously deplete the oxygen within the plume (Segar and Berberian, 1977), with subsequent loss of marine life.

The accumulation of benthic deposits by sludge dumping is probably the most serious environmental effect of ocean disposal. Although data are scarce, there seems little doubt that sludge deposits in all dumping areas have had some effect on the benthic ecosystems (National Academy of Sciences, 1978).

In the New York Bight, for example, where 50 years of sludge dumping has resulted in a 50-km^2 area of dead sea, with the bottom covered by an essentially anaerobic black ooze (Lerner and Wood, 1971), deep currents move the mass up and down the Hudson Submarine Canyon. Lear (1976), found that in Philadelphia, where barge dumping has been going on only since 1973, a shift in benthic populations has occurred, with more resistant species such as nematodes becoming more numerous. Metals in the sludge seem to be taken up by the shellfish. Accumulations of sludge on the continental shelf may cause disease in marine organisms such as shell erosion (Young and Pierce, 1975) and fin erosion (Mahoney *et al.*, 1973).

CONCLUSIONS

Around the turn of the century, a disagreement of principle developed between public health (sanitary) engineers and public health physicians as to the proper method of water treatment. The physicians wanted to treat every sanitary discharge into any watercourse and thus have water fit to drink flowing in all streams. The logic was that in that case water filtration plants were not necessary.

Prominent engineers such as Whipple, Hazer, Eddy and others argued that discharges should be treated only to the extent that they not pose a health hazard and that the assimilative capacity of watercourses be utilized to their fullest. Drinking water, they argued, should always be treated, and the presence of residual pollution would not impair this treatment.

Obviously, cold engineering logic prevailed in this case, mainly because the engineers' solution was the less expensive of the two alternatives.

In retrospect, however, we are no longer totally sure as to the wisdom of this approach. Should streams be used as extensions of wastewater treatment plants? Is there social good in clear lakes and rivers, or must good be always measured in monetary terms?

Such questions apply equally well to the present arguments on ocean disposal. In the absence of a demonstrable detrimental effect, should ocean disposal of sludge not be allowed? Or is the absence of detrimental effect merely a reflection of our inability to measure the right parameters? And indeed, so *what* if there is a gooey sludge mousse in the bottom of New York Bight?

What this boils down to is a question of ethics and social values. Ocean disposal of sludge is a prime example of the conflict of such values. Land disposal of sludge, the subject of the next chapter, is another unresolved problem requiring similar value judgments.

REFERENCES

Brooks, N. H. (1960). "Diffusion of Sewage Effluent in an Ocean Current," in *Waste Disposal in the Marine Environment*, E. A. Pearson, Ed. (New York: Pergamon Press, Inc.).

Burchett, M. E., G. Pchobanaglous and A. J. Burdoin. (1967). "A Practical Approach to Submarine Outfall Calculations," *Public Works* 98:5.

Camp, T. R. (1963). *Water and Its Impurities* (New York: Van Nostrand Reinhold Company).

Gunnerson, C. G. (1958). "Sewage Disposal in Santa Monica Bay," *San. Eng. Div. ASCE* 84:SA1.

Kehr, R. W., and C. T. Butterfield. (1943). "Notes on the Relation between Coliforms and Enteric Pathogens," in *Public Health Reports* (Washington, D.C.: U.S. Public Health Service).

Lear, D. W. (1976). Testimony on Committee on Merchant Marine and Fisheries, U.S. Congress, as reported in National Academy of Sciences (1978).

Lerner, J., and R. T. Wood. (1971). "Solid Waste Problems; Proposals and Progress in the Tri-State Region," Tech. Report 4251-2430, Tri-State Regional Planning Commission, New York.

Mahoney, J. B., F. H. Midlige and D. C. Devel. (1973). "A Fin Rot Disease of Marine and Euryhaline Fishes in the New York Bight," *Trans. Am. Fish. Soc.* 102:3, 596.

McNown, J. S., and E. Hsu. "Application of Conformal Mapping to Divided Flow," *Proc. Conf. on Fluid Dynamics*, Ann Arbor, MI (from Rawn *et al.* 1960).

National Academy of Sciences. (1978). *Multimedium Management of Municipal Sludge, Vol. IX* (Washington, D.C.: Commission on Natural Resources, Natural Resources Council).

Pearson, E. A. (1956). "An Investigation of the Efficacy of Submarine Outfall Disposal of Sewage and Sludge," State Water Poll. Contr. Board Pub. No. 14 (Sacramento, CA).

Pomeroy, R. (1960). "The Empirical Approach for Determining the Required Length of an Ocean Outfall," in *Waste Disposal in the Marine Environment*, E. A. Pearson, Ed. (New York: Pergamon Press, Inc.).

Rawn, A. M., F. R. Bowerman and N. H. Brooks. (1960). "Diffusers for Disposal of Sewage in Sea Water," *J. San. Eng. Div., ASCE* 86:SA2.

Segar, D. A., and G. A. Berberian. (1977). "Oxygen Depletion in the New York Bight Apex; Causes and Consequences," *Proc. Am. Soc. of Limnology and Oceanography*,Vol. 2 (Lawrence, KA: Allen Press).

Young, J. S., and J. B. Pierce. (1975). "Shell Disease in Crabs and Lobsters from New York Bight," *Mar. Poll. Bull.* 6(7):101.

PROBLEMS

1. An ocean outfall with no ports (just end of pipe) emits 100 m^3/min of effluent. The current is uniform and steady at 0.5 m/sec. Estimate the dilution 1 km downcurrent.

2. A wastewater containing 1 x 10^6 coliforms per 100 ml is discharged into the ocean at a rate of 25 m^3/min, at a depth of 30 m from a 100-m-long diffuser. Estimate the coliform count in the surf zone.

3. A 1 m in diameter diffuser lays flat on the ocean floor so that the static head on the end of the pipe is 30 m. The sharp-edged ports, each 5 cm in diameter, are spaced at 3-m intervals. If the $\Delta s/s = 0.025$, find the required length of diffuser.

4. A 100-m-long diffuser is discharging 200 m^3/min of raw primary sludge into the ocean. The diffuser is 1,500 m from shore, at a depth of 40 m.

The current is onshore at 4 m/min. Estimate the coliform concentration at the surf, using Figure 10-3. Assume the coliform concentration of raw sludge as 100×10^6 coliforms/100 ml.

CHAPTER 11

ULTIMATE DISPOSAL ON LAND

> Paris casts twenty-five millions of francs annually into the sea; an approximate amount given by the estimates of modern science. Science knows now that the most fertilizing and effective manures is the human manure . . . .Do you know what these piles of ordure are, those carts of mud carried off at night from the streets, the frightful barrels of the nightman, and the fetid streams of subterranean mud which the pavement conceals from you? All this is a flowering field, it is green grass, it is the mint and thyme and sage, it is game, it is cattle, it is the satisfied lowing of heavy kine, it is perfumed hay, it is gilded wheat, it is bread on your table, it is warm blood in your veins.–Victor Hugo in *Les Miserables*

One does not usually think of sludge in such emotional terms. Yet the concept of land disposal of human excrement as an integral part of our total ecology and benefit to mankind has been known for many centuries. People in China and other eastern countries have used human excrement as the major source of fertilizer since the beginnings of their recorded history. Western civilizations have not needed to be quite so inventive, and in Europe the use of human wastes as a benefit in farming dates back only to the middle of the last century.

In the mid-1800s Edinborough, Scotland, installed one of the earliest citywide drainage systems. The effluent ran into a covered stream called Foul Burn and then through farmland to the sea. The farmers along the way began to experiment with this effluent and soon discovered its benefits. In fact, the rights to the sewage came into dispute and had to be settled in court (Ridgeway, 1970).

As cities grew and population densities increased, sludge disposal became an increasingly serious problem. Successes with land disposal of wastes prompted the first Royal Commission on Sewage Disposal (1857-1865) to conclude that "The right way to dispose of town sewage is to apply it continuously to land" (Benarde, 1973). This advice was heeded, and sewage treatment plants in England are still called "sewage farms."

In the United States, disposal of sewage effluent or digested sludge on farmland has not been used widely due partly to the availability of inexpensive and convenient inorganic fertilizer. The usual method of disposing wastewater sludge in the U.S. used to be to haul it by truck to some hole in the ground and forget it. Only recently have some towns experimented with spreading sludge on land and using the soil as a sludge assimilator.

In this chapter, the disposal of "stabilized" (see Chapter 3) sludge is discussed first, followed by descriptions of two alternative land disposal techniques–chemical fixation and deep-well injection–which do not require prior stabilization.

BENEFITS OF SLUDGE DISPOSAL ON LAND

The benefits of sludge addition to soil is graphically illustrated in Figure 11-1, where the growth of marigolds is plotted against the fraction of sludge in the soil. The three curves are typical of numerous similar test results with other plants. In all three test series the best results were obtained with about 70% (by weight) or 30% (by volume) of sludge. This emphasizes the point that sludge is a soil additive and fertilizer, and not, in itself, a good soil for growing crops.

Data such as the corn and potatoes results shown in Table 11-1 have been widely published. There is little doubt as to the benefits accrued from sludge addition to farmland.

Sludge is especially beneficial in improving very poor soils, such as stripmine spoils. Lejcher and Kunkle (1973) showed that sludge application on stripmine spoils improved the physical conditions for germination and growth after 18 months. The pH of the soil, with an application of 135 dry tonnes/ha, increased from 2.3 to 6.3.

METHODS OF LAND DISPOSAL

Sludge can be spread on land in a number of ways, the simplest being driving a tank truck over a field and allowing the sludge to dribble out the back. This, however, causes operational problems since the sludge does not cover the field evenly and the truck is often required to drive over previously applied sludge. A number of variations have evolved which lessen these problems. A "T-Bar," for example, is a simple pipe manifold for spreading the sludge out in a wider swath. Similarly, a splash plate on the end of the exit pipe spreads out the sludge. As the hydraulic head decreases during the application run, the driver reduces speed so as to more evenly apply the sludge.

A major evolution of trucks for liquid sludge disposal are big-wheeled (for traction) subsod disposal vehicles. The pressurized injection system

Table 11-1. Corn and Potato Yield Response to Applications of Sewage Sludge on a Hubbard Sandy Loam (Larson *et al.*, 1974)

Corn		Potatoes	
Sludge Application Rate (tonne/ha)	Grain Yield (kg/ha)	Sludge Application Rate (tonne/ha)	Potato Yield (kg/ha)
0	727	0	18,500
81	10,100	112	44,200
159	10,600	225	53,100
320	11,600	450	66,600

Note: Municipal sludge from sand drying bed; 24% solids, 1.3% N, 1.8% P.

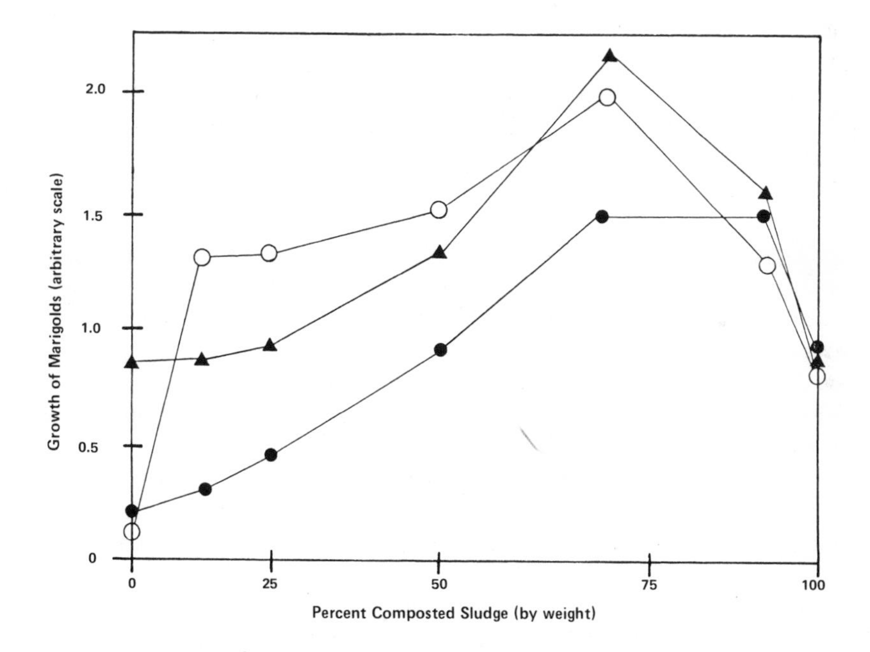

Figure 11-1. Composted sludge has significant agricultural value (Walker, 1975).

eliminates the problems of uneven application and tracking sludge back to the highway.

Another variation is pressurized spraying from the side of the truck. This has been modified in several places to pressurized spraying through irrigation lines, thus freeing the truck to make more trips to the plant and increasing its efficiency. These "big gun" sprinklers have nozzle diameters of about 2 cm (0.75 in.) and can be stationary or travelling.

If dewatered (not liquid) sludge is applied, the method usually used is spreading by a tractor and disking in the sludge. Alternatively, dewatered sludge can be transported by truck, reslurried at the site, and the liquid sludge applied as usual.

POTENTIAL PROBLEMS WITH LAND DISPOSAL

Although significant benefits can be accrued by disposing of sludge on land, some potential problems must be considered. The adverse environmental effects which sludge disposal can cause can be classified as:

- offense to senses (mainly odor)
- transmittal of toxins (pathogens or chemicals) into water supplies
- detrimental effect on the growth of crops
- transfer of pathogens or the translocation of chemicals to crops which are consumed by animals or people

The first point is discussed in Chapter 3 under stabilization, and little else need be added here.

The transfer of undesirable chemicals or organisms into water supplies depends mainly on the location of the disposal site and the topography and geology of the ground. When sludge disposal sites are planned, soil analyses and subsurface geology must be considered. In areas where limestone is prevalent, for example, fissures can form which could carry liquids long distances, essentially untreated or unfiltered.

At sufficiently high application rates sludges can cause damage to crops. This damage can be as suppression of seed germination, stunting of growth or toxicity.

Problems with the inhibition of seed germination are usually due to ammonia in the sludge. Aeration of digested sludge will destroy this inhibitory effect, as will spreading the sludge on the ground a few days before seeding (Hinsley and Sosewitz, 1969).

There are many materials that might cause stunted growth or other symptoms of toxicity. Of greatest concern seem to be potentially toxic elements such as heavy metals which enter municipal sludges from both industrial wastes and domestic sources. In some cases, the contribution of toxins from nonindustrial sources has exceeded 50%. Some of the chemicals most likely to cause problems are zinc, copper and nickel, with cobalt and boron as two other potential plant toxins.

Tables 11-2 and 11-3 are two collections of data on heavy metal concentration in sludges. Two points should be made: (1) there is a great variability in metal concentrations, and (2) most of the high metal ion sludges are from

Table 11-2. Typical Metals in Municipal Sludges (mg/kg dry sludge) (EPA, 1976)

	Range	Mean	Median
Ag	nd[a] - 960	225	90
As	10 - 50	9	8
B	200 - 1,430	430	350
Ba	nd - 3,000	1,460	1,300
Be	nd	nd	nd
Cd	nd - 1,100	87	20
Co	nd - 800	350	100
Cr	22 - 30,000	1,800	600
Cu	45 - 16,030	1,250	700
Hg	0.1 - 89	7	4
Mn	100 - 8,800	1,190	400
Ni	nd - 2,800	410	100
Pb	80 - 26,000	1,940	600
Sr	nd - 2,230	440	150
Se	10 - 180	26	20
V	nd - 2,100	510	400
Zn	51 - 28,360	3,483	1,800

[a]nd = not detected.

Table 11-3. Metal Content of Digested Municipal Sludges (Chaney and Giordano, 1976).

Element (ppm)	Purely Domestic[a]	Controlled Municipal[b]	Observed Maximum
Zn	750	2,500	50,000
Cu	250	1,000	17,000
Ni	25	25	8,000
Cd	5	1,000	3,410
Pb	150	10	10,000
Hg	2	10	100
Cr	50	1,000	30,000

[a]Observed in sludges from newer suburban communities.

[b]Typical of sludges from communities without excessive industrial waste sources or with adequate source abatement.

industrial areas. Pretreatment of industrial wastes can result in substantial cuts in heavy metal concentrations in large industrial cities. EPA's estimates for the potential of pretreatment in one community are shown in Table 11-4.

The second danger in the application of high levels of such toxins on land is that these toxins can be translocated to crops or to animals by ingestion, and then to humans. Similarly, pathogenic organisms can cling to food crops and be transmitted to the public.

The types of materials in sludge which can be of public health concern are heavy metals, organic toxins such as PCBs and nitrogen.

Heavy Metals

The uptake of various heavy metals by potatoes, lettuce and corn is shown in Table 11-5. Note that lead and boron are poorly translocated. Cadmium seems to be picked up only by lettuce, as is zinc. Such data point out an important variable in evaluating the risks of applying sludge–the type of plant grown must be considered in deciding the maximum application rate. To date such standards have not been proposed.

The susceptibility of various plants to damage by heavy metals is shown in Table 11-6. Again, the crop to be grown should be considered when setting the application rate. In fact, it is at times feasible to plant a less-tolerant species alongside the main crop and use this as an inexpensive early warning system. Should the monitors show distress, thorough tests on the main crop would be necessary.

Hansen and Tjell (1973) have correctly pointed out that standards for the sludge application rate must take into account the total body burden of potentially toxic materials. For example, the World Health Organization recommends the maximum daily intake of lead and cadmium at 450 μg and 50 to 70 μg, respectively. In developed countries, the level of lead intake is already at 200 μg and cadmium at 30 μg per day. Of this intake fully 73% of the lead and 60% of the cadmium comes from foods. Any substantial increase in the levels of these two toxins in agricultural soil could easily result in excessive total intake.

It is also true, however, that the input of toxins to land is from routes other than sludge. For example, Hansen and Tjell have calculated the balance of lead and cadmium in soil and found that a substantial fraction of these two toxins comes from the atmosphere. The annual increase of Pb and Cd in Danish soils is 0.6% and 0.55%, respectively. If sludge is applied to farmland, the accumulation rate jumps to 3.7% and 5.8%. These increases are clearly intolerable; if sludge utilization on land is to be continued in the future, the removal of toxins prior to disposal will be necessary.

Table 11-4. Effect of Industrial Pretreatment on the Concentration of Metals in Buffalo, NY, Sludge (EPA, 1977)

Metal	Percent Reduction in Metals	Metal Concentration (ppm)	
		Before	After
Cd	50.0	100	50
Cr	59.1	2,540	1,040
Cu	79.0	1,570	330
Pb	66.4	1,800	605
Ni	63.5	315	115
Zn	84.0	2,275	364

Table 11-5. Uptake of Heavy Metals by Potatoes, Lettuce and Corn (Larson *et al.*,1974)

Sludge Applied (tonne/ha)	Elements (ppm)					
	Mn	Zn	Cu	Cd	Pb	B
			Potatoes (tubers)			
0	9	24	8.6	0.12	0.4	6
112	6	31	12.8	0.11	0.4	7
225	7	41	15.8	0.21	0.4	6
450	7	53	19.0	0.23	0.6	6
			Lettuce (leaf)			
0	133	21	1.6	0.61	1.1	23
112	84	94	5.4	1.28	1.4	27
225	77	155	8.1	1.72	0.7	28
450	79	225	11.9	2.67	0.8	28
			Corn (grain)			
0	12	41	2.0	0.02	0.2	3
112	8	47	2.9	0.02	0.2	3
225	9	48	0.5	0.03	0.2	3

Note: Sludge added contained 1,070 ppm Zn, 245 ppm Cu, 545 ppm Mn, 515 ppm Pb, 64 ppm Cr, 24 ppm Ni, 7.4 ppm Cd and 14 ppm B.

The Special Case of Cadmium

Cadmium is a highly toxic element which plants readily pick up from the soil, but which seldom has a toxic effect on the plants. Cadmium accumulation in vegetation seldom shows any sign of damage.

Several incidents involving cadmium poisoning have occurred (Flick, 1971; Kobayashi, 1971). The episodes in Japan gave cadmium poisoning its

Table 11-6. Relative Tolerance to Metal Toxicity (Chaney, 1974)

Very Sensitive:	beet crops (chard, sugarbeet, red beet); kale, mustard; turnip; tomato
Sensitive:	beans; cabbage; collards; other vegetables
Moderately Tolerant:	corn; small grains; soybean
Tolerant:	most grasses (fescue, ryegrass, Bermuda grass)
Very Tolerant:	ecotypes of grasses

common name—the Itai-Itai disease—from the Japanese equivalent of "ouch-ouch," since people suffering from chronic cadmium poisoning are extremely sensitive to touch. These tragic experiences, and the cumulative effect of cadium in the body, have prompted great concern and conservative guidelines.

The source of cadmium in a sludge is generally presumed to be from industrial sources. Surprisingly, however, a substantial fraction of Cd comes from the water supply (Martin, 1977). Table 11-7 lists some approximate fractions of Cd as found in Pittsburgh sludges. The conclusion from these studies is that a simple prohibition of Cd discharges by industry will not solve the Cd problem and that effective removal of Cd at water and wastewater treatment plants may be necessary.

Table 11-7. Source of Cadmium in Pittsburgh's Wastewater (Martin, 1977)

Source of Cadmium	kg/day	%
Industry	1.1	46.6
Urban Runoff	0.01	0.4
Waste Oil	0.01	0.4
Domestic Waste	0.04	1.7
Water Supply	1.0	42.4
Infiltration	0.2	8.5
		100.0

Other Toxins

Little is known on the fate of such chemicals as polychlorinated biphenyls (PCBs) and chlorinated hydrocarbon pesticides when sludge is applied on land. This is a potentially serious problem, since some sludges have been reported to have PCB levels of up to 450 mg/kg (National Academy of Sciences, 1978). Commonly, PCB levels are between 3 and 30, and 0.3 and 2.2 mg/kg, respectively (Farrell and Saletto, 1973). Because of the extreme toxicity of PCBs (PCBs have been shown to cause reproductive failures, skin lesions and liver cancer), and its potential biomagnification (magnifications as high as 2.7×10^{14} have been observed) (Massachusetts Audubon Society, 1976), EPA has suggested a water quality standard for PCBs of one part per trillion (0.000001 ppm). This standard would effectively eliminate all ocean disposal of sludges.

Nitrogen

A major concern with the application of sludge on land is nitrogen management. Nitrogen in sludge exists as either organic N, ammonia N or nitrate N. The first two are not of public health concern, although ammonia can inhibit seed germination. High concentration of nitrates in drinking water, however, can cause a disease in infants known as "blue babies," or methemoglobinemia, in which the nitrates are reduced to nitrites and inhibit the transfer of oxygen. In essence, infants suffocate due to a lack of oxygen, thus the common name for this disease.

Pathogenic Organisms

The problem of disease transmission has been a significant factor in preventing wider application of this sludge disposal method. It was thought for many years that anaerobic digestion effectively destroys all pathogens; unfortunately, this is not ture.

Viruses and bacteria seem to cling to solids and are concentrated in the sludge removal process (Lund 1970; Lecre *et al.*, 1970). This is especially true when phosphorus removal is practiced by alum addition. The flocs formed sweep out almost all of the viruses in the water (Scott and Horlings, 1974). In one study (Subrahmanyan 1974) many enteric (animal viruses were isolated from sludge, including poliovirus, coxsackie and hepatitis. The various strains of coxsackie are responsible for a number of illnesses ranging from the common cold to pericarditis, meningitis and encephalitis. Hepatitis virus A (infectious hepatitis) and B (serum hepatitis) at present cannot be isolated, but as a detection technique becomes available this will no doubt become a major parameter.

Secondly, although some bacteria such as coliforms are destroyed during digestion, *Salmonella typhosa*, for example, can survive in large numbers (Lecre *et al.*, 1970), and 20 days of digestion will destroy only 92% of *S. typhosa*. Helminth ova require at least a 30-day detention time for destruction (Chang, 1965). Poliovirus seem to survive digestion but other viruses seem to be less hardy (Foster and Engelbrecht, 1973).

Although there is no epidemiological evidence to suggest that the use of digested sludge on farmland can cause disease in humans or animals, health authorities remain skeptical. Once on the soil, pathogens can contaminate edible crops or water from runoff and percolation. Some pathogens such as *Shigella* and *Salmonella* only survive about a week, but others such as *Ascaris* ova can survive longer than 35 days (Rudolfs *et al.*, 1951). Tubercle bacilli have been known to survive more than two years in nature (Muller, 1959).

In some countries, raw sewage is used to irrigate farmland. There the problems of disease are understandably acute. Freytag (1967) concluded that although *E. coli* disappeared 14 days after sewage was placed on a vegetable garden, the operation is still questionable from the health standpoint.

STANDARDS FOR APPLICATION OF SLUDGE ON LAND

The rates of sludge application are usually expressed in terms of tons of dry solids per acre (or tonnes/ha) either on an annual basis or as the total lifetime load for the site. The maximum application rates for a given disposal location depend on a number of factors, including:

1. soil
2. crops
3. topography
4. potential for water contamination
5. weather
6. potential for odor production
7. method of application

The characteristics of the sludge include its concentration of

1. pathogenic organisms
2. heavy metals
3. nutrients (primary nitrogen)
4. other toxins

The objective of land application standards is to incorporate these variables into a fair and enforceable statement.

The major concern at present seems to be with the seeping of toxins and nitrates into ground- or surface waters. Accordingly, the maximum allowable application rates are calculated on the basis of either heavy metals or nitrogen, and the lowest of the two is used in practice.

It should be emphasized that there is little agreement on allowable sludge application rates.

Garrigan (1977) has shown, for example, that the maximum allowable annual sludge application rate for a typical sludge and application can vary from 0.6 to 20 ton/ac, depending on whom one believes. The lifetime application can vary from a measly 9 to a whopping 408 ton/ac. This variation illustrates the sorry state of the art in evaluating the environmental impact of sludge application onto land.

Standards Based on Heavy Metals

There are at this writing no universally accepted standards for application based on heavy metals. A number of researchers and governmental agencies have made suggestions, but it will take a long time to resolve this issue.

Chumbly (1971) has proposed that no application of sludge be greater than 250 ppm zinc equivalent for soils with pH > 6.5. He defines zinc equivalent as

$$\text{Zn Equivalent (ppm)} = [\text{Zn}] + 2[\text{Cu}] + 8[\text{Ni}]$$

where the concentrations of Zn, Cu and Ni are in ppm or mg/kg. The underlying logic is that copper seems to be twice as toxic to plants as zinc, and nickel about eight times as toxic. The restriction on soil pH is because at a pH below 6.5, most metals are dissolved and are thus available to the plants. At pH > 6.5, most of the metals appear as oxides and hydroxides. All standards for heavy metals suggested to date carry the restriction of pH > 6.5. If the pH of the soil (plus applied sludge) is less than 6.5, it must first be increased by lime addition before sludge is applied.

The zinc equivalent seems to be a reasonable first guideline for sludge application but since it ignores other metals and soil conditions, it is an obviously inadequate standard. Leeper (1972) suggested that any single toxic metal not exceed 5% of the cation exchange capacity. The cation exchange capacity (CEC) of a soil is its ability to tie up cations (such as heavy metals). A low CEC, as measured in the laboratory by ammonium ion adsorption, suggests that metal ions will be available to plants for uptake and for leaching into the groundwater.

Chaney (1974) recommends a maximum allowable concentration of various metals. These are listed in Table 11-8. The restriction on cadmium is listed as 15 mg/kg or 1% zinc because plants will preferentially pick up zinc to cadmium. If there is sufficient zinc, cadmium will not be readily absorbed. With too much zinc, the plant will begin to suffer damage. Without zinc, however, cadmium will be absorbed and since Cd is not as harmful to plants as it is to animals (including people), its presence cannot be easily detected.

The maximum allowable sludge application rate for soils with a pH $\geqslant 6.5$, was suggested by Chaney (1974) as

$$\text{Total Sludge Application (dry ton/ac)} = \frac{16{,}300\ (\text{CEC})}{[\text{Zn}] + 2[\text{Cu}] + 8[\text{Ni}] - 300}$$

Note that the denominator is the "zinc equivalent" as defined above. The 300 is an adjustment to compensate for the extra cation exchange capacity added by the organics in the sludge itself. The factor in the numerator is a conversion factor, which reflects the basic premise that the zinc equivalent should not exceed 5% of the CEC.

Wisconsin has adopted a similar approach, but the toxic elements addition is limited to 10% of the soil CEC instead of the 5% in the above formula, Wisconsin's recommended application rate can be calculated as

$$\text{Total Sludge Application (dry ton/ac)} = \frac{32{,}500\ (\text{CEC})}{[\text{Zn}] + 2[\text{Cu}] + 4[\text{Ni}]}$$

where the heavy metal concentrations are in $\mu g/g$ (or mg/kg).

The CEC is difficult to measure in the laboratory, and there is as yet no standard procedure. The procedure used by agricultural scientists is the ammonium acetate method (Chapman, 1965). For a simplified procedure, acceptable for rough analyses and student use, see Vesilind (1978). CEC values for various soils are shown in Table 11-9.

Table 11-8. Suggested Maximum Heavy Metal Concentrations in Sludge (Chaney, 1974)

Metal	Maximum Concentration (mg/kg)
Zn	2,000
Cu	1,000
Ni	200
Cd	15 or 1% of Zn
Pb	1,000
Hg	10
Cr	1,000
B	100

Table 11-9. Cation Exchange Capacity of a Wide Variety of Surface Soils from Various Parts of the United States (meq/100 g of dry soil)

Soil Type	Exchange Capacity	Soil Type	Exchange Capacity
Sand		Silt Loam	
Sassafras (NJ)	2.0	Delta (MI)	9.4
Plainfield (WI)	3.5	Fayette (MN)	12.6
Sandy Loam		Spencer (WI)	14.0
Greenville (AL)	2.3	Dawes (NE)	18.4
Sassafras (NJ)	2.7	Carrington (MN)	18.4
Norfolk (AL)	3.0	Penn (NJ)	19.8
Cecil (SC)	5.5	Miami (WI)	23.2
Coltneck (NJ)	9.9	Grundy (IL)	26.3
Colma (CA)	17.1	Clay and Clay Loam	
Loam		Cecil clay loam (AL)	4.0
Sassafras (NJ)	7.5	Cecil clay (AL)	4.8
Hoosic (NJ)	11.4	Coyuco sandy clay (CA)	20.2
Dover (NJ)	14.0	Gleason clay loam (CA)	31.5
Collington (NJ)	15.9	Susquehanna clay (AL)	34.2
		Sweeney clay (CA)	57.5

Example 11-1. A sludge analysis shows the following heavy metal concentrations: Zn = 500 mg/kg, Cu = 600 mg/kg, Ni = 50 mg/kg. This sludge is to be applied to a silty loam with a CEC of 20. On the basis of heavy metals (only) what is the maximum application rate?

$$\text{Total Sludge Application (dry ton/ac)} = \frac{16{,}300\,(20)}{(500) + 2(600) + 8(50) - 300}$$

$$\cong 60 \text{ ton/ac}$$

The answer to the above example is considerably higher than common application rates. The reason for this is that for most domestic sludges, heavy metals are not the limiting factor in application rates, as discussed later.

Maximum loadings of heavy metals have been suggested by the USDA and tentatively adopted by the EPA. These are shown in Table 11-10.

Chaney (1974), based on the guidelines of cadmium levels not exceeding 1% zinc, suggests that the maximum cadmium concentration would be 2 kg/ha for the life of the disposal site, or 0.1 kg/ha/yr.

Ontario guidelines work out to about 1.6 kg/ha for the life of the area (Garrigan, 1977).

Table 11-10. Total Amount of Sludge Metals Allowed on Agricultural Land (EPA, 1977)

	Soil Cation Exchange Capacity (meq/100 g)		
	0-5	5-15	>15
Metal	Amount of Metal (kg/ha)		
Pb	500	1,000	2,000
Zn	250	500	1,000
Cu	125	250	500
Ni	50	100	200
Cd	5	10	20

Note: Soil pH >6.5.

EPA has suggested that although annual cadmium loading levels of 1 to 2 kg/ha have not been shown to be harmful, the concern over potential effects will prompt a move toward more strict control of adding cadmium to croplands (EPA, 1977).

Toward that end, the proposed EPA regulations for cadmium show a progressively restrictive standard. To the end of 1981, the annual allowable cadmium application is 2 kg/ha, which drops to 1.25 kg/ha until the end of 1985, and then to 0.5 kg/ha. The maximum lifetime application is 5 kg/ha for soils with a CEC less than 5 meq/100 g, 10 kg/ha for soils with a CEC between 5 and 15 meq/100 g, and 20 kg/ha for soils with CEC exceeding 15 meq/100 g. A further restriction is that sludge containing cadmium in excess of 25 mg/kg should not be applied to sites where tobacco, leafy vegetables or root crops are grown. (*Federal Register*, 1978).

The question of permissible cadmium loadings is yet to be decided.

Standards Based on Nitrogen

As noted above, nitrogen is considered a potential pollutant because of the possible seepage of excess nitrate into the groundwater. The solution to this problem is to add exactly the nitrogen from sludge that is removed by the plants, plus whatever is volatilized into the atmosphere or is lost due to denitrification, or considered an allowable concentration in the groundwater.

The proposed procedure for estimating maximum allowable sludge application rates requires the calculation of a nitrogen balance on the proposed plot. The balance can be expressed nitrogen applied (in sludge) = (ammonia volatilized) + (nitrogen to groundwater) + (nitrogen loss in harvesting crops) + (denitrification)

The intent is to minimize nitrogen loss to groundwater. This equation ignores the fixation of nitrogen from the air and any other addition and assumes a steady-state condition.

The plant-available nitrogen includes any mineralized (inorganic) nitrogen in the soil, the inorganic fraction of nitrogen in the sludge, and the nitrogen mineralized during the growing season (about 15 to 20% of organic nitrogen during the first season, and 3% for the subsequent seasons). (Larson *et al.*, 1974). The release of inorganic nitrogen from organic nitrogen as estimated by the EPA is shown in Table 11-11.

Table 11-11. Release of Residual Nitrogen During Sludge Decomposition in Soil (EPA, 1977)

Years After Sludge Application	Organic N Content of Sludge (% of dry solids)						
	2.0	2.5	3.0	3.5	4.0	4.5	5.0
	lb N Released per ton Sludge Added						
1	1.0	1.2	1.4	1.7	1.9	2.2	2.4
2	0.9	1.2	1.4	1.6	1.8	2.1	2.3
3	0.0	1.1	1.3	1.5	1.7	2.0	2.2

Volatilization of ammonia varies with crops, temperature, soil, etc. Experience has shown that as much as 50% of ammonia nitrogen can be volatilized if the sludge is not immediately incorporated into the soil (EPA, 1977).

Denitrification, $NO_3^- \rightarrow N_2\uparrow$, will reduce available nitrogen. The rate of this process is more rapid than nitrification and can be enhanced by proper application techniques.

Some typical values of various forms of nitrogen are shown in Table 11-12, and the utilization of nitrogen by various crops is listed in Table 11-13.

Example 11-12. If corn is to be grown in a sandy loam soil, estimate the application rate of a sludge which has the following nitrogen analysis: organic N = 20,000 ppm, ammonia N = 1,500 ppm, nitrite N = 5 ppm, nitrate N = 50 ppm.

From Table 11-13, the nitrogen uptake is about 200 lb/ac (224 kg/ha). Assume the nitrogen mineralized during the first year is 20% of the organic nitrogen and that the volatilized fraction is 50% of the ammonia nitrogen. Also assume that no N flows into the groundwater, and no denitrification occurs. The balance is

$$N_{\text{applied}} = N_{\text{volatilized}} + N_{\text{to groundwater}} + N_{\text{used by plants}} + N_{\text{denitrification}}$$

We need to calculate N_{applied} as kg/ha.

$N_{\text{volatilized}}$ depends on the ammonia nitrogen. Since 20% of the organic is mineralized (assume to ammonia) the total available ammonia N is 1,500 ± 0.2(20,000) = 5,500 ppm, and the volatilized ammonia is 0.5(5,500) = 2,250 ppm. If the total application of sludge is X kg/ha, the total N volatilized is 2.25×10^{-3} (X) kg/ha.

The N to groundwater and loss due to nitrification are zero.

The N used by the corn is 224 kg/ha.

The total N applied is the product of dry solids application rate, X (kg/ha) times the fraction of total nitrogen. The latter is the sum of all forms of nitrogen, or 21,555 ppm, or 22.155×10^{-3} (X) kg/ac.

Table 11-12. Concentrations of Carbon and Nitrogen in Sewage Sludge on Dry Solids Basis (Sommers, 1977)

	Sample				
Component	Type[a]	Number	Range	Median	Mean
Organic C, %	Anaerobic	31	18 - 39	26.8	27.6
	Aerobic	10	27 - 37	29.5	31.7
	Other	60	6.5 - 48	32.5	32.5
	All	101	6.5 - 48	30.4	31.0
Total N, %	Anaerobic	85	0.5 - 17.6	4.2	5.0
	Aerobic	38	0.5 - 7.6	4.8	4.9
	Other	68	<0.1 - 10.0	1.8	1.9
	All	191	<0.1 - 17.6	3.3	3.9
NH_4-N, ppm	Anaerobic	67	120 - 67,600	1,600	9,400
	Aerobic	33	30 - 11,300	400	950
	Other	3	5 - 12,500	80	4,200
	All	103	5 - 67,600	920	6,540
NO_3-N, ppm	Anaerobic	35	2 - 4,900	79	520
	Aerobic	8	7 - 830	180	300
	Other	3	---	---	780
	All	45	2 - 4,900	140	490

[a]"Other" includes lagooned, primary, tertiary and unspecified sludges.
"All" signifies data for all types of sludges.

$$22.155 \times 10^{-3} (X) = 2.25 \times 10^{-3} (X) + 0 + 224 + 0$$
$$\text{and } X = 11253 \text{ kg/ha}$$
$$= 11.2 \text{ tonne/ha}$$

Note that the solids loading calculated above is considerably lower than the permitted loadings based on heavy metals.

Table 11-13. Annual Nitrogen, Phosphorus and Potassium Utilization by Crops (EPA, 1977)

Crop	Yield	Nitrogen	Phosphorus	Potassium
			lb/ac	
Corn	150 bu	185	35	178
	180 bu	240	44	199
Corn Silage	32 ton	200	35	203
Soybeans	50 bu	257[b]	21	100
	60 bu	336[b]	29	120
Grain Sorghum	8,000 lb	250	40	166
Wheat	60 bu	125	22	91
	80 bu	186	24	134
Oats	100 bu	150	24	125
Barley	100 bu	150	24	125
Alfalfa	8 ton	450[b]	35	398
Orchard Grass	6 ton	300	44	311
Brome Grass	5 ton	166	29	211
Tall Fescue	3.5 ton	135	29	154
Bluegrass	3 ton	200	24	149

[a]Values above are for the total aboveground portion of the plants. Where only grain is removed from the field, a significant proportion of the nutrients is left in the residues. However, since most of these nutrients are temporarily tied up in the residues, they are not readily available for crop use. Therefore, for the purpose of estimating nutrient requirements for any particular crop year, complete crop removal can be assumed.

[b]Legumes get most of their nitrogen from the air, so additional nitrogen sources are not normally needed.

A strong word of caution is necessary here. The mechanics of nitrogen use and application in agricultural land are not well understood. Farmers, for example, apply two or three times the necessary fertilizer to their crops without apparent harm. Most of this fertilizer is inorganic and has the greatest immediate potential for groundwater pollution. It thus seems unlikely that the meager application rates as calculated above are of much value, except for comparative purposes.

Guidelines on Topography, Permeability and Proximity of Watercourses

Surface water contamination is prevented by applying sludge only to fairly level surfaces far removed from flowing watercourses and areas not susceptible to flooding. The maximum allowable slope is a function of soil permeability. Bell (1971) suggests that the minimum acceptable soil permeability is 10^{-5} cm/sec (0.015 in./hr). Land with slopes greater than 3% should not be used if the soil permeability is less than 10^{-4} cm/sec (0.15 in./hr). Suggested guidelines for acceptable distances from watercourses are shown in Table 11-14.

Table 11-14. Maximum Sustained Slope vs Minimum Distance to Watercourses (Bell, 1971)

	Minimum Distance to Watercourse	
Maximum Sustained Slope	For Sludge Application During May to Nov., Inclusive	For Sludge Application During Dec. to Apr., Inclusive
0 to 3%	200 ft	600 ft
3 to 6%	400 ft	600 ft
6 to 9%	600 ft	No sludge to be applied
Greater than 9%	No sludge to be applied unless special conditions exist	No sludge to be applied

European Standards for Sludge Application

Sludge application guidelines developed in other parts of the world are surprisingly similar to U.S. values. For example, in Finland the annual maximum sludge application is 4 tonne/ha (Hansen and Tjell, 1978). Norway allows 50 tonne/ha of sludge for the lifetime of the site, and 10 tonne/ha for 5 yr. Sweden has stricter standards; 1 tonne/ha annual application, with repeated applications strongly discouraged. In Denmark, the rate is variable depending on the nitrogen balance and Cd concentration, similar to the U.S. guidelines. In all Scandinavian countries the maximum allowable levels of heavy metals are also specified. The allowable concentration of lead in the sludge, for example, is 1,200 mg/kg in Denmark and Finland, and 300 mg/kg in Norway and Sweden. The cadmium level is 30 mg/kg in Denmark and Finland, and 15 mg/kg in Norway and Sweden.

The concentration of heavy metals in sludges in other parts of the developed world does not differ greatly. For example, the metals in sludge from Scandinavia are shown in Table 11-15.

In the Netherlands, farmers are advised to use only sludges containing no more than 2,000 mg/l Zn, 500 mg/l Cu, 50 mg/l Ni and 10 mg/l Cd on a dry weight basis, and to apply not more than 2 tonne/yr/ha on arable land and no more than 1 tonne/yr/ha for grassland. In all, no more than 2,000 tonne/ha should be used on arable land and 100 tonne/ha on grassland (de Haan, 1975).

IS LAND DISPOSAL FEASIBLE?

Counterbalancing all the disadvantages to sludge disposal on land is an impressive list of advantages. One important advantage is that this is truly a final disposal–the sludge returns to the environment. Not only is it final but it is also beneficial–something not often achieved in pollution control.

Only limited capital investment is required to enable the sludge to be piped or trucked to farmland, and operating expenses can be quite low, especially if wet sludge is moved by pipeline. Disposal of wet sludge has an additional disadvantage because supernatant would not have to be returned to the plant where it might create an unnecessary load and result in operational problems. Lyman *et al.* (1972) proposed that the ultimate disposal of sludge should fulfill four requirements:

1. It should not pollute air or water.
2. It should be economical.
3. It should conserve organic matter for beneficial purposes.
4. It should provide a permanent solution to the disposal problems of treatment plants.

Although this is a demanding list, sludge disposal on farmland meets all these requirements. It can be stated without much qualification that, before a disposal method is selected in all cases, sludge disposal on land should be seriously considered and its costs compared to those incurred with other methods.

DISPOSAL OF SLUDGE INTO LANDFILLS

Local municipalities and state regulatory agencies seem to have a fear of sludge disposal in landfills. Most states require special permission to do this, and many insist on monitoring wells and other environmental safeguards.

This concern does not seem reasonable in most cases. Municipal refuse, as delivered to the landfill, is about 20% moisture. Its field capacity (the

Table 11-15. Metal Characterization of Municipal Sludges in Scandinavia (mg/kg) (Hansen and Tjell, 1978)

Metal mg/kg	Denmark		Finland		Norway	Sweden	
	Range	Median	Range	Median	Range	Range	Median
Cr	25 - 3,434	36	20 - 3,700	46	20 - 1,640	6 - 67,000	80
Mn			60 - 1,500	350		36 - 5,100	
Co	3 - 49	5	8 - 95	23		2 - 180	
Ni	16 - 320	20	18 - 1,300	52	10 - 550	9 - 3,700	51
Cu	151 - 2,234	241	50 - 3,000	160	230 - 2,180	27 - 5,300	560
Zn	1,316 - 16,418	1,731	210 - 5,300	920	290 - 2,920	90 - 18,000	1,567
Cd	5 - 55	7	1.6 - 710	6	2 - 37	1 - 350	7
Hg	3 - 33	5	0.3 - 16	3	1 - 14	0 - 110	
Pb	251 - 3,709	314	55 - 1,600	150	44 - 639	18 - 5,100	180

[a]The number of treatment plants investigated in Norway was relatively few. Therefore, only the range is given.

moisture content at which water can no longer be held and seepage begins) is about 60% moisture. There is, in short, a large reserve of moisture-holding capacity in refuse; as a result it does not seem reasonable to require sludge dewatering before disposal in a landfill.

In fact, the addition of sludge could have a beneficial effect on the anaerobic microbial decomposition occurring within the landfill. An accelerated decomposition will allow for a more rapid use of landfill areas, and will facilitate the production and recovery of methane from landfills. At the present time, such recovery is only economical for deep landfills (200 feet), but an increased methane production rate may make even recovery from shallow (50 feet) landfills economically feasible.

Of course, all the concerns voiced previously on the land disposal of sludge apply to disposal in solid waste landfills as well, and the proper precautions must be taken. It does not seem reasonable, however, to unfairly restrict the disposal in landfills because of potential environmental problems.

CHEMICAL FIXATION

Chemical fixation (cementation) is a technique by which the waste material is mixed with a chemical slurry which will set into a solid, thus capturing the waste within the solid structure. Cementation materials used in the past are Portland cements, lime-based mortars, plasters and epoxies. Several proprietary processes are offered for which the specific fixative agents are not publicly known.

Typically, a waste sludge (or other material) is pumped out of a lagoon or other holding basin into a mixing chamber, and back into another lagoon where the mixture will set.

The most important concern, from the environmental perspective, is the leaching of toxins from the fixed solids. The probable danger of this occurring can be measured using a simple test in which the solidified material is placed in a long chromatography tube, packed down and distilled water filtered through. Based on such tests, some chemical fixation techniques have been shown to produce almost no toxins such as heavy metals, cyanide or phenols (Conner, 1973).

The most serious disadvantage of chemical fixation seems to be the cost. With some imagination, it is generally possible to devise a less expensive alternative to chemical fixation. In some cases, however, this is the only feasible method of eliminating a potentially hazardous sludge or liquid waste problem.

DEEP-WELL INJECTION

An alternative land disposal method is to inject the waste through deep wells into the ground. This technique was first developed by the petroleum companies which pumped brine taken from the gas and oil wells back into the formations from which the brine had originally been extracted. This not only solved a waste disposal problem, but also restored the formations to equilibrium pressure.

The idea of deep-well injection has, however, been applied to disposal of poisonous, hazardous and/or radioactive wastes which would otherwise be very expensive to treat.

Although the basic principle of deep-well injection is to fill subterranean caverns with waste, essentially forever, the problem is that the geological structures necessary for such containment simply do not exist. Wastes injected into the ground have, in many cases, appeared as eruptions or leaks, sometimes miles from the disposal site. Far more serious is the possibility that a leak would not be detected and the waste material would finally end up in a water supply or cause other environmental damage.

From the philosophical standpoint, it does not make sense to place a hazardous material in a situation where it may cause severe damage and from where its extraction would be difficult or impossible.

CONCLUSIONS

Problems with the ultimate disposal of sludges have occurred at many wastewater treatment plants. Most of these problems are, unfortunately, due to the lack of adequate foresight by planners and decision-makers, and sometimes due to benign neglect by consulting engineers.

A few years ago, a large metropolitan plant on one of the Great Lakes suddenly found itself without a means of getting rid of the sludge. As the politicians haggled and the engineers perspired, the plant was filling up with solids. The digesters, primary clarifiers and all tanks which held water were full.

The only solution for the operation was to have periodic accidents. About once a week, someone (unknown) would accidentally leave a valve open and, before it was discovered, half a digester would be discharged into the lake.

This unfortunate but true incident is illustrative of the altogether too common lack of planning and design in the ultimate disposal of wastewater residues.

REFERENCES

Bell, R. M. (1971). *Sewage Sludge Disposal.* Report by Ontario Ministry of the Environment, Ottawa.

Benarde, M. A. (1973). "Land Disposal and Sewage Effluent: Appraisal of Health Effects of Pathogenic Organisms," *J. Am. Water Works Assoc.* 65:6.

Chaney, R. L. (1973). "Crop and Food Chain Effects of Toxic Elements in Sludges and Effluents," *Proceedings,* Recycling, Municipal Sludges and Effluents on Land, Champaign, IL.

Chaney, R. L. (1974). *Recommendations for Management of Potentially Toxic Elements* (Beltsville, MD: USDA).

Chaney, R. L., and P. M. Giordano (1976). "Microelements as Related to Plant Deficiencies and Toxicities," in *Soils for Management and Utilization of Organic Wastes and Wastewaters,* L. F. Elliott and T. J. Stevenson, Eds. (Madison, WI: Soil and Science Society of America).

Chang, S. L. (1965). Discussion of a paper by H. Liebman, *Adv. Water Poll. Res.* 2:279.

Chapman, H. D. (1965). "Cation Exchange Capacity," in *Methods of Soil Analysis,* C. A. Black, Ed. (Washington, D.C.: American Society of Agronomy Inc.).

Chumbley, C. G. (1971). "Permissible Levels of Toxic Metals in Sewage Used on Agricultural Land," A.D.A.S. Advisory Paper No. 10.

Conner, J. R. (1973). "Ultimate Disposal of Liquid Wastes by Chemical Fixation," in *Proceedings of the 29th Industrial Waste Conference* (West Lafayette, IN: Purdue Research Foundation).

Dallaire, G., and N. Godfrey (1972). "Chicago Reclaiming Strip Mines with Sewage Sludge," *Civil Eng.* 42:9.

de Haan, S. (1975). "Discussion: Land Application of Liquid Municipal Wastewater Sludges," *J. Water Poll. Control Fed.* 47(11):2707.

Dotson, G. K., R. B. Dean and G. Stern (1973). "Cost of Dewatering and Disposing of Sludge on Land," *Am. Inst. Chem. Eng. J.* 69.

Doyle, C. B. (1967). "Effectiveness of High pH for Destruction of Pathogens in Raw Sludge Filter Cake," *J. Water Poll. Control Fed.* 39:8.

Environmental Protection Agency (1976). *Municipal Sludge Management,* U.S. EPA 430/9-76-009 (Washington, D.C.: U.S. EPA).

Environmental Protection Agency (1977). *Process Design Manual for Sludge Treatment and Disposal,* U.S. EPA 625/1-74-006 (Washington, D.C.: U.S. EPA).

Evans, J. O. (1973). "Soils as Sludge Assimilators," *Compost Sci.* 14:6.

Farrell, J. B., and B. V. Saletto (1973). "The Effect of Incineration on Metals, Pesticides, and Polychlorinated Biphenyls in Sewage Sludge" *Ultimate Disposal of Wastewaters and Their Residuals* (Raleigh, N C : Water Resource Res. Inst.).

Farrell, J. B., J. E. Smith, S. W. Hathaway and R. B. Dean (1974). "Lime Stabilization of Primary Sludges," *J. Water Poll. Control Fed.* 39.8.

Federal Register (1978). "Solid Waste Disposal Facilities," 43(25), February 6.

Flick, D. F. (1971). "Toxic Effects of Cadmium: A Review," *Environ. Res.* 4:71.

Foster, D. H., and R. S. Engelbrecht (1973). "Microbial Hazards in Disposing of Wastewater in Soil," in *Recycling Treated Municipal Wastewater and Sludge Through Forest and Cropland,* W. E. Sopper and L. T. Kardos, Eds. (University Park, PA: Pennsylvania State University Press).

Freytag, B. (1967). "Hygienic Aspects of the Utilization of Sewage Sludge in Agriculture," *Schr. Reih Kratoriums Kultibauw,* No. 16.

Garrigan, G. A. (1977). "Land Application Guidelines for Sludges Contaminated with Toxic Elements," *J. Water Poll. Control Fed.* (12):2380.

Hansen, J. Aa., and J. C. Tjell (1973). "Toxicological Problems in Sludge Disposal on Farmland," Danish Technological University, Lyngby (in Danish).

Hansen, J. Aa., and J. C. Tjell (1978). "Guidelines and Sludge Utilization Practice in Scandinavia," *Proceedings on the Conference on Utilization of Sewage Sludge on Land* (Medmenham, England: Water Research Centre).

Hinsley, T. D., and B. Sosewitz (1969). "Digested Sludge Disposal on Crop Land," *J. Water Poll. Control Fed.* 41:5.

Jorgensen, S. E. (1970). "Analytical Examination of Anaerobic Sludge," *Vatten* 26 (in Swedish).

Kampelmacher, E. H., and L. M. van Noorle Jansen (1972). "Reduction of Bacteria in Sludge Treatment," *J. Water Poll. Control Fed.* 44:2.

Kobayashi, J. (1971). "Relation between the 'Itai-Itai' Disease and the Pollution of River Water by Cadmium from a Mine," *Proceedings,* 5th International Water Pollution Research Conference, San Francisco, CA.

Kugel, G. (1972). "Pasteurization of Raw and Digested Sludge," *Water Res.* 6:651.

Larson, W. E., R. H. Susag, R. H. Dowdy, C. E. Clapp and R. E. Larson (1974). "Use of Sewage Sludge in Agriculture with Adequate Environmental Safeguards," *Proceeding, Sludge Handling and Disposal Seminar,* (Ottawa: Environmental Protection Service).

Lecre, H., A. Perchet, G. Savage, S. Andrieu and R. Nguematcha (1970). "Microbiological Aspects of Sewage Treatment," *Proceedings,* 5th International Conference on Water Pollution Research, San Francisco, CA.

Leeper, G. W. (1972). "Reactions of Heavy Metals with Soils with Special Regard to Their Application in Sewage Wastes" Contract No. DACW 73-73-C-0026, U. S. Army Corps of Engineers.

Lejcher, T. R., and S. H. Kunkle (1973). "Restoration of Acid Spoil Banks with Treated Sewage Sludge," in *Recycling Treated Municipal Wastewater and Sludge through Forest and Cropland,* W. E. Sopper and L. T. Kardos, Eds. (University Park, PA: The Pennsylvania State University Press).

Lund, E. (1970). "Observations on the Virus Binding Capacity of Sludge," *Proceedings,* 5th International Conference on Water Pollution Research, San Francisco, CA.

Lyman, B. T., B. Sosewitz and T. D. Hinsley (1972). "Liquid Fertilizer to Reclaim Land Produce Crops," *Water Res.* 6:545.

Martin, E. J. (1977). "Control at Sources of Heavy Metals in Sludge," *Proceedings,* National Conference on Composting of Municipal Resources and Sludges, Rockville, MD.

Massachusetts Audubon Society (1976). *Critical Document for PCB's* (Washington, D.C.: National Technical Information Service).

Muller, G. (1959). "Tuberculin Bacteria in Sludge from Mechanical Biological Wastewater Treatment," *Städtehyg.* 5:46 (in German). (From Foster and Engelbrecht, 1973).

National Academy of Sciences (1978). *Multimedium Management of Municipal Sludge* (Washington, D.C.: National Research Council).

Ridgeway, J. (1970). *The Politics of Ecology* (New York: E. P. Dutton & Co.).

Rudolfs, W., L. Falk and R. Ragotzkie (1951). "Contamination of Vegetables Grown in Polluted Soil," *Sew. Works J.* 23:253.

Scott, D. S., and H. Horlings (1974). "Removal of Phosphates and Metals from Sewage Sludges," *Proceedings, Sludge Handling and Disposal Seminar* (Ottawa: Environmental Protection Service).

Smith, J. E. (1973). EPA Technology Transfer Seminar, Atlanta.

"Solid Waste Disposal Facilities" (1978). *Federal Register* 43(25), February 6.

Sommers, L. E. (1977). Chemical Composition of Sewage Sludges and Analysis of their Potential Use as Fertilizers," *J. Environ. Qual.* 6:225.

Sosewitz, B. (1973). "Sludge for Land Reclamation," *Reporter APWA* 40:5.

Swanwick, J. D. (1972). "Theoretical and Practical Aspects of Sludge Dewatering," *Proceedings,* 2nd European Sewage and Refuse Symposium, Munich.

Subrahmanyan, T. P. (1974). "Virological Investigations on Sludges from Selected Ontario Sewage Plants," in *Proceedings, Sludge Handling and Disposal Seminar* (Ottawa: Environmental Protection Service).

Vesilind, P. A. (1978). "Sludge Treatment and Disposal Laboratory Manual," Duke Environmental Center, Duke University, Durham, NC.

Walker, J. M. (1975). "Sewage Sludges–Management Aspects for Land Application," *Compost Sci.* 16(2), March/April.

Water Pollution Control Federation (1973). "Sludge Disposal," *J. Water Poll. Control Fed.* 45:10.

Wolfel, R. M. (1967). "Liquid Digested Sludge to Land Surfaces–Experiences at St. Mary's and Other Municipalities in Pennsylvania," presented at the 39th Conference, Water Pollution Control Association of Pennsylvania.

PROBLEMS

1. A recent news item reported that domestic wastewater sludge contains \$200 worth of gold per dry ton of sludge. Using the percent of gold (1978–\$190 per ounce), estimate the concentration of gold in sludge. Why isn't there a gold rush?

2. An industrialist once argued that pretreatment for cadmium removal really did not make much sense, since drinking water already contains about 42% of the cadmium found in wastewater. Do you agree?
3. A sludge has the following heavy metal concentrations:

	mg/kg
zinc	1,800
copper	500
nickel	120
cadmium	9

   If the soil on which it is to be disposed of is a loam with some silt in it, what application rate might be appropriate?
4. Consider the information given in the problem above. What other information would you require before it would be possible to establish an application rate?
5. A raw sludge which has 15,000 mg/l organic nitrogen, 1,000 mg/l ammonia nitrogen, no detectable nitrite, and 25 mg/l of nitrate, is to be applied to a sandy soil where soybeans are grown. What would be a reasonable application rate in terms of nitrogen limits?

CHAPTER 12

REGIONAL SLUDGE MANAGEMENT*

In sparsely populated areas, where one wastewater treatment plant serves a community, it is usually reasonable to treat and dispose of the sludge locally. By contrast, metropolitan and suburbanized areas may have many plants located within short distances of each other. Such densely populated regions usually share another problem and that is the absence of obvious and inexpensive ultimate disposal sites. In some areas, such as the Washington, D.C. suburban area, Los Angeles County, Chicago, Allegheny County (Pittsburgh), others in the U.S. and many more overseas, the treatment plants are run by one authority, and thus the ultimate disposal problem becomes a regional one. In this chapter, the problems of regional sludge management are discussed and a mathematical procedure for evaluating regional sludge management options is presented.

OPTIONS IN REGIONAL SLUDGE MANAGEMENT

If the reader is familiar with the first eleven chapters of this book, he no doubt recognizes that the options for sludge management are literally infinite. Within a wastewater treatment plant, the operator can vary the operation so as to produce many different sludges of different characteristics in different quantities. Any general scheme for regional sludge management must therefore recognize the existence of a vast array of decision variables which can be grouped into four general areas:

- wastewater processes
- inplant sludge processing

*This chapter is coauthored by Dr. Yakir Hasit, Department of Environmental Engineering, Cornell University, Ithaca, NY.

- sludge transport
- ultimate disposal

Obviously, changing the treatment system (*e.g.*, from activated sludge to lime precipitation) will significantly influence the sludge handling and disposal options. Within the plant, sludge processing such as thickening, digestion and dewatering all affect the last two variables; transport to the disposal area or an intermediate stop, and the final disposal of the sludge.

Graphically represented in Figure 12-1 is a generalized scheme showing two treatment plants producing several sludges. There are a number of processing alternatives (with one alternative being no processing). The sludges can be transported to several ultimate disposal sites, or they can be transported from one plant to another where they may be processed and then disposed of. The development of the most economical solution requires summing the costs for each alternative and comparing overall costs. Obviously, this is not a simple problem, and its solution requires first that the costs for each of the processes be defined and, second, that an objective function be developed with the proper constraints so that computer-aided optimization schemes can be used.

It is convenient to consider the optimization of regional sludge management in terms of two different models: (1) the determination of disposal facilities for sludges generated within existing plants and (2) the determination of sludge handling processes, transportation modes and disposal facilities for sludges generated in expanding or proposed plants. In the first case, the purpose is to minimize transportation and offplant disposal costs, while in the second case the purpose is to minimize inplant handling costs in addition to the previously mentioned costs.

THE TRANSPORTATION MODEL

The first type can be formulated as the *transportation model* found in operations research literature (Hillier and Lieberman, 1974; Wagner, 1969). Each plant is a source and each landfill, farm application site, ocean disposal site, stripmined land and so on is a sink. The objective is to assign each plant to one or more disposal sites in such a manner that the transportation and disposal costs are minimized. This objective, which can be expressed as a mathematical function (called the objective function), is subject to two sets of constraints: (1) the waste transported from each plant to all disposal sites must be equal to the waste generated at that plant, *i.e.*, all the waste has to be transported; (2) the waste transported from all sources to each disposal site must be less than or equal to the capacity of the disposal site. The first set of constraints contains one equation for each source and the second set of

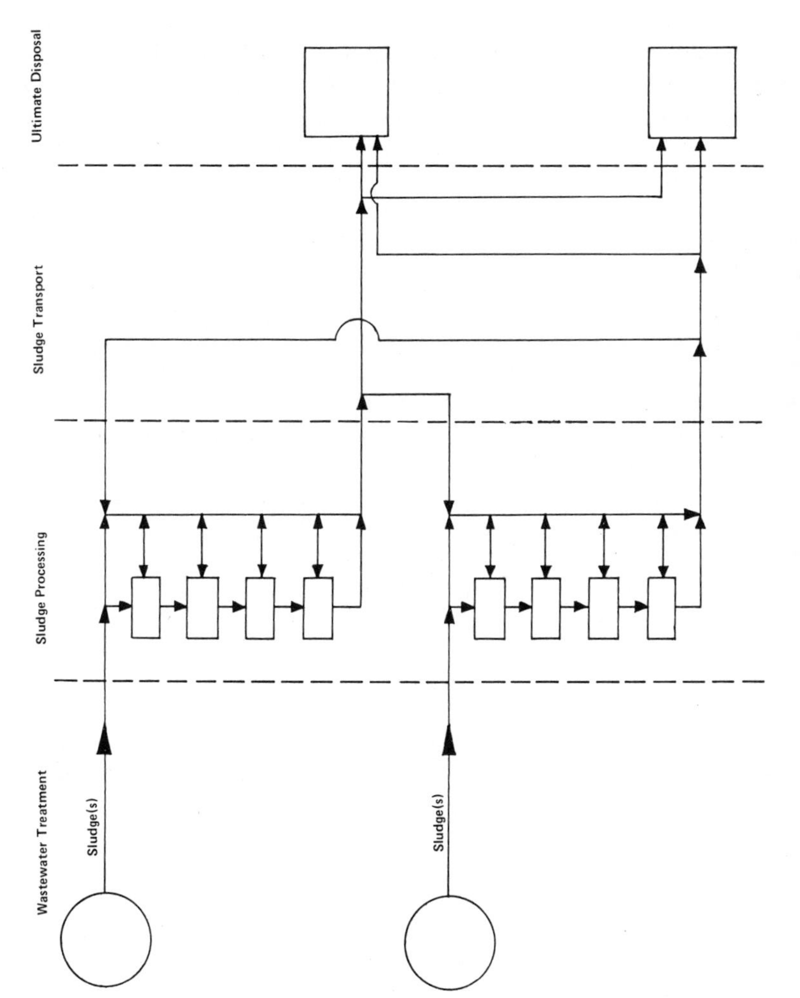

Figure 12-1. Generalized options for regional sludge management.

constraints contains one equation or inequality for each sink. If more than one type of sludge is generated at a plant, then the sources have to be further divided to represent each type.

For the model to be feasible the total capacity of all disposal sites considered together must be larger or equal to the total amount of sludge produced by all the plants per unit time. Also it is important to know that linking all sources to all sinks is sometimes not possible. For example, due to regulations, sources generating liquid sludges may not utilize landfills, or resources generating raw sludges may not use farm application sites. There may also be sociopolitical reasons which may prohibit the use of some disposal facilities by some plants.

Sometimes a third set of constraints may be necessary. If there are limitations to the waste flow between sources and sinks, such as the number of trucks that can operate between two points or the amount of a certain type of sludge a disposal site can accept, then restrictions on the flows have to be added as the third set of constraints.

If the sludge disposal systems include incinerators, then the problem can be formulated as a *transshipment model* (Hillier and Lieberman, 1974), where sludges can be transshipped to and from incinerators. In this case, the mass going through the incinerators is reduced.

A simple example, which shows the mathematical formulation, is presented below. For a detailed presentation refer to Hasit (1978).

Example 12-1: There are three wastewater treatment plants (A,B,C), one incinerator (N), two land application sites (X,Y) and one landfill (L) in a region (Figure 12-2). The regional authority wants to minimize the sludge transportation and disposal costs for the region. The following tables represent the relevant data.

Sludge Production

Plant	Sludge Generated (dry ton/day)	Type of Sludge	Solids (%)	Volatiles (%)
A	1	Digested Mixed	8	40
B	4	Digested Mixed	40	45
C	2	Raw Primary	30	60

Unit Costs

Solids (%)	Transportation ($/dry ton/mile)	Land Disposal ($/dry ton)	Landfill Disposal ($/dry ton)	Incineration ($/dry ton)
8	1.50	10.00	---	---
30	0.65	4.00	8.00	12.00
40	0.50	6.00	5.00	12.00
100 (ash)	0.20	---	3.00	---

Capacities of Facilities

Facility	(wet ton/day)
N	9
X	15
Y	10
L	12

Distances (miles)

From/To	N	X	Y	L
A	–	15	10	–
B	10	8	15	12
C	20	–	–	18
N	–	–	–	3

The final cost table is computed by multiplying the unit transportation costs with the one-way distances and adding the processing costs at the destination. It is assumed that unit costs remain unchanged regardless of the amount of sludge processed or transported.

Final Costs ($/dry ton)

	N	X	Y	L
A	--	32.50	25.00	--
B	17.00	10.00	13.50	11.00
C	25.00	--	--	19.70
N	--	--	--	3.60

As mentioned previously the purpose is to find the least costly solution such that all the restrictions on the system will be satisfied.

The *objective function*, Z, is the mathematical expression which requires the minimization of transportation and processing costs, that is:

$$\text{Minimize } Z = 32.50\, W_{AX} + 25.00\, W_{AY} + 17.00\, W_{BN} + 10.00\, W_{BX} +$$

$$13.50\, W_{BY} + 11.00\, W_{BL} + 25.00\, W_{CN} + 19.70\, W_{CL} + 3.60\, W_{NL}$$

where W_{ij} = dry weight of solids (tons) transported from i to j and handled at j

The objective function is subject to a few sets of *constraints*. The first set ensures that all the sludge generated at a source will be processed.

$$W_{AX} + W_{AY} = 1 \text{ (sludge generated at A)}$$

$$W_{BN} + W_{BX} + W_{BY} + W_{BL} = 4 \text{ (sludge generated at B)}$$

$$W_{CN} + W_{CL} = 2 \text{ (sludge generated at C)}$$

The second set ensures that the sludge received at each facility will not exceed the facility's capacity. Note that since the capacity of each plant is given in wet ton/day, the daily sludge produced by each plant must also be expressed in wet tons.

$$\frac{W_{BN}}{0.40} + \frac{W_{CN}}{0.30} \leqslant 9 \text{ (capacity of N)}$$

$$\frac{W_{AX}}{0.08} + \frac{W_{BX}}{0.40} \leqslant 15 \text{ (capacity of X)}$$

$$\frac{W_{AY}}{0.08} + \frac{W_{BY}}{0.40} \leqslant 10 \text{ (capacity of Y)}$$

$$\frac{W_{BL}}{0.40} + \frac{W_{CL}}{0.30} + \frac{W_{NL}}{1.0} \leqslant 12 \text{ (capacity of L)}$$

The third set of constraints (in this case only one) relates the amount of sludge going into the incinerator to the amount leaving the incinerator. The mass reduction taking place at the incinerator due to the destruction of volatile solids is reflected in the constraints.

$$0.55\, W_{BN} + 0.40\, W_{CN} = W_{NL} \text{ (mass balance for the incinerator)}$$

Note that the coefficients are the ratios of fixed solids to total solids.

The solution of this system of equations and inequalities is possible by linear programming (LP) which is a mathematical optimization technique. As the scope of this book does not cover optimization techniques, the interested reader may refer to any text on operations research or systems analysis for a background on LP (Hillier and Lieberman, 1974; Wagner, 1969; Murty,

1976). Using any linear programming computer program, the solution to the above problem can be found to be:

$$W_{AX} = 0.2 \text{ dry tons/day}$$
$$W_{AY} = 0.8 \text{ dry tons/day}$$
$$W_{BX} = 4.0 \text{ dry tons/day}$$
$$W_{CL} = 2.0 \text{ dry tons/day}$$
$$W_{BN} = W_{BY} = W_{BL} = W_{CN} = W_{NL} = 0$$

The total cost for the system is \$105.9/day or \$38,644/year (Figure 12-3).

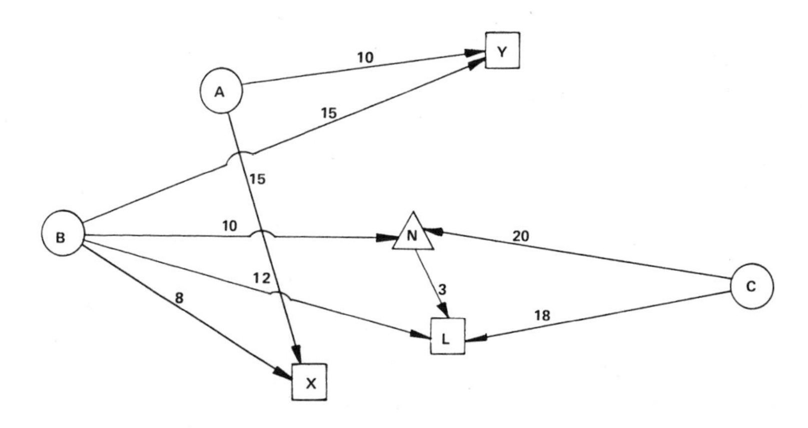

Figure 12-2. Flow diagram for Example 12-1, showing possible flows and distances (in miles) between any two sites.

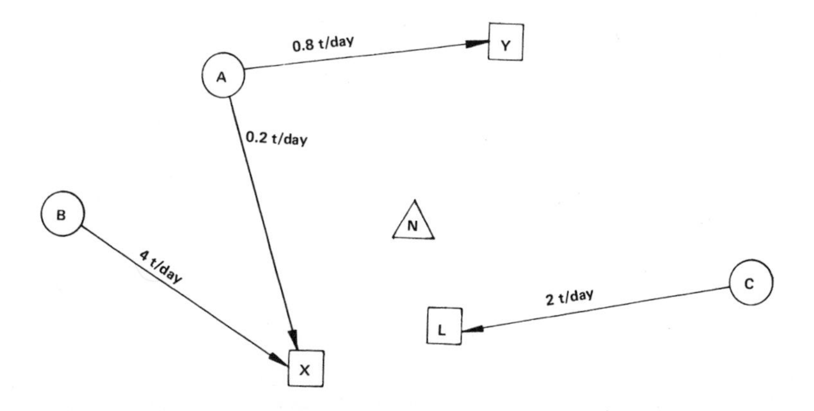

Figure 12-3. Solution for Example 12-1.

THE MIXED INTEGER PROGRAMMING MODEL

The determination of inplant sludge handling processes along with disposal sites is a more complex problem than the one just presented. In this case interactions among various processes must also be considered. The complexity of the problem can best be illustrated by Figures 12-4 and 12-5 which are typical flow charts for primary and waste activated sludges. In these figures it is assumed that the anaerobic digesters are high-rate secondary units and thus do not need thickening after digestion while the aerobic digesters have no settling tanks and thus may have thickeners before or after digestion. It is also assumed that sand beds and land application sites accept only digested, and not raw, sludges.

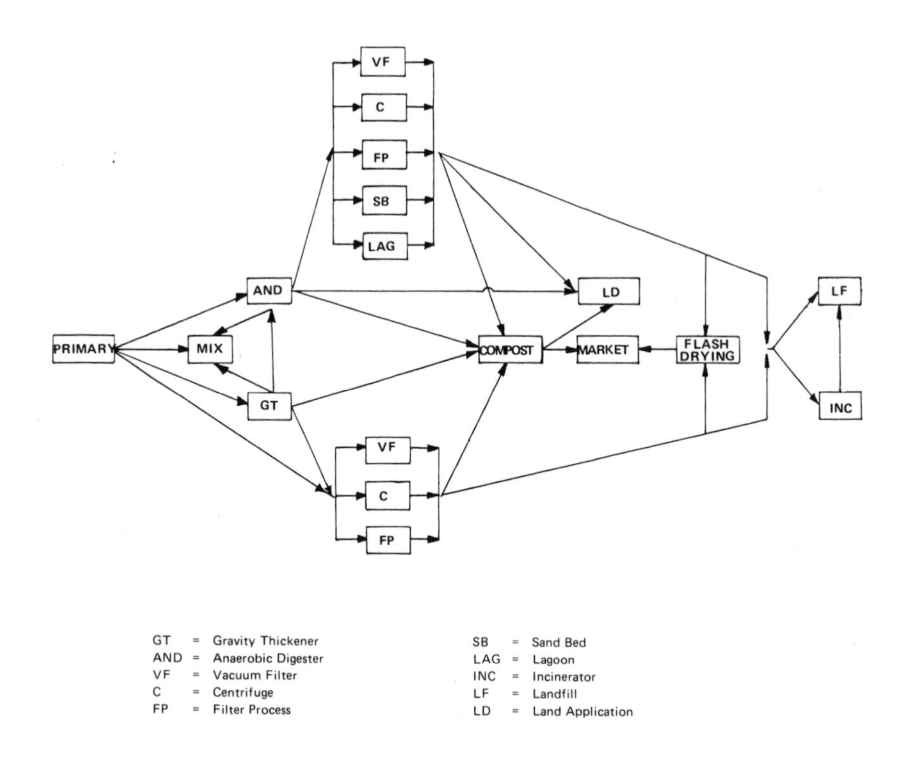

Figure 12-4. Typical flow chart for primary sludges (based on Dick and Simmons, 1974 and Sverdrup & Parcel, 1977).

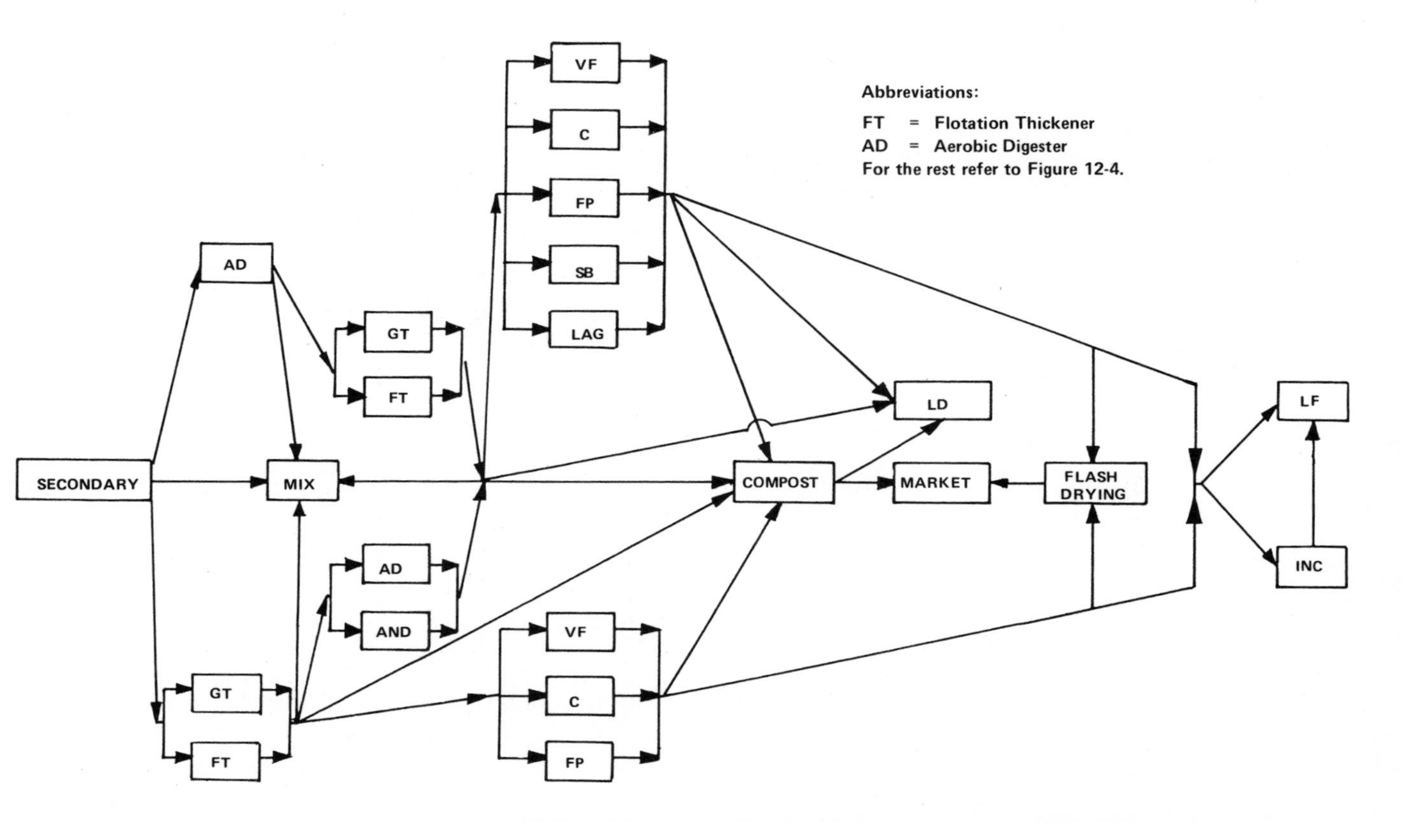

Figure 12-5. Typical flow chart for waste activated sludges (based on Dick and Simmons, 1974, and Sverdrup & Parcel, 1977).

The objective in this case, too, is to minimize the processing, transportation and disposal costs; however, costs must further be divided into capital (amortized) and operation and maintenance (O & M) costs in order to account for existing and proposed facilities. For existing facilities only O & M costs are considered, while for new or expanding facilities (or units) total costs (capital plus O & M) are considered.

This model also consists of sources, intermediate processes and ultimate sites. Each treatment plant is a main source and, depending on the type of plant, the source may further be divided into several sources. For example, a water treatment plant has only one type of sludge (ignoring filter backwash), while a tertiary plant may have three types of sludge, namely: primary, secondary and tertiary (or chemical). Thus a tertiary plant has three sources.

Intermediate processes are stabilization, thickening, dewatering and combustion (conditioning can be considered as part of thickening or dewatering processes). Intermediate processes can receive flows from sources or from other intermediate processes, subject to certain constraints. For example, an incinerator cannot receive a flow from a primary or secondary clarifier but it can receive a flow from a vacuum filter or a sand bed. Ultimate processes include landfills, land application sites and so on. They can receive flows only from intermediate processes and not from sources, assuming that raw sludges, including water plant sludges, cannot be disposed of without prior treatment. Ultimate processes have no flows out.

The objective function for an expanding single plant can be represented by:

> Minimize [(existing intermediate process O & M costs + proposed intermediate process total costs + existing ultimate process O & M costs + proposed ultimate process total costs) + (transportation costs for hauling sludge from intermediate processes to ultimate processes)].

For a group of plants the objective function would expand into:

> Minimize [(existing intermediate process O & M costs + proposed intermediate process total costs + existing ultimate process O & M costs + proposed ultimate process total costs) + (transportation costs for hauling sludge from sources to intermediate processes in other plants + transportation costs for hauling sludge among intermediate processes in separate plants + transportation costs for hauling sludge from intermediate processes to ultimate processes)].

This objective function is subject to the following constraints.

1. Source balance equations ensure that all waste generated at a source must be processed. There must be one equation for each type of sludge at each source.

2. Process capacity equations ensure that for existing processes the total amount of sludge sent to these processes must not exceed their handling capacities. For proposed processes, these equations determine the capacities of the units and place upper limits on them, if necessary. As is usually the custom, if a capacity is given as the surface area of a unit (thickeners, sand beds, etc.) or as the volume (digesters, etc.), then the loading rates of the units must be incorporated. For an existing ultimate process, k, this constraint would be:

$$\sum_{j} \sum_{m} \frac{W_{jkm}}{L_{kjm}} \leqslant P_k \text{ for all k} \tag{12-1}$$

where W_{jkm} = dry weight of sludge type m hauled from intermediate process j to ultimate process k

L_{jkm} = loading rate of sludge type m hauled from intermediate process j to ultimate process k

P_k = capacity of ultimate process k

If the ultimate process is farm application, then L_{kjm} can be given in terms of dry ton/ac/yr, the value depending on the type of sludge, the sludge's nitrogen and/or heavy metals content and the characteristics of the soil. As the weight of the sludge would be given as dry ton/yr, then the capacity unit would be acres. The same argument holds for other types of processes also. Typical loading rates are given in Table 12-1.

3. Process mass balance equations relate the weight of dry solids entering a process to the weight of dry solids leaving that process. The solids capture, the reduction in mass that occurs in some processes (digestion, combustion), or the increase in mass that sometimes occurs due to conditioning (especially in filter presses) must be incorporated.

These three sets of constraints must be present in each model; in fact they are similar to the ones presented in the previous model, the only difference being that the transshipment points are incinerators in the first case, while the transshipment points can be any of the intermediate processes in the latter case.

Furthermore, depending on the system under consideration, there may be other constraints unique to each system. For example, the nitrogen or heavy metals content may limit the extent of land application. Also there may be upper and lower limits to the WAS/primary sludge ratio when blending them. There may also be limits in the number of trucks, barges or railroad cars that can be used to transport the sludge.

If the assumption made in Example 12-1 holds, that is, that costs remain linear with capacity and thus do not exhibit any economies of scale, then

Table 12-1. Typical Percent Solids Out, Loadings, Yields and Cycle Times for Different Types of Sludge and Processes

Process	Type of Sludge	Percent Solids Out	Loading ($lb/ft^2/day$)	Remarks
Gravity Thickener	Raw primary	10 - 15 (WPCF, 1969)	18 - 25 (WPCF, 1977)	Without conditioners
	WAS	3 - 5 (WPCF, 1969)	4 - 6 (WPCF, 1977)	Without conditioners
	Primary + WAS	5 - 7 (WPCF, 1969)	6 - 10 (EPA, 1977)	
	Lime	19 (Bond and Straub, 1974)	-	
	Aerobic digested WAS	-	5 - 10 (WPCF, 1977)	
			Loading ($lb/ft^2/hr$)	
Flotation Thickener	WAS	3 - 5 (WPCF, 1977)	0.4 - 1 (WPCF, 1977)	Without conditioners
	WAS	4 - 6 (WPCF, 1977)	(1.5 to 2) x (0.4-1) (WPCF, 1977)	With conditioners
	Primary + WAS	6 - 7.5 (Bond and Straub, 1974)	-	
			Loading	
Anaerobic Digester (High-Rate Secondary)	Primary	6 - 12 (WPCF, 1977)	Choose loading from Figure 27-3 in WPCF, 1977, or use an equivalent diagram from another source	Solids retention time 10 - 20 days
	Primary + WAS	6 (WPCF, 1977)		
Aerobic Digester	WAS	2 - 3 (WPCF, 1977)	Choose loading from Figure 26-2 in WPCF, 1977, or use an equivalent diagram from another source	
			Loading ($lb/ft^2/yr$)	
Sand Bed	Primary	35 - 50 (Liptak, 1974)	27.5 (EPA, 1974)	All biological sludges are digested

	Primary + WAS	28 - 50 (Liptak, 1974)	15 (EPA, 1974)	
	Tertiary	–	22 (EPA, 1974)	
Centrifuge (Solid Bowl)	Raw primary	28 - 35 (WPCF, 1977)		For dewatering only
	Digested primary	28 - 35 (WPCF, 1977)		
	Raw (primary + WAS)	18 - 24 (WPCF, 1977)		
	Digested (primary + WAS)	15 - 30 (EPA, 1974)		
	WAS	12 - 15 (WPCF, 1977)		
	Alum	15 - 25 (WPCF, 1977)		
	Lime	45 - 60 (WPCF, 1977)		
			Yield ($lb/ft^2/hr$)	
Vacuum Filter (Rotary)	Raw primary	25 - 30 (WPCF, 1977)	5 - 10 (WPCF, 1977)	Values from (WPCF, 1977) are with conditioners
	Digested primary	28	3 - 7 (WPCF, 1969)	
	Raw WAS	12 - 18 (WPCF, 1977)	1 - 2 (WPCF, 1977)	
	Digested WAS	20	2 - 5 (WPCF, 1969)	
	Raw (primary + WAS)	16 - 24 (WPCF, 1977)	2 - 5 (WPCF, 1977)	
	Digested (primary + WAS)	20 - 24 (WPCF, 1977)	3 - 5 (WPCF, 1977)	
	Low lime	25 - 30 (WPCF, 1977)	3 - 6 (WPCF, 1977)	
	High lime	30 - 40 (WPCF, 1977)	5 - 10 (WPCF, 1977)	
	Polyelectrolyte	25 - 38 (WPCF, 1977)	8 - 10 (WPCF, 1977)	
			Cycle Time (hr)	
Filter Press	Raw primary	40 - 50 (WPCF, 1977)	1.5 - 3 (WPCF, 1977)	Loading rates are 10-20% of those given for vacuum filters (WPCF, 1977)
	Digested primary	45 - 50 (EPA, 1974)	1.5 - 2 (WPCF, 1969)	
	WAS	24 - 40 (WPCF, 1977)	2 - 6 (WPCF, 1977)	
	Raw (primary + WAS)	30 - 40 (WPCF, 1977)	2 - 4.5(WPCF, 1977)	

Table 12-1, continued

Process	Type of Sludge	Percent Solids Out	Cycle time (hr)	Remarks
	Digested (primary + WAS)	35 - 45 (WPCF, 1977)	2 - 4 (WPCF, 1977)	
	Low lime	35 - 40 (WPCF, 1977)	2 - 4 (WPCF, 1977)	
	Alum	25 - 40 (WPCF, 1977)	1 - 4 (WPCF, 1977)	
			Loading (ft^2/capita)	
Drying Lagoons	Digested primary	-	1 (EPA, 1974)	For arid climates
	Digested WAS	-	3 - 4 (EPA, 1974)	For rainy climates
	-	-	2.2 - 2.4 $lb/ft^3/yr$ (WPCF, 1977)	
		Loading		
Land Application	Liquid on orchards	11 } $yd^3/ac/yr$ (WPCF, 1969)		
	Liquid on grass	10 - 20 } $yd^3/ac/yr$ (WPCF, 1969)		
	Liquid on vegetation	20 - 60 } $yd^3/ac/yr$ (WPCF, 1969)		
	Liquid	10 - 20 dry ton/ac/yr (EPA, 1974)		Slight limitations on soils
	Liquid	less than 10 dry ton/ac/yr (EPA, 1974)		Moderate limitations on soils

linear programming can be used again to find a cost-optimal solution. However, the problem becomes more complicated when there are economies of scale. First, the cost functions become concave leading optimization procedures to local optimal solutions rather than global ones. (The solution may not be the least expensive alternative, and is dependent on the conditions at the start of the program). Second, the optimization technique may require modifications. In Figure 12-6, the solid line, which is a concave cost function, can be linearly approximated by the dotted line and consequently expressed as:

$$\text{Total cost} = \begin{cases} b + cW & \text{when } W > 0 \\ 0 & \text{when } W = 0 \end{cases} \tag{12-2}$$

where W = the amount of sludge processed
b = the fixed charge
c = the unit cost of processing

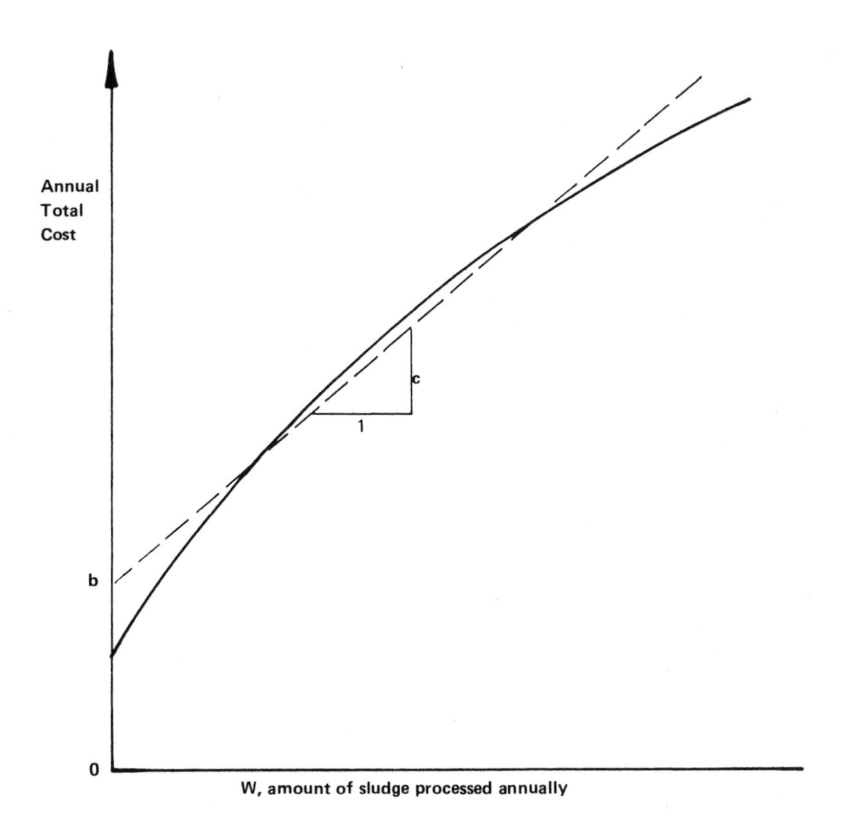

Figure 12-6. A concave cost function.

The presence of fixed charges introduces integer variables and thus necessitates the use of mixed integer programming (MIP) instead of linear programming. This can be observed by rewriting Equation 12-2 as:

$$\text{Total cost} = bi + cW$$

where $i = 0$ when $W = 0$ and $i = 1$ when $W > 0$. Thus i has to be introduced into the model as an integer variable.

If it is necessary to seek solutions with discrete sizes for processes like vacuum filters and centrifuges, as they are manufactured only in certain sizes, then additional integer variables have to be introduced into the model. Other cases where integer variables may be used are, for instance, when the possibility of putting thickeners either before or after aerobic digesters are being considered. To enable the model to represent the problem correctly, integer variabies which will permit the planning of thickeners only before or after digestion, but not at both locations, have to be included.

Though MIP is an extension of LP, due to the presence of integer variables, it is more complex and its algorithms are computationally less efficient than LP algorithms. For full details about cost functions, variables, objective function and constraints refer to Hasit (1978) and for MIP refer to Murty (1976), Hillier and Lieberman (1974) and IBM (1973).

Due to the details required to give a complete account of the model, a simplified version of a case study, as analyzed by Hasit (1978), is presented to convey an idea of the planning possible with MIP.

Example 12-2. A regional government is investigating the feasibility of centralization by comparing two options: the cost-optimal management of individual plants and the cost-optimal management of the whole region with centralized facilities. Figure 12-7 presents the planning region which has five wastewater (four expanding and one proposed) and two water (one existing and one proposed) treatment plants, one incinerator and two landfills. Two plants (WW2 and WW4) are being considered as regional facilities. Possible flows among different sites are shown on the figure. The types and amounts of sludge produced in the plants are given in Table 12-2. The existing and proposed processes of each plant are presented in Table 12-3. With the cost functions and the loading rates used, and the assumptions made (see Hasit, 1978), the flow chart shown in Figure 12-8 gives the regional cost-optimal solution. The annual cost for the region with centralization is \$1,076,000 (June 1978 dollars) while it is \$1,255,000 without centralization. Thus regionalization would reduce annual costs by about \$180,000 or 14%.

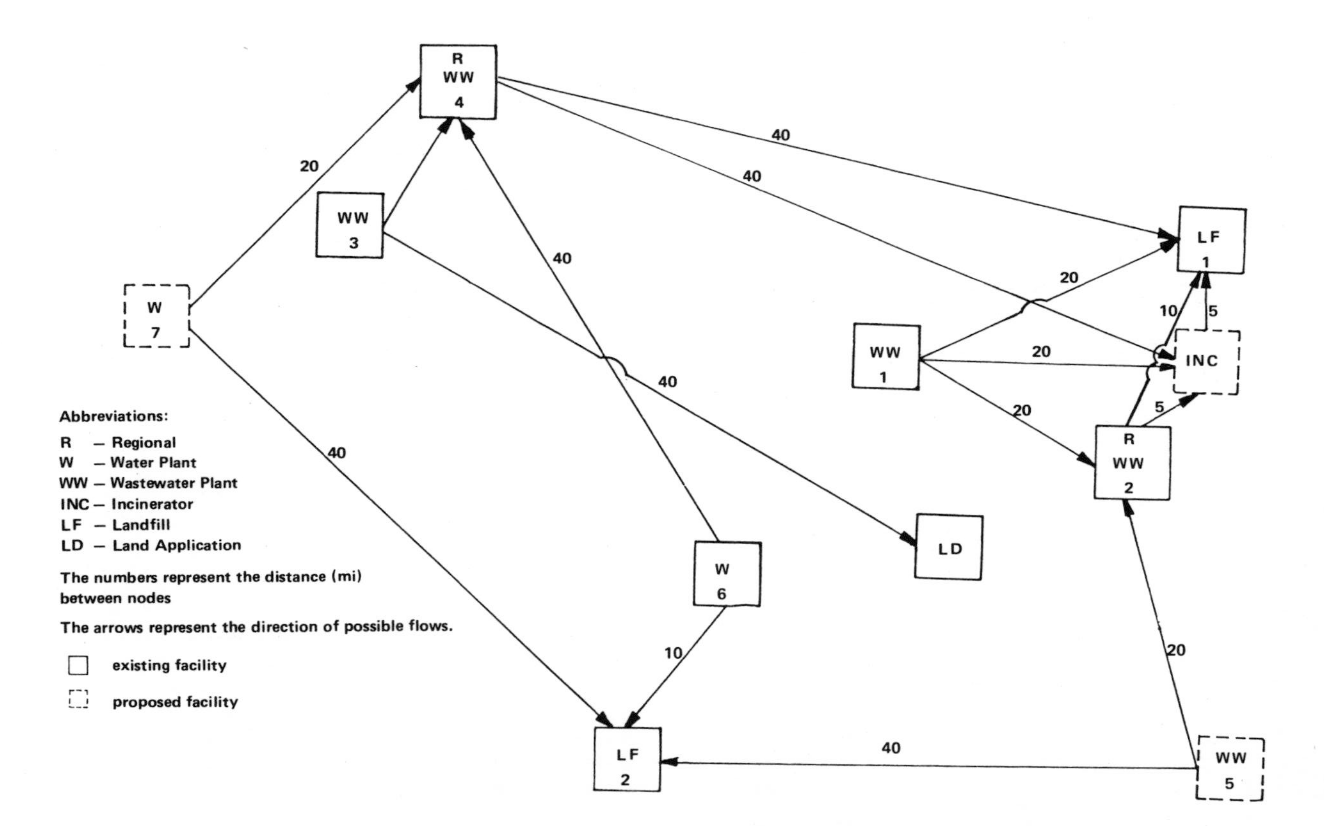

Figure 12-7. Schematic representation of the regional case study.

Table 12-2. Types and Amounts of Sludge Produced in the Plants

Plant	Plant Capacity (mgd)		Type of Sludge	Percent Solids	Sludge Flow (gpd)		Sludge Dry Weight (lb/day)	
	Present	Future			Present	Future	Present	Future
I	2	4	Primary	5	6,000	12,000	2,500	5,000
II	12	18	Primary	4	36,000	54,000	12,000	18,000
			WAS	1	48,000	72,000	4,000	6,000
III	0.5	1	WAS	1	9,500	19,000	800	1,600
IV	3	5	Primary	4	9,000	15,000	3,000	5,000
			WAS	1	–	20,000	–	1,700
V	–	2	Primary	5	–	6,000	–	2,500
			WAS with alum	1	–	12,000	–	1,000
VI	20	20	Water alum	1.8	24,600	24,600	3,700	3,700
VII	–	14	Water alum	2	–	23,200	3,900	3,900

Table 12-3. Status of Various Processes in the Plants
(p = planned; e = existing)

Process	Status	Optimal Solution (a = accepted; r = rejected)	
		Regional with Centralization	Local without Centralization
Plant I (expanding wastewater plant)			
Gravity thickening	p	a	a
Anaerobic digestion	e	a	a
Anaerobic digestion	p	r	r
Vacuum filtration	p	r	r
Centrifuging	p	r	r
Filter pressing	p	r	r
Sand bed drying	e	r	r
Sand bed drying	p	r	r
Lagoon drying	p	a	a
Plant II (expanding wastewater plant)			
Gravity thickening	e	a	a
Gravity thickening	p	a	a
Flotation thickening	p	r	r
Aerobic digestion	p	r	r
Anaerobic digestion	e	a	a
Anaerobic digestion	p	a	a
Vacuum filtration	p	r	r
Centrifuging	p	r	r
Filter pressing	p	r	r
Sand bed drying	e	a	a
Sand bed drying	p	a	a
Plant III (expanding wastewater plant)			
Gravity thickening	p	r	a
Flotation thickening	p	r	r
Aerobic digestion	e	r	a
Aerobic digestion	p	r	r
Vacuum filtration	p	r	r
Filter pressing	p	r	r
Sand bed drying	e	r	r
Plant IV (expanding wastewater plant)			
Gravity thickening	p	a	a
Flotation thickening	p	r	r
Anaerobic digestion	e	r	a
Anaerobic digestion	p	r	a
Vacuum filtration	p	a	r

Table 12-3, continued

Process	Status	Optimal Solution (a = accepted; r = rejected)	
		Regional with Centralization	Local without Centralization
Centrifuging	p	r	r
Filter pressing	p	r	r
Sand bed drying	e	r	r
Sand bed drying	p	r	a
Plant V (planned wastewater plant)			
Gravity thickening	p	a	r
Flotation thickening	p	r	r
Aerobic digestion	p	r	r
Anaerobic digestion	p	r	a
Vacuum filtration	p	r	r
Centrifuging	p	r	r
Filter pressing	p	r	r
Sand bed drying	p	r	a
Plant VI (existing water plant)			
Filter pressing	p	r	r
Sand bed drying	p	r	a
Lagoon drying	e	a	r
Plant VII (planned water plant)			
Vacuum filtration	p	r	r
Centrifuging	p	r	r
Filter pressing	p	r	r
Sand bed drying	p	r	a
Incinerator	p	r	r
Landfill (1)	e	a	-[1]
Landfill (2)	e	r	-[1]

[1] In the local system, the decision to reject or accept a landfill is trivial, because the one giving lower transport and disposal costs can immediately be chosen. However, this may not be apparent if there are capacity limitations on either of them, or if the decision is between a landfill (which accepts only dewatered sludge) and a land application site which can accept liquid sludge. In the latter case, the trade-off is between the extra cost of dewatering and the extra cost of transporting liquid sludge in addition to the ultimate disposal costs.

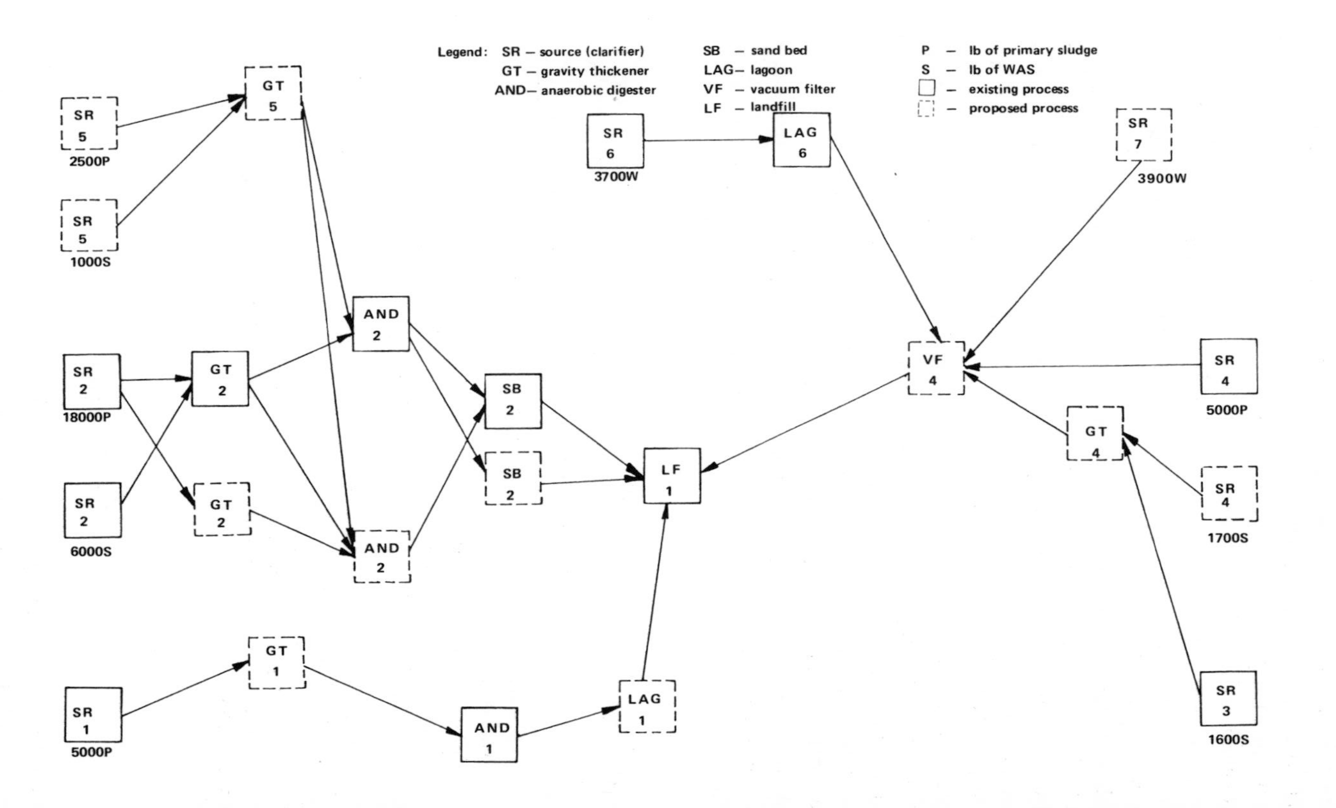

Figure 12-8. Flow diagram for regional problem [for clarity only accepted processes and the amount of sludge generated at the clarifiers (pounds) are included in the diagram].

This solution, however, cannot be generalized and in different situations centralization may cause higher or lower costs or even no savings at all. The status of each process, *i.e.*, its acceptance or rejection, in each option is also presented in Table 12-3. Notice that in plants WW1 and WW2 the status of the processes remain unchanged in both options, while in the other plants some processes that are accepted (or rejected) in one option are rejected (or accepted) in the other option.

In the regional option, the incinerator and one of the landfills are rejected and WW2 acts as a regional facility for WW5, and WW4 for WW3, W6 and W7. Also note that though WW1 has the choice of using the facilities of WW2, it is more cost-effective to build a new gravity thickener in the plant than to haul it to WW2. It must be remembered, though, that the optimal solution is a regional one; that is, the total cost is minimum for the region as a whole and not for each individual plant. Considered separately, some plants (particularly regional plants) cost more to operate than would be the case without regionalization (*i.e.*, they would process their own sludge only). However, due to the elimination of some processes, the savings at other plants more than compensate for the extra cost of regional plants. The reader is reminded again that no generalizations can be made from this example and each problem is unique.

Some typical computational statistics for the regional option are given in Table 12-4.

Table 12-4. Computational Statistics for Example 12-2[a]

CPU time (sec)	342
No. of constraints	255
No. of continuous variables	533
No. of integer variables	58
Matrix density = $\frac{\text{no. of nonzero coefficients}}{\text{total no. of coefficients}}$	0.011
No. of iterations for MIP solution	2,494
Total no. of iterations	6,219

[a]The computer program used was Mathematical Programming System Extended with Mixed Integer Programming (MPSX/MIP), (IBM, 1973), and the computer was an IBM 370/168.

CONCLUSIONS

Municipal sewage sludge is a residual which exists as a direct result of a human psychological and economic need to live in organized communities.

As land space is allocated to various community activities, the ability of the land to assimilate wastes is reduced; remote wastewater treatment becomes necessary with the inevitable production of sludge. The majority of this book is devoted to the analysis and solution of the sludge problems of a single wastewater treatment plant or a single community. By contrast, this last chapter presents a means of planning for sludge management on a regional basis, with many communities or a single community with many wastewater treatment plants.

It is recognized that the information presented in this chapter is inadequate for the practicing engineer or planner. Accordingly, a users' manual, describing the use of MIP in analyzing and optimizing regional sludge management alternatives has been prepared and is available from the Department of Civil Engineering, Duke University, Durham, NC 27706.

REFERENCES

Bond, Richard G., and Conrad P. Straub, Eds., (1974). *CRC Handbook of Environmental Control, Volume IV, Wastewater: Treatment and Disposal* (Cleveland: CRC Press, Inc.)

Dick, Richard I., and David L. Simmons (1974). "Integration of Unit Operations in Wastewater Residuals Management," in *Williamsburg Conference on Management of Wastewater Residuals,* J. L. Smith and E. H. Bryan, Eds., November 13, 14, Williamsburg, VA.

Environmental Protection Agency (1974). *Process Design Manual for Sludge Treatment and Disposal*, EPA 625/1-74-006.

Hasit, Yakir (1978). "Optimization of Municipal Sludge Handling and Disposal Systems," Ph.D. dissertation, Duke University, Durham, NC.

Hillier, Frederick S., and Gerald J. Lieberman (1974). *Operations Research*, 2nd ed. (San Francisco: Holden-Day, Inc.).

IBM (1973). *Mathematical Programming System Extended (MPSX) Mixed Integer Programming (MIP) Program Description*, SH20-0908-1, 2nd ed., White Plains, NJ.

Liptak, Bela G., Ed. (1974). *Environmental Engineers' Handbook, Volume 1, Water Pollution* (Radnor, PA: Chilton Book Company).

Murty, Katta G. (1976). *Linear and Combinational Programming* (New York: John Wiley & Sons, Inc.).

Sverdrup & Parcel and Associates, Inc. (1977). *Sludge Handling and Disposal Practices at Selected Municipal Wastewater Treatment Plants*, EPA-430/9-77-007, MCD-36, U.S. Environmental Protection Agency.

Wagner, Harvey M. (1969). *Principles of Operations Research* (Englewood Cliffs, NJ: Prentice-Hall, Inc.).

Water Pollution Control Federation (1977). *Wastewater Treatment Plant Design,* Manual of Practice No. 8.

Water Pollution Control Federation (1969). *Sludge Dewatering,* Manual of Practice No. 20.

EPILOG

The famous American linguist and writer H. L. Mencken, in his treatise *The American Language*, observes that many of the newer words in our language have been formed as a combination of sounds which in themselves convey a picture or a meaning. For example "crud" started out as C.R.U.D., Chronic Recurring Unspecified Dermititis, a medical diagnosis for American soldiers stationed in the Philippines in the early 1900s. The word has a picture without a definition. Try smiling, and in your sweetest, friendliest way say "crud." It just can't be done. It always sounds . . . well . . . cruddy.

A combination of consonants which Mencken points out as being particularly ugly is the "sl" sound. Scanning the dictionary for words starting with "sl" produces slimy, slither, slop, slovenly, slug, slum and, of course, sludge.

Is it possible that some of our sludge disposal problems could be caused by this unfortunate name? It was noted in the previous chapter that a catchy trade name is necessary if stabilized sludge is to be marketed. Perhaps the generic name should also be changed to something which more accurately reflects the value of this material.

This suggestion is only half in jest. The problems with sludge disposal would be more humorous if they weren't so serious. In fact, sludge disposal problems are but a small part of our environmental crisis. Unfortunately, we are still trying to live an open-ended lifestyle, not recognizing that all species must be in closed-cycle harmony with all other living and nonliving things of the worldwide ecosystem. If imbalances and perturbations occur, the species either adapt, die out or destroy the ecosystem. I recognize that perhaps the transition from unwise sludge disposal to ecosystem catastrophe is a bit strained. The point of this discussion is, however, to elucidate the proper *philosophy* of a sludge management. Sludge, in my opinion, should be thought of as a waste product with value (no contradiction) and a material which should find its proper *use*, not just a method of disposal.

If this philosophy makes sense, such problems as heavy metals in sludge are no longer controversial. Heavy metals at high concentrations simply do not belong in sludge. Similarly, deep-well injection and ocean disposal would be contrary to wise residuals management, and decisions on the possible use of these disposal techniques should be tempered with ecological concerns.

Some would accuse me of idealism. And yes, we ivy-covered professors do think idealistic thoughts, and are paid for it! But there are precious few ideals and social tenets in our Western society that didn't start out as idealistic speculation and ephemeral theorizing. Slavery was not abolished by plantation owners. The idea that slavery is a social abomination arose first as a theoretical idea, and only through pain and bloodshed was this idea translated to social realization.

One can easily become pessimistic when viewing the fate of humanity and how the present social systems just do not match with what is obviously the solution, the recognition of ecosystem balance and stability.

But I do not want to conclude this book on an unnecessarily sour note. Sludge treatment and disposal have come a long way in the last few years. All the signs are positive, and people in all countries are beginning to re-examine the techniques and methods of residuals management. The trend in sludge incineration seems to have reversed, the need for pretreatment of industrial wastes is gaining momentum, and regulatory agencies are looking with increased disfavor on ocean disposal and deep-well injection, especially for hazardous materials. There are all good signs.

To some people, sludge is still an ugly four letter word. To others (and to me) it is a residual of our society which must be treated and used to the maximum benefit of mankind, recognizing that mankind is but one species on this wonderful planet.

APPENDIX

ABBREVIATIONS

ac = acres
BOD = biochemical oxygen demand
BTU = British thermal unit
C = degrees Celsius
CA = centrifugal acceleration
COD = chemical oxygen demand
CST = capillary suction time
cal = calories
cfm = cubic feet per minute
d = day
DCG = dual cell gravity filter
est = estimated
F = degrees Fahrenheit
ft = foot
g = gram
gal = gallon
gpm = gallons per minute
ha = hectare
hp = horsepower
hr = hour
J = joule
K = degrees Kelvin
kcal = kilocalories
kg = kilogram
km = kilometer
kW = kilowatt
kWh = kilowatt hours
l = liter
lb = pound
lb_m = pounds mass
lb_f = pounds force
m = meter
mg = milligrams
mgd = million gallons per day
mi = mile
min = minute
ml = milliliter
mm = millimeter
N = newton
Pa = pascal
ppm = parts per million
psi = pounds per square inch
rad = radian
rpm = revolutions per minute
sec = second
SS = suspended solids
SVI = sludge volume index
t = tonne (metric)
ton = 2000 lb
tonne = 1,000 kg
VSS = volatile suspended solids
W = watt
wt = weight
yr = year
μm = micrometer

INDEX